ENCYCLOPAEDIA
OF
ASIA

NAUTCH GIRLS OF HAIDARABAD.

ENCYCLOPAEDIA
OF
ASIA

Compendium of Geography
with Maps and Illustrations

Volume II

Southern and Western Asia

By

A.H. KEANE, F.R.G.S.

DISCOVERY PUBLISHING HOUSE PVT. LTD.
NEW DELHI-110 002

Published by:

DISCOVERY PUBLISHING HOUSE PVT. LTD.
4383/4B, Ansari Road, Darya Ganj
New Delhi-110 002 (India)
Phone : +91-11-23279245; 23253475; 43596065
E-mail : discoverybooksindia@gmail.com
discoverypublishinghouse@gmail.com
namitwasan9@gmail.com
web : www.discoverypublishinggroup.com

Reprinted: 2021

ISBN: 978-81-7141-405-5 (Set)

Encyclopaedia of Asia

Printed at:
Infinity Imaging Systems
Delhi

CONTENTS

SOUTHERN ASIA

(AFGHANISTAN, INDIA, INDO-CHINA, MALAY PENINSULA)

CHAPTER I

AFGHANISTAN AND BALUCHISTAN (KABUL AND KELAT)

CHAPTER II

THE INDIAN EMPIRE

CHAPTER III

INDO-CHINA AND MALACCA

WESTERN ASIA: MUHAMMADAN STATES

(TURKEY IN ASIA, ARABIA, AND PERSIA)

CHAPTER IV

ASIA MINOR

CHAPTER V

THE EUPHRATES AND TIGRIS BASIN

CHAPTER VI

SYRIA AND PALESTINE

CHAPTER VII

ARABIA

CHAPTER VIII

PERSIA

LIST OF ILLUSTRATIONS

SOUTHERN ASIA

AFGHANISTAN, INDIA, INDO-CHINA, AND THE MALAY PENINSULA

CHAPTER I

AFGHANISTAN AND BALUCHISTAN (KABUL AND KELAT)

1. *Boundaries—Extent—Area.*

ALTHOUGH now connected politically with British India, the countries forming the subject of this chapter belong geographically to the Iranian world. They were even for many centuries comprised within the Persian monarchy, from which they are now separated by little more than conventional frontiers. Nevertheless the valley of the Hari-rud, the Sistan depression, and the change of direction in the mountain system of West Makran, offer a sufficiently defined physical parting-line between the western and eastern divisions of the Iranian plateau. The eastern section, stretching thence to the Indus valley and bounded on the south by the Arabian Sea, northwards by the Hindu-Kush and its little-known western extensions to the Hari-rud, forms a quadrangular mass about 600 miles long north and south, and 550 broad east and west, with an area of some 400,000 square miles.

Of this area about 170,000 square miles are comprised in the southern division forming the Khanate of

Kelat, and 230,000 in the northern division forming the Amirate of Kabul, States more commonly known as Baluchistan and Afghanistan respectively. But since 1873 a large tract, about 70,000 square miles in extent, lying beyond the natural limits of the Iranian plateau, has been recognised as politically belonging to the Amir of Kabul. This tract, known as Afghan Turkestan, lies mainly between the northern scarp of the plateau and the Upper Oxus, the boundary here following the left bank of the river from its source on the Pamir to Khojah Saleh. The northern frontier line runs thence across the Dasht-i-Chul desert a little north of Maruchak on the Murgh-ab river to Zulfikar on the Persian frontier. Afghanistan has thus a total area of 300,000 square miles, with a population variously estimated at from 4,000,000 to 6,000,000. The boundary between Kabul and Kelat is scarcely anywhere clearly determined, but may be said to follow the 30th parallel from Persia to within 30 miles of Quetta, whence it runs north-east to an undetermined point on the frontier of British India.

One of the arrangements arrived at between the Amir and Sir Mortimer Durand in 1894 had reference to a more accurate determination of the frontier towards British India. In the Khaibar section the line west of the Sitsobi Pass, at the head of the Bara valley, will continue to follow the crest of the hills to the Safed-Koh, overlooking the Kuram valley. But east of Landi Kotal the boundary line has been drawn to the Kabul river a few miles below Fort Dakka and Lalpura. From this point to the southern extremity of Chitral, now included in British territory, the boundary coincides with the range of hills between the Mohmand country and Bajaur. In the Kuram valley the line is traced from the Sita Ram peak of the Safed-Koh along the crest of the hills between the Kuram and Khorb rivers as far as the latter stream. The Waziri border has not yet been settled; but in the Zhob district the line has been drawn from Kandar Domandi at the confluence of the Kandar and Gomal rivers along the Kakar border to Chaman, and thence southwards along the Shorawak frontier, the section from Shorawak to the Helmand being reserved for future settlement. In the unsettled Hindu-Kush border-lands the arrangement is satisfactory. Here Chitral and the neighbouring districts are henceforth recognised as lying within the undivided sphere of British influence, and to England is left the absolute control of all the approaches to India on the south-east side of the Pamirs. Thus British India becomes the guardian of all the passes over the Hindu-Kush and Lahori ranges, which here include the Dora and Nuksan above Chitral, the Ishtragh above Mastuj, the Baroghil, Shundr, and Darkot above Yasin and Gilgit. In return for these concessions the Amir's annual allowance is raised from £75,000 to £112,000, and he is permitted the free import of arms from India.

2. *Relief of the Land: Highlands—Hindu-Kush—Paropamisus—Safed-koh—Suliman Mountains—Hala and Coast Ranges—Desert.*

The eastern section of the Iranian plateau rises from the central Hamun depression towards the highlands, by which it is enclosed on the south, east, and north, and which in the north-east gather to a head in the Hindu-Kush, connecting the whole system and the tableland itself with the Pamir and great Central Asiatic plateau.

The recent surveys of the Afghan highlands, covering an area of nearly 30,000 square miles, have shown that while the southern ranges are more elevated, the Hindu-Kush, at least in its western section, is a far less formidable barrier between India and Central Asia than had been supposed. "Throughout the whole length of it visible from the Kabul plain, it is by no means an imposing range. No part of it is snow-covered, except for a few months in winter; there are no grand peaks, no magnificent altitudes. Previous estimates of its general altitude must be reduced by from 1000 to 2000 feet at least. . . . It is crossed by mountain paths at intervals along its whole length, from the Irak Pass leading to Bamian to the Khawak Pass, east of which the Hindu-Kush rises into a really formidable mountain chain, increasing gradually eastwards till we arrive at peaks of truly Himalayan proportions. The Tirich Mir, at the Nuksan Pass, is fixed now at 25,000 feet, and others have been seen not far west which cannot differ by many thousand feet. Still, so far as the Koh-i-Daman or the plains of Kabul are concerned, the line of the Hindu-Kush is hardly a defensible, and is certainly a most undesirable, military frontier" (Capt. T. H. Holdich).

At its north-east end the Hindu-Kush is crossed by the Baroghil Pass, leading from India, Chitral, and Kashmir to the Upper Oxus valley, Kashgar, and Yarkand. To the south-west of the Tirich Mir stretches the still little known Kafiristan section of the system, where, however, at least one pass, the Apaluk mentioned by Major Raverty, leads to the Oxus basin.

From the south-west corner of the Pamir the Hindu-Kush runs mainly south-west to about the 68th meridian, whence it is continued for over one hundred miles westwards by the Koh-i-Baba. Before the survey of Captains Talbot and Maitland in 1885, this range, apparently the Paropamisus of the ancients, was one of the least-known highland regions on the globe. Its three western ramifications—the Tirband-i-Turkestan, Safed-koh, and Siah-koh—are now known to run nearly parallel through the Hazarajat and Zamindawar to Herat and the Hari-rud valley, whence they are continued by the Daman-i-koh system north-westwards through the Little and Great Balkans to Krasnovodsk on the Caspian. Between the Koh-i-Baba and Herat they throw off numerous spurs running almost uniformly north-east and south-west, and forming longitudinal valleys, which drain through the Helmand and other rivers to the Hamun depression.

Much light has been thrown on the orography of the extreme north-east by recent exploration, and its leading features may now be fairly traced. From the angle formed by the converging Hindu-Kush and Mustagh ranges spring a number of lofty spurs separating the head-streams of the Gilgit River. One of these, with many peaks over 20,000 feet high, forms the water-parting between the Chitral and Gilgit basins, and is crossed by the Darkot and Moshabar Passes. Just south of the 36th parallel a remarkable transverse range runs from the Indus at Bunji nearly to Chitral, throwing off a succession of spurs between the Kandia, Swat, Panjkora, and Chitral (Kunar) river valleys. Here the peaks diminish from nearly 20,000 feet to between 4000 and 7000 as we proceed southwards to the Kabul river. This transverse range, supposed by Major Tanner to be the Hindu Roj of the Afghans, is an important feature in the physical geography of the Hindu-Kush, as it separates the comparatively rainless tracts of Gilgit, Hunza, and Yasin from the well-watered southern valleys of Panjkora, Kashkar, and Swat.

From the junction of the Hindu-Kush and Koh-i-Baba an important spur, running eastwards between the Helmand and Ghorband basins, sweeps round the head-waters of the Arghand-ab to the north of Ghazni, and thence trending north-east follows the 34th parallel as the Safed-koh ("White Mountains") between the Kabul and Kuram river basins, eastwards to the plains of

Peshawar. From this range, which culminates with Mount Sikaram (15,620 feet), the whole system of the Suliman Mountains projects southwards between the Iranian plateau and the valley of the Indus. At right angles with Mount Sikaram runs the Peiwar range, a well-wooded spur crossed by the Peiwar Pass, the scene of General Roberts's signal victory over the Afghans on 2nd December 1878. The main range of this complicate system runs from near the Shutargardan Pass (10,900 feet) southwards to the great Kund Peak, where it branches off into a number of minor spurs, ultimately merging in the East Baluchistan highlands, which continue to skirt the Indus valley to the coast. Besides the main chain forming the watershed between the Helmand and Indus, it is now ascertained that a continuous system of parallel ranges runs from the gorge of the Gomul to about the 30th parallel.

South of the Gomul Pass run two main ranges nearly 12,000 feet high, which include several remarkably parallel ranges, increasing in number southwards, till no less than twelve distinct ridges are observed where the Nari river pierces the whole system. Many other streams or torrents rising on the eastern slopes of the Inner Sulimans, when swollen with the rains or melting snows, penetrate across the intervening ridges down to the Indus. These *darahs*, or river gorges, afford easy access at many points from India to the Afghan uplands, so that the whole frontier from Peshawar to Jacobabad is now found to be traversed by a large number of "excellent natural roads and passes" (Holdich).

Between the Gomul and Kuram (Kurmah) valleys lie the Waziri highlands, and the Safed-koh skirting the Kuram river on the north maintains a uniform level of 12,500, culminating eastwards in a double-peaked mountain 14,680 feet high. North of the Safed-koh project three important spurs, one east of the Logar river traversed by the ill-omened Khurd Kabul defile, another (the Karkacha ridge) washed by the Tezin and Surkh-ab rivers, and a third springing from the intersection of the meridian of 70° 45′ with the main range, and dividing the Khaibar from the Bazar valley.

Towards Baluchistan the most prominent and important range is the Khoja Amran, running nearly north and south between the Pishin valley and the Kandahar country, and forming in this direction the present political frontier of Afghanistan. It culminates with the Khoja Amran Peak, and is crossed in the north by the Psha Pass, in the centre by the Khojak (8000 feet), in the south by the less known but easier Gwaja; through the Khojak the Indo-Afghan railway is carried to Chaman Fort, within 65 miles of Kandahar.

North of the Shal district the Khoja Amran ramifies northwards into the Toba and Surkh-ab ridges, the latter enclosing the Pishin valley on the north-east and sloping gently towards Shal (Quetta). Eastwards the hills fall more abruptly, and here the chief approaches from India are through the famous Bolan and Mula river gorges.

The Suliman system, which culminates with the Takht-i-Suliman (11,298 feet), and which has many other peaks, such as the Takatu between Pishin and Quetta, Chapar and Kalipat farther east, and several others ranging from 10,000 to 12,000 feet, has a mean width of about 150 miles between the Indus and the desert. The whole distance from Sukkur on the Indus to Kandahar through the Bolan and Khojak Passes is 410 miles, of which 140 are comprised in the alluvial riverain tract and the Kachi desert as far as Sibi, which is still only 700 feet above sea-level. Beyond the Khojak Pass, which is 90 miles from Kandahar, the land again falls rapidly towards the central desert, so that the true highlands between the southern end of the Bolan Pass and the Khoja Amran range[1] between the Pishin valley and the Kandahar district are about 180 miles wide.

The southern section between Baluchistan and the Lower Indus has no general native name, but is variously known to Europeans as the Brahui or Hala range.[2] This highland region, which is politically included in the territory of the Khan of Kelat, is approached from the Indus valley through short steep watercourses to a height of 1200 feet. The main ridge, running north and south, throws off various branches east and west. Eleven such offshoots occur between Kelat (6700 feet) and Khozdar (3800) at the foot of the Mula Pass, forming in a tract scarcely 100 miles long as many as thirteen upland plains at various elevations. From Khozdar the route surveyed by Bellew in 1872 descends towards the coast through the Purali valley, and towards India through the dangerous Mula Pass.

The Baluchistan southern highlands run mainly east and west parallel with the coast from the Indus delta to the Persian frontier, where they change abruptly to the south-west. The intervening valleys ascend successively inland to a height of 2500 feet, and are often of great length. One of them runs from the Khelat hills uninterruptedly westwards nearly to Bampur in Persia, and 70 miles south of it is another stretching for 250 miles westwards to Kasr-Kand also within the Persian frontier, where all these valleys are closed in by the intricate

[1] The Khoja Amran has no general native name, and the term Khoja (properly Khwaja) is merely the name of a peak in the Gwaja Pass at its southern end. Khojak also is rather the name of the river, the bed of which forms the pass, than the pass itself.

[2] The *Hala* seems to be properly only a small ridge running from Kelat southwards to the Baghwana River, lat. 28° to 29° N. lat., and 66° 30′ E. long.

highland system of West Makran. "No difficulty exists for wheeled traffic from one end to the other of these two valleys" (Major Lovett). Farther inland a third parallel range, the Wushuti or Mue Mountains, stretches along the border-land of the two states at a distance of about 280 miles from the coast.

Most of the inner space enclosed between the northern, eastern, and southern highlands consists of an extensive sandy plateau, at a mean elevation of perhaps 3000 feet above the sea, and sinking everywhere towards the central Hamun depression. Except along the river banks, this region may be regarded as waste; and south of the Helmand, where there seem to be no more rivers, the desert formation is complete. It begins at the foot of the Khoja Amran range, and stretches thence almost uninterruptedly along the Afghan and Baluch border-lands eastwards to Sistan and Persia. No European has yet ventured across this almost impassable wilderness, which still remains nearly a blank on our maps. Seen from the neighbourhood of Kandahar, it presents the appearance of endless undulating sand-hills rolling up from the far south.

Similar desert tracts are found within the uplands themselves—as, for instance, the Kachi desert below Sibi, 90 miles long, and now traversed by a railway, and the Dasht-i-Be-daulat ("Desolate Plain," or, more exactly, "the plain without wealth") in the very heart of the highlands above the Bolan Pass and south of Quetta, 200 square miles in extent.

3. *Hydrography: Inland and Seaward Drainage—The Hari-rud—Helmand and Kabul Basins.*

The East Iranian drainage system is threefold—two inland to the Hamun, Aralo-Caspian and some smaller depressions, one to the Indian Ocean either directly or through the Indus. Afghanistan belongs to all three, but mainly to the Hamun basin, while Baluchistan drains almost exclusively seawards.

Afghan Turkestan is comprised entirely within the Aralo-Caspian basin, all its rivers flowing from the northern slopes of the Hindu-Kush and Paropamisus to or towards the Oxus and Aral or Caspian. Here we again meet with the same undeveloped or partially dried-up water system which was found prevailing in Arabia and Persia, and which forms such a striking feature of the great Central Asiatic tableland. In the east the Kokcha and Kunduz still reach the Upper Oxus, but as we proceed westwards we find that all the rivers flowing north fail to reach either the main stream or either of the great inland seas. Thus the Dehas-rud (Balkh), rising in the Koh-i-Baba, gets no farther than Mazar-i-Sherif, where it takes the name of Band-i-Barbari, and runs dry in the Siyagird district after a course of over 180 miles; the Sar-i-pul is lost in the sands beyond Shibarghan; the Murgh-ab, after irrigating the Merv oasis, disappears in the Karakum desert, and the same fate overtakes the Hari-rud (Tajand) after skirting the Daman-i-koh on its way to the Caspian.

The Hari-rud, or river of Herat, has its source in a deep valley 9500 feet above the sea at a point where the Koh-i-Baba ramifies into the Siah-koh and Safed-koh. It flows thence rapidly through an unexplored region down to the town of Obeh, where its waters are largely diverted into irrigating rills. Its course lies thence westwards to Herat and Ghorian, where it turns abruptly northwards along the Persian frontier to its junction with the Keshef-rud above Sarakhs. The united streams now take the name of the Tajand, whose course has been described at p. 479.

The southern slopes of the Hindu-Kush within the Afghan frontier all drain to the Indus through the Kabul river, which also receives on its right bank several streams from the Koh-i-Baba. Thus the north-eastern portion of Afghanistan is comprised in the Indus basin, to which also belongs the eastern slope of the great watershed of the Suliman, as well as all the intervening outer parallel ridges. But nearly all the land west of this parting line, and south of the northern scarp of the plateau, an area of about 200,000 square miles altogether, drains to the great Hamun depression. Of this vast basin the chief stream is the Helmand, which flows from the west side of the Pughman range through a deep channel in the Hazarajat south-westwards to within 40 miles of Girishk, where it enters the plains which merge southwards with the Baluchistan desert. Here it is largely utilised for irrigation purposes, and at Girishk is crossed by the great caravan route from Kandahar to Herat. It then sweeps southwards through the fertile Garmsel country, beyond which it turns north-west to the Hamun or Sistan swamp. The Helmand, which has a course of about 700 miles, is never without an abundant supply of water, but in winter after the floods it comes down with great rapidity, sometimes overflowing its banks in consequence of the neglected state of the old embankments. Its chief tributaries are the Arghand-ab, Tarnak, and Dori, whose united stream joins it from Kandahar a few miles below Girishk. West of the Helmand the Kash-rud, Farrah-rud, and Harut all flow from the Ghor highlands in nearly parallel beds southwards to the Hamun swamp.

To the same system belongs the lagoon Abistada, the only other body of water in East Irania deserving the name of lake. It lies over 7000 feet above the sea some 60 miles south-west of Ghazni, and is fed by the river of that name. It is about 17 miles by 15 in extent, and, although hitherto supposed to be a closed basin, there is little doubt that during the floods it overflows to the Arghasan, a tributary of the Arghand-ab. Its water is brackish and very shallow, nowhere exceeding 5 or 6 feet in depth.

The crest of the water-parting between the Helmand and Kabul basins is marked by the Sher-i-Dahan ("Lion's Mouth") Pass, crossed by the road going south to Ghazni. Rising at a height of about 8400 feet above the sea, the Kabul flows mainly east by Kabul and Jelalabad, to the Indus at Attock. During its rapid course of about 250 miles it receives from the Hindu-Kush the Swat, Kunar, (Chitral), Alingar, Tagao, Panjshir, and Ghorband; from the Safed-koh the Logar, Surkh-ab, Bara, and Tira. Of the northern affluents the most important is the Kunar, which flows from the Baroghil Pass through the Chitral valley for nearly 300 miles down to the main stream, a few miles below Jelalabad. The Yarkhun or Mastuj, as its upper course is called, is often represented as rising in a Lake Karambar Sar, figuring on the maps as also the source of the Karambar or Ashkaman River of Gilgit. But it was always doubted whether any basin in this region could have a double outflow, and M. Dauvergne, who returned to India by the Baroghil Pass in 1889, found that there are two distinct lakes separated by a low rocky divide, the Gazkul, source of the Yarkhun (Kunar), and, a few hundred yards farther east, the Karambar Sar, properly Ishky-kul, source of the Karambar or Gilgit river.

South of the Kabul river are the important Gomul and Kuram basins, the former of which covers an area of perhaps 13,000 square miles between the western and eastern Suliman ranges, along which the great trade route from Central Asia to India passed for centuries. The

Kuram, which rises on the eastern slopes of the great water-parting between the Indus and Helmand basins, is joined on its course to the former river by numerous affluents from the Safed-koh on the north, and from the hilly country of the Mangal tribes on the west and south.

The Lower Indus receives no important stream from Baluchistan, which seems to be almost as riverless a country as Arabia itself. To its inland drainage belongs the Lora, which rises with several head-streams on the east slope of the Khoja Amran, and after watering the Pishin valley, escapes through the Tang gorge in the Tang range south-westwards to the Hamun Lora morass in 29° 30′ N. lat., 65° E. long. Its lower course, like those of Bale and other streams flowing in the same direction, still remains to be traced. On the Makran coast the only noteworthy river is the Dasht or Bhingwur, which is supposed to rise far inland, and to make its way through all the intervening ranges and valleys to the sea at Gwattar Bay in 61° 40′ E. long. But here scarcely any perennial streams seem to exist, and few of them flow through regular or well-defined beds.

4. *Natural and Political Divisions: Wakhan—Badakhshan—Afghan Turkestan, Afghanistan Proper—Kafiristan—Baluchistan—Kelat—Makran.*

Afghanistan as at present constituted comprises three physically distinct regions—the northern slopes of the escarpment forming geographically a part of Western Turkestan, the basin of the Helmand embracing most of the central plateau, and the eastern highlands mostly included in the Indus basin. But to these natural regions the political divisions correspond in part only. Since the foundation of the modern Afghan State by Ahmad Shah in 1747, comparatively little progress has been made towards moulding it into one political system. So heterogeneous are its ethnical components, so inaccessible many of the highland tracts, and so persistently upheld is the old tribal organisation of the dominant Afghan race itself, that in many places the Amir's authority is merely nominal, in others openly defied, in some never yet recognised. Protracted internecine feuds between the rival branches of the royal Durani (including the Barakzye) tribes, combined with several disastrous foreign wars, first with the Sikhs of the Panjab and then with the British Government of India, have added to the confusion to such an extent that disintegration rather than consolidation of empire has lately seemed imminent. As it is, the Amir has been fain to sacrifice external independence, and to accept a somewhat indefinite position of subordinate relationship to the Suzerain of India.

In the north, Afghan Turkestan, comprising that portion of the land included in the Aralo-Caspian basin, possesses a certain ethnical as well as physical unity, for here the bulk of the people belong to the Usbeg branch of the Turki stock. It is administratively divided into a number of provinces corresponding with the old Usbeg khanates, all of which have completely lost their autonomy.

In the extreme north-east is the alpine territory of Wakhan, which consists of two upland valleys traversed by the Sarhad (Panja), the chief southern head-stream of the Oxus. On either side the valleys are hemmed in by lofty mountains, those to the south forming the northern section of the Hindu-Kush, here crossed by some difficult passes, the easiest of which are the Dora and Baroghil (12,000 feet) leading to Chitral and Gilgit. The chief resources of the people are derived from their flocks, mainly sheep and the Tibetan yak. The land is too elevated and sterile for tillage, but yields a pasturage like that of the Pamir, possessed of peculiar fattening qualities. In this alpine region the lowest hamlet is 8000, and Sarhad, the highest, no less than 11,000 feet above the sea, or higher than the loftiest peaks of the Pyrenees. Yet a little pulse and barley are grown in a few sheltered glens. When Lieutenant John Wood, discoverer of the source of the Oxus, visited Wakhan in 1838, he estimated the population at about 1000; but Forsyth, thirty-five years later, raised the number to 3000, which agrees with the Russian estimate. The mir or chief, who, like so many others in this region, claims descent from Alexander the Great, resides in Kila-Panja ("Five Forts"), on the Oxus, and close to the Pamir. In Wood's time he was almost independent; but since then has become tributary through Badakhshan indirectly to Kabul.

Badakhshan, adjoining Wakhan on the west, comprises the valley of the Kokcha and the little-known tracts enclosed on one side in the great northern bend of the Oxus, and stretching on the other to Kafiristan. Besides the Kokcha, it is watered by the Wardoj, and both streams unite a few miles above Faizabad, the capital, before joining the Oxus. In the upper parts the crops are often nipped by summer frosts. But lower down the more favoured sites yield wheat, barley, mulberries, walnuts, pistachios, and pulse. The country is exposed to earthquakes, one of which in 1832 was very destructive to life, and was felt as far south as Lahore. Badakhshan is noted for its mineral wealth, salt, sulphur, iron, and especially the ruby[1] and lapis lazuli,[1] prominently men-

[1] These rubies, which are of a delicate rose colour, were formerly known as *balais* or *balash*, a corrupt form of Badakhshan, which Marco Polo calls Balacian. The lapis lazuli takes its name from the district of Lajurd or Lazurd, whence both the words *lazuli* and *azure*.

tioned by Marco Polo. The lapis lazuli mines, which lie close under the crest of the Hindu-Kush, have been fully described by Wood.

West of Faizabad the road diverging to the right through Rustak crosses the Oxus to Kulab, Karateghin, and other Trans-Oxian districts. The gold-washings in a small stream between Rustak and the Oxus yielded a revenue of about £100 in 1874. The main road beyond this point still runs westwards over the Lattaband Pass, through Talikhan down to Kunduz. Here the descent from the Badakhshan highlands to the marshy plains of Turkestan, here scarcely 500 feet above sea-level, is attended by a marked change of climate, that of Kunduz being excessively hot and unhealthy especially in summer.

Kunduz is watered by the river of like name, which rises with several head-streams in the Koh-i-Baba. Beyond the town of Kunduz it joins the Oxus below Hazrat-Imam. Here are extensive undulating plains yielding good pasturage, and tenanted by nomad Usbegs and Hazaras. From Talikhan and Kunduz there diverge to the left routes which lead over the Sir-alang and Khawak Passes to Kabul. But here the chief highway is the historical route which passes through Bamian and Heibak, joining the Badakhshan road at Khulm or Tashkurgan. This route was traversed for some distance by the British troops with horse artillery in 1839. Near Bamian are two gigantic idols, one of which is said to be 100 feet high, cut in bold relief in the face of the cliff skirting the road. They stand in deep niches, and are clothed in flowing drapery. These idols and caves are generally supposed to be of Buddhist origin, but all memory of the time and hands by which they were executed has long perished. Here also are the stupendous ruins of Ghulgulch destroyed by Chingiz Khan, besides many other remains, which have been fully described by Masson.

Adjoining Kunduz is the smaller but not less populous khanate of Khulm. It occupies a vital position in the heart of the ancient Baktriana, the converging point of all the natural highways from India, Persia, and Central Asia. Here are the ruins of Baktra and of its successor Balkh, now supplanted by the modern towns of Siyagird, Mazar-i-Sherif, and Khulm. The country has been largely encroached on by the desert, and the Khulm river, flowing from the Kara-koh hills, now no longer reaches the Oxus. In the plains the river of Balkh, here called the "Band-i-Barbari," or "Dyke of the Barbarians," is soon absorbed in irrigation works in the gardens interspersed amidst the vast ruins and flourishing towns of this historic land.

West of Balkh are the four petty Usbeg States of Akcha, Shibarghan, Sar-i-pul with Andkhui, and Maimana, lying mostly between the outer spurs of the Paropamisus and the sands by which they are now cut off from the Oxus. This tract is very fertile and well watered by the streams from the mountains, but it is also proverbially unhealthy. Nevertheless here are the populous towns of Shabirkhan and Andkhui, lying close to the Russian frontier.

Of these khanates Andkhui alone has retained a certain measure of independence. All the rest are absolutely controlled, and even administered, by Kabul, though the old geographical and political divisions are still preserved. The village of Gurzivan and the Darzat valley in the hills south of Sar-i-pul have also lost their autonomy, though still retaining the empty titles of khanates.

The Usbeg inhabitants of these districts are not called upon to render military service; but, according to the authority of Grodekov, which, however, is not above suspicion, they are so heavily taxed that they are impatiently awaiting the arrival of the emancipating Russians.

On the southern slopes of the Hindu-Kush, bordering eastwards on Kashmir, south-eastwards on the Panjab, are the territories of Kafiristan, Gilgit, Chitral, Swat, and Chilas, which were hitherto conventionally supposed to belong to Afghanistan. But Kafiristan has now (1896) been effectively occupied by the Afghans, while the other petty states have since 1893 been brought under the British rule. The change has been caused partly by internal disorders, partly by the movements of the Russians in the Pamir region, which have obliged the Indian authorities to establish a strong frontier in that direction.

Of Kafiristan, "Land of the Infidel," next to nothing was known before Mr. G. T. Robertson's expedition of 1889-90. In 1885 Sir W. Lockhart had crossed from Chitral into the Upper Bashgul valley, returning by another route after a stay of a few days in that district on the north-east frontier. Later M'Nair visited the Kalash territory, which, however, belongs to Chitral and forms no part of Kafiristan proper. Even Robertson's journeys were confined to the eastern and central parts, comprising the whole of the Bashgul basin and the Viron (Presun) valley. But during a residence of over a year in the country he was able to collect much accurate information regarding its physical features, hydrographic system, climate, vegetation, and inhabitants.

Kafiristan is bounded on the east by Chitral and the Chitral (Kunar) river valley, on the south and west by Afghanistan proper, and on the north by the Hindu-Kush, which is here crossed by the Mandal and some other passes leading into Badakhshan at altitudes of over 15,000 feet. Towards Chitral the passes are lower, one falling to 8400 between Utzun and Gurdesh, but still difficult and blocked for two or three months in winter. The surface is extremely rugged and mountainous, being disposed in a number of deep, narrow, tortuous river valleys, "into which a varying number of still more difficult, narrower, and deeper valleys, ravines, and glens pour their torrent waters. The hills which separate the main drainage valleys, one from the other, are all of considerable altitude, rugged and toilsome. As a consequence, during the winter Kafiristan is practically converted into a number of separate communities, with no means of intercommunication."[1] All the drainage is to the Kabul river, either directly through the Alingar (Kao) and others, or indirectly through the Chitral, which receives on its right bank the Bashgul just above Arun (Arundo), and the Presun with its Kti and Wai affluents at Chigar Sarai. Too little space is left for much tillage between the river banks and the enclosing hills. Hence most of the vegetation is arborescent—pomegranates and other fruit-trees, horse-chestnuts, olives, and evergreen oaks flourishing at the lower elevations, dense pine and cedar forests between 5000 and 8000 or 9000 feet, the willow, birch, juniper, cedar, with wild rhubarb, up to 13,000 or 14,000, the limit of all vegetation except coarse grasses and mosses. The whole region abounds in wild romantic scenery, presenting great diversity according to the different altitudes and aspects of the land. "In many places where the tortured water foams and lashes itself against the rocks, the river almost assumes the nature of a cataract and is indescribably beautiful. Tree-trunks encumber the waterway, jam themselves against the rocks, pile up in picturesque confusion, or hurry round and round in the swirl of a backwater" (*ib.*).

Mr. Robertson is of opinion that the present inhabitants are mainly sprung from the old Indian population

[1] Robertson, *Geo. Journal*, Sept. 1894, p. 197.

of Eastern Afghanistan, who rejected Islam and took refuge in the eleventh century in the almost inaccessible Hindu-Kush valleys, enslaving or partly blending with the aborigines. These aborigines are still represented by the Presuns, Jazhis, Wai, Arams, and others, while the bulk of the conquering intruders (Katirs, Kams, Madugals, etc.) are collectively known as *Siah-Posh*, "Black-clad," from the sombre colour of their clothes. Hence, although nearly all have a dark complexion like that of the average Panjabi, there are two distinct types, one high with regular features, "purely Aryan," the other low with flattish nose, coarse features, and hair worn nearly down to the eyebrows, giving them a singularly forbidding appearance. The women also, who do all the hard work, are for the most part debased and unlovely. There are three linguistic groups: Siah-Posh, Wai (with Ashkun ?), and Presun, the last being absolutely distinct, and not only unintelligible to the other Kafirs, but so difficult that none of them can ever learn the Presun language. Apart from a few Muhammadans towards the frontiers, all are still pagans and polygamists, with from one to four or five wives, who on the death of their master revert to his family, and are either sold or kept by his surviving brothers. They are a brave, freedom-loving people, capable of much self-devotion, and naturally most intelligent, but quarrelsome and excessively covetous. There are, however, no blood feuds, and murderers or homicides are made outcasts for ever or till payment of a ransom, which is so heavy that it is very rarely paid. The tribal system still prevails, each tribe comprising numerous clans, whose affairs are arranged by consultation between the several *jasts* or headmen, who are not hereditary but chosen for their wealth, valour, or other personal qualities.

A southern branch of the Siah-Posh Kafirs, or "Black-

Clad Infidels" as they are called by the surrounding Muhammadans, are the Safis and Chagnans, whose domain reaches down to the Kabul river. Masson describes them as a straightforward manly race.

In Afghanistan proper the political divisions are often far less distinctly defined than in its outlying Turkestan possessions. Some regions in the Hindu-Kush, such as Kafiristan and Swat, as well as nearly the whole of the northern highlands between Bamian and Herat, besides many tracts in the Suliman Mountains, had never acknowledged the Amir's authority, and had retained their independence until the recent partition. Elsewhere, as in the districts of Herat and Kandahar, and even in Kabulistan itself, the tribal organisation still largely prevails, so that the limits of the provinces are scarcely anywhere carefully laid down, and it becomes impossible to speak of provincial administration in the ordinary sense of the term. Hence, instead of taking the various provinces separately, it will be more convenient to deal with them in connection with the chief towns round which they are grouped.

In Baluchistan, although more order has recently been introduced, a similar state of things still largely prevails. The Khan of Kelat, who may be said to have frankly accepted the suzerainty of the Kaisar-i-Hind (Empress of India), is nominal ruler of the whole land. But his authority has often been confined to the Kelat district itself, and is still challenged by many of the tribal chiefs, especially towards the Persian frontier. The natural divisions of the country, the eastern and southern highlands merging inland with the desert, are grouped in seven recognised provinces: Sarawan and Katch-Gandava, including the Mari and Bugti country on the north-east; Kelat, between these two; Jhalawan on the east; Lus on the south-east; Makran, comprising the southern coast region; Kohistan, or the "highlands" of the west.

Most of the land is still practically unknown. The north-eastern section lying between the Indus and the Pishin valley, along the Afghan border, and thence southwards to Kelat, has been thoroughly surveyed, and a military station has even been established by the British at Quetta, above the Bolan Pass, and overlooking the Pishin valley. The south coast has also been carefully surveyed by the Admiralty, and somewhat farther inland by the English Telegraph Staff; while the country has been crossed, chiefly from east to west, by Grant, Pottinger, Ferrier, Goldsmid, Bellew, Lovett, and a few other explorers during the present century. Still, most of the interior has never been visited, and the sandy plains stretching beyond the hills towards the Hamun depression remain a blank on our maps. Elsewhere the highland formation everywhere predominates, although in the south the parallel ridges are separated by long and almost level valleys reaching from the Persian frontier to the eastern uplands. This southern region, from the sea to the desert, is usually spoken of collectively as Makran; but the term should properly be restricted to the strip of land between sea and the first parallel ranges. Here the geographers of Alexander placed the Ichthyophagi, or "Fish-eaters," apparently a mere translation of the local name.

The country is almost entirely occupied by pastoral tribes under semi-independent sirdars and chiefs. Hence the so-called provinces are not administrative divisions in the ordinary sense, and should be more properly called territories. Besides those above mentioned there are several others current amongst the natives as applicable to particular cantons, especially in Makran and Kohistan. Here there are several semi-independent chiefs, of whom the most powerful was, till recently, the Khan of Kej, in central Makran. But the native ruler was, some ten years ago, replaced by a direct nominee of the Khan of Kelat, and although the change was at first followed by disturbances, it has had the effect of somewhat consolidating the Khan's authority, and thus barring the further progress of Persia in this direction.

The Khan or Mir of Kelat, who belongs, not to the Baluch, but to the Brahui stock, concluded a treaty in 1877 with England, in virtue of which he has become a feudatory of the Empress of India. The right had already been secured of occupying at pleasure the mountain passes between Kelat and Afghanistan. But the new treaty places the whole country at the disposal of the British Government for all military and strategical purposes. In return the Khan has acquired a certain prestige amongst the tribal chiefs and sirdars, who no longer seriously question his supremacy, and his subjects have begun to enjoy the blessings of peace.

5. *Climate.*

In Afghanistan the prevailing climatic conditions are dryness combined with great extremes of temperature. Snow lies on the ground for three months in the Kabul and Ghazni districts, and many of the peaks from the Hindu-Kush to Kelat rise above the snow-line. But so much depends on elevation that Jelalabad, 2000 feet above the sea, is scarcely colder than India, while the winters are almost as severe as those of Russia on the neighbouring Kohistan uplands. The summer heats, on the other hand, are everywhere intense, more so, in some places, even than in Bengal. At Kabul (6500 feet) the glass rises to 90° and 100°, and in Kandahar even higher. Yet the country is on the whole decidedly salubrious, in this respect presenting a marked contrast to the fever-stricken lowland districts of Afghan Turkestan.

In Baluchistan also intense heats are followed by almost equally intense colds, the snow lying for two months on the ground even in the Shal and other valleys. The Kej district and some other parts of Makran are said to be the hottest places in the whole of Asia. Even in March Major Lovett registered "125° F. in the shade in the neighbourhood of Kej." On the other hand, Pottinger found it so cold in February at Kelat that water poured on the ground froze instantaneously. Owing to its proximity to the ocean, Baluchistan receives on the whole more moisture than Afghanistan. The dry season lasts from March till September, but rain or snow falls intermittently throughout the winter, and often heavily in February and March. Unfortunately most of it is precipitated on the outer ranges, leaving little for the deserts of the interior, where the sultry heats are intensified by fierce sand-storms.

6. *Flora and Fauna: The Karez Irrigation System.*

Bare, treeless mountains, sandy and absolutely unproductive plains, fertile valleys and riverain tracts, producing enormous quantities of magnificent fruits and vegetables, besides cereals of various kinds, are the prevailing features almost everywhere from the Upper Oxus to the Indian Ocean, and from Persia to the Indus valley. In the north, however, the southern slopes are often clothed with forests of walnut, birch, oak, and conifers, the latter growing to a height of 10,000 feet. In Afghanistan the asafœtida covers extensive tracts, and here the most productive districts are those of Herat, Kandahar, the Lower Helmand, the valleys of the Kabul and Logar rivers, and the Koh-i-Daman. Wheat, maize, and rice are the staples of food; the vine and many other fruits are indigenous; cotton, sugar, and tobacco thrive in the well-watered low-lying tracts, and the melon and many other vegetables arrive at astonishing perfection. The apples, the grapes, the pomegranates of Afghanistan are celebrated throughout India.

In Baluchistan wheat, barley, rice, cotton, pulse, madder, indigo, and tobacco are cultivated; the date-palm prevails in Makran; magnificent fruits and vegetables are grown in the valleys. Asafœtida abounds, and of forest trees the chief are the plantain, walnut, sycamore, wild fig and olive, mulberry, tamarisk, and mimosa.

The "karez," or underground system of irrigation peculiar to the Iranian plateau, is well suited to this region, and extensively practised. "The soil being naturally open and porous, composed of water-worn stones embedded in a sandy soil, which, however, having a large admixture of lime, hardens at a short distance below the surface into an impermeable conglomerate, it is easy to understand how flowing water may in many places be found 20 or 30 feet from the surface, while on the surface itself for miles round there is nothing but an arid plain. The water thus found is led gradually towards the surface through the karez. A series of wells are dug at intervals of 15 to 25 yards, and connected below by an underground passage, through which the water runs till at last it reaches the surface and is utilised for irrigating the fields" (Capt. R. Beavan).

In East Irania wild animals are comparatively scarce. Lions and leopards of a small type haunt the upper valleys of the Hindu-Kush, where are also met the wolf and two species of bear. The so-called Angora cat is indigenous in Kabulistan, and in the plains the dromedary or one-humped camel is the chief beast of burden. Here the horse is far inferior to the Turkoman breed. The ass exists in the wild and domestic state, but sheep and goats form everywhere the chief resources of the pastoral tribes.

In Baluchistan the lion, tiger, leopard, and wolf are occasionally met, the jackal, wild dog, fox, wild goat, and

WILD ASSES.

ass more frequently. There is a distinct species of gazelle (*Gazella fuscifrons*), and both species of camel occur, the Baktrian or two-humped on the uplands, the dromedary on the plains, where it is highly valued for its speed by the marauding tribes. Serviceable horses are bred in the north and west, but those of Makran are small, weak, and spiritless. In Baluchistan is found the curious Uromastix lizard, one of the most remarkable animals in the world. It looks at a distance somewhat like a rabbit in appearance and size, but is really a sort of diminutive saurian, called by the Persians buz-miji, or goat-sucker, from its peculiar habit of bleating like a kid to attract the goats, whose teats it then sucks. The Makran coast abounds in fish, where it still forms the staple food of its ichthyophagous inhabitants.

7. *Inhabitants: The Afghans—The Brahuis, Baluchis, and Luri.*

East Irania presents a greater complexity of races even than Persia itself. For to nearly all the elements contained in the west must here be added at least three others—the Galcha of the Hindu-Kush, the Hindu of the large towns, and the Brahui of Kelat; this last being distinct in speech, not only from all the others, but from all other known linguistic groups. The subjoined table comprises all the races in the region, classed according to their most probable ethnical affinities:—

<table>
<tr><td rowspan="14">ARYANS</td><td rowspan="6">Galcha Branch.</td><td>Wakhis</td><td rowspan="2">Hindu-Kush (northern slopes).</td></tr>
<tr><td>Badakhshis</td></tr>
<tr><td>Siah-Posh Kafirs</td><td rowspan="3">Hindu-Kush (southern slopes).</td></tr>
<tr><td>Safis</td></tr>
<tr><td>Chagnans</td></tr>
<tr><td>Kohistanis</td><td>Hills north of Kabul.</td></tr>
<tr><td rowspan="5">Iranic Branch.</td><td>Afghans</td><td>Kabul; Suliman Mountains, Kandahar, Helmand basin; Herat.</td></tr>
<tr><td>Tajiks</td><td>Herat; most towns and settled districts.</td></tr>
<tr><td>Baluchis</td><td>Baluchistan lowlands; Makran.</td></tr>
<tr><td>Sistanis</td><td>Lower Helmand; Hamun.</td></tr>
<tr><td>Kurds</td><td>Baluch Kohistan.</td></tr>
<tr><td rowspan="3">Indic Branch.</td><td>Hindkis</td><td>Most large towns.</td></tr>
<tr><td>Lassis</td><td>Prov. Las, So. Baluchistan.</td></tr>
<tr><td>Jats
Luris</td><td>Makran chiefly.</td></tr>
</table>

MONGOLO-TATARS	*Mongol Branch.*	Hazarahs . .	N. highlands between Bamian and Herat.
		Aimaks . .	
	Turki Branch.	Usbegs . .	Afghan Turkestan.
		Turkomans .	Herat, Maimana, and Andkhui.
		Kizil-Bashis .	Kabul chiefly.
	?	Brahuis . .	Mainly East Baluchistan highlands.

Of these various peoples four only possess a decided political or social preponderance in their respective areas —the Usbegs in Afghan Turkestan, the Afghans in Afghanistan, the Brahuis and Baluchis in Baluchistan. The Usbegs, here represented by the Kateghan family, differ in no material respect from their kinsmen of the adjoining khanates of Bokhara and Khiva, and will therefore be more conveniently dealt with in the chapter devoted to that region.

THE AMIR.

The Afghans, commonly known in India as Pathans, are all Sunnis in religion, but are socially still in the tribal state, a fact that is not sufficiently taken into account in estimating the political situation of the country. There is an Afghan *race*, one in physical type, speech, religion, and culture; but there is, strictly speaking, no Afghan *nation* possessing a distinct sense of its unity as a whole, with common political sentiments and aspirations. Such common sentiments are scarcely felt beyond the several great sections into which the race continues to be divided. The Duranis, the Ghiljis, the Waziris, the Afridis, the Mangals, Momands, Jusafzais, and others, form so many States, as it were, within the

State, each with its own separate interests, and each capable of combining rapidly for some common tribal object, but all incapable of acting in concert for a common national object, except under a strong ruler, such as Ahmad Khan, or the present Amir, Abdur-Rahman. When Ayub Khan of Herat moved in 1881 against Abdur-Rahman, the people of the intervening Kandahar district refused to pay revenue, not through any love of the Amir, but through indifference to the claims of the rivals for supreme authority. For both Abdur-Rahman and Ayub are chiefs rather of the Durani tribe than of an Afghan nation. And the Duranis themselves are regarded by other almost equally powerful sections merely as usurpers of the sovereignty, their usurpation dating only from the death of Nadir Shah in 1747, when their chief, Ahmad Khan, took advantage of the disorders in Persia to raise the royal standard in Kandahar. Ahmad endeavoured to give a national importance to his tribe, not only by changing its name from Abdali to Durani,[1] but also by associating with it some other sections, such as the Jusafzais, Momands, Afridis, Shinwaris, Orakzais, and Turkolanis, under the common designation of Bar-Duranis. But the attempt failed, these sections still retaining their tribal integrity, and refusing to be fused into a common Afghan nationality.

In the Durani tribe there are several sections, among which are included the two royal branches—the Suddozais and the Barakzais. It was to the Suddozais that Shah Suja belonged, who was placed on the throne by the British in 1839, after the first Afghan war. It is to the Barakzais that the equally well-known Dost

[1] Derived not, as is often stated, from the supposed custom of wearing a pearl (*durr*) in their right ears, but from the title of Durr-i-Dúrán ("Pearl of the Age"), adopted by Ahmad when he assumed the royal power.

Muhammad and his successors on the throne of Kabul belong.

The sections themselves are divided into a multiplicity of minor branches, septs, and clans,[1] offering still further obstacles to a general amalgamation of the whole race. And the race itself is everywhere opposed to other races speaking different languages, such as the Tajiks, Hindkis, Usbegs, Siah-Posh Kafirs, Hazaras, and Aimaks, which, although numerically inferior, possess greater national cohesion, and which in some cases have been able to maintain their independence.

But for these untoward circumstances the Afghan race, by its warlike spirit and remarkable physical vitality, might seem destined to subdue the surrounding peoples. But their national resources have hitherto for the most part been frittered away in internecine broils and struggles for the local independence of individual chiefs and tribes.

Yet it would be a mistake to suppose that the Afghans are absolutely incapable, under proper conditions, of turning from turbulent to peaceful ways. Although surrounded by hostile and marauding tribes, the Povindahs of the Suliman inner ranges have for ages occupied themselves with tillage, stock-breeding, and trade. These itinerant and warlike dealers, who claim descent from a goatherd of Ghor in the days of the Ghaznevid Mahmud, follow their industrious pursuits in the face of extraordinary difficulties. In the summer they pitch their tents on the plains near Kalat-i-Ghilzai and Ghazni, where they pay £60 to the Amir's government for grazing rights, and where the women and children remain under a sufficient guard, while the men are away trading at Samarkand, Bokhara, Herat, or Kabul. In the autumn

[1] Usually termed *zais* or *khels*, as in Barakzai, Abdur-Rahmanzai, Ali Khel, Utman Khel, etc.

they repair to the Indian plains through the Gomul route, fighting their hereditary foes, the Waziris, on the way, and encamping on the Derajat plains. From this point the men again disperse towards Multan, Lahore, Benares, retailing their raw silk, druggets, clothes, saddlery, horses, saffron, dried fruits, and other wares. In April the Povindahs reassemble for the return journey, and ascend the pass towards Kandahar and Ghazni.

BALUCH WOMAN.

Many other promising elements of future progress exist in the land, such as the Kakar and Tajik agriculturists, the Hindki traders, met with in every large town, and even the despised Kizil-Bashis of Kabul.

In Baluchistan the ruling race are not the Baluchis, but the Brahuis, who are moreover both the aboriginal and the most numerous element. Hence the term Baluchistan, unknown in the country itself, is altogether inappropriate, though it may now be too late to substitute the expression Brahuistan, as some geographers have proposed. The Brahuis, whose racial and linguistic affinities still remain an unsolved problem, are predominant in all the eastern highlands; the reigning Khan and most of the chiefs and nobles are of Brahui stock. But they are no longer independent rulers, having by recent treaties recognised the sovereignty of the British raj. The

Baluchis still dwell mainly in the lowlands, and form the rural population both in the direction of India and Persia. Both races are Muhammadans, the Brahuis like the Afghans being Sunnis, the Baluchis like their Persian kinsmen Shiahs for the most part. There can be little doubt that the Baluchis penetrated eastwards originally from Karman, and they are still predominant in the adjoining districts of Makran and Sistan. Bellew describes them all in two words—needy and hungry. They are true nomads, migrating, like many Afghan tribes, with their families and flocks from the uplands to the lowlands. But some few are settled in villages.

Distinct both from the Brahuis and Baluchis are the Luri, a sort of gipsies of Indian origin scattered in single families all over the country. They are generally met with as strolling minstrels, potters, tinkers, ropemakers, weavers of mats, and pedlars. They owe no lands, never cultivate the soil, and are regarded as outcasts by the rest of the people. Each troop has a "king," and Pottinger noticed their "marked affinity to the gipsies of Europe."

8. *Topography: Khulm—Mazar-i-Sherif—Balkh—Herat—Kandahar—Kabul.*

In Afghan Turkestan the chief places are—1. Tashkurgan (New Khulm), at the junction of the Bamian and Badakhshan routes, where the Khulm river emerges from the mountains. It is three miles in circumference, and its houses are built of clay or adobe. The inhabitants are chiefly Tajiks, Kabulis, and Hindkis, trading in livestock, cottons, leather ware, fruits, and melons. Four miles north of Tashkurgan are the ruins of Old Khulm, which was abandoned by a former Khan, because the water supply was liable to be cut off. 2. Mazar-i-Sherif,

50 miles west of Tashkurgan, capital of Afghan Turkestan and residence of the Governor-General. It is surrounded by well-cultivated fields and orchards, and in 1885, when Captain Talbot passed through it, the population (Usbegs, Afghans, and Tajiks) was rapidly increasing, although still inferior to that of Tashkurgan. Six miles farther west is the military cantonment of Takhtapul, the arsenal of which has been removed to Mazar. But Mazar is chiefly noted for its mosque, held in great veneration for a tomb supposed to be that of Ali; and for the shrine of Hazrat Shah, a famous Moslem "saint." 3. Baktra and Balkh, both now ruins on the Dehas or Balkh River, a few miles west of Mazar. Baktra, capital of the Græco-Baktrian monarchy, was one of the oldest cities in Central Asia, and its successor Balkh still bears the title of "Mother of Cities." It was the chief town in Afghan Turkestan till 1872, when a terrible outbreak of cholera caused the seat of government to be removed to Mazar, and in 1878 Balkh was an insignificant village, whose former greatness was attested only by numerous canals and miles of ruins. Here are buried the travellers Moorcroft and Guthrie. 4. Andkhui, on the verge of the desert due west of Balkh, a large but proverbially unhealthy place, of which the Persians say that with its salt water, its scorching sands, venomous flies and scorpions, "it is a real hell on earth." 5. Maimana, on a plain near the foot of the Koh-i-Baba, noted for its excellent horses and textiles woven of wool and camel's hair. Previous to 1874 Maimana was a very large place, with a population estimated at 60,000. But in that year it was besieged and nearly destroyed by the Afghans, who massacred 18,000 of its inhabitants. Since then it has somewhat revived, and must always enjoy a certain importance from its position at the junction of the routes from Herat and Kabul. Captain Talbot describes it as

two-thirds the size of Herat, with a large covered bazaar, but badly built with mean houses irregularly distributed over the space enclosed by the walls.

In Afghanistan the three cities of Herat, Kandahar, and Kabul stand out conspicuously as at once the chief centres of power and population, as well as the most important strategical points in the country. They occupy the three angles of a triangle, whose base crosses the northern scarp of the plateau, and whose apex lies nearly in the centre of the State. Thus Herat and Kabul at the west and east ends of the base respectively are separated by intervening impassable highlands occupied by the hostile and semi-independent Mongolo-Tatar Hazaras and Aimaks. Hence the route from one to the other is deflected southwards to the apex, where Kandahar thus occupies the key of the whole position. North of the scarp is the Turkoman country, now entirely absorbed in the recently organised Russian Trans-Caspian territory. From this direction the plateau can be approached in the east only by the difficult "Gate of Bamian," in the west by the easy Tajand and Murgh-ab valleys. Here, therefore, the importance of Herat becomes obvious. And this circumstance itself enhances the importance of Kandahar, which bars the direct and only route from Herat to India, and which lies on the flank of the not impossible route through Bamian and Kabul to India. It is satisfactory to know that under these circumstances the railway is already completed from the Indus to Quetta above the Bolan Pass, and thence through the Gwaja Pass to Chaman Fort on the main route for Kandahar.

The city of Herat lies in the well-watered valley of the Hari-rud, or Upper Tajand, which is extremely fertile, and capable of furnishing supplies for an army of occupation of 150,000 men. This, coupled with its lofty ramparts and fortifications, and its central position as the converging point of routes from the Caspian, Mashhad, Merv, Bokhara, and India through Kandahar, has invested it with a strategic importance which has earned for it the title of the "key of India." In Pottinger's time it was

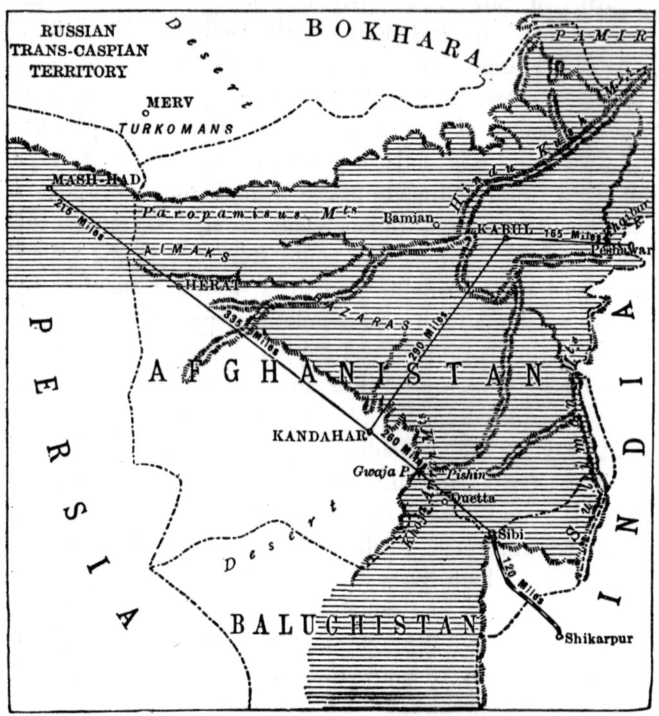

the great emporium of trade in Central Asia, and though its many vicissitudes of fortune and innumerable sieges have caused its population to fluctuate excessively, it still contained 50,000 inhabitants in 1878, a motley gathering of Afghans, Indians, Tatars, Turkomans, Jews, Tajiks, and others. Carpets of soft texture and brilliant colour

are here produced, and the district is noted for its excellent horses, wheat, water, and grapes, of which there are no less than seventeen varieties.

The road from Herat to Kandahar lies through the fertile but little cultivated Zamindawar country, peopled by the Duranis as far as the fort of Girishk, near the fords of the Helmand. Although its fortifications are slight, the strong position of Girishk on the main route and in the vicinity of supplies has at all times invested it with strategic importance. About midway between it and Kandahar is Khushk-i-Nakhud, mournfully signalised in 1880 by the defeat of General Burrows at the hands of Ayub Khan.

Kandahar, the chief city of the south, lies in a level cultivated plain about 7 miles wide, bounded by low hills between the Arghand-ab and Tarnak Rivers. It forms an irregular oblong of about 3 miles circuit, enclosed by a substantial baked-mud wall 27 feet high. Towards the north end is the citadel, shut off by a massive mud wall, and to the west the tomb of Ahmad Shah, within an octagonal structure surmounted by a golden dome. The population is variously estimated at from 50,000 to 80,000, of whom the Hindkis are the wealthiest class. During the British occupation they have always developed a profitable trade with Bombay through Shikarpur and Karachi, but at other times are subject to heavy imposts. They import British produce, such as silks, calicoes, muslins, chintzes, merinoes, woollen and broad cloths, cutlery, needles, paper, besides indigo, spice, sugar, drugs, and other Indian produce. Their exports consist of madder, asafœtida, wool, dried fruits, tobacco, raw silk, besides such Persian goods as carpets, copper utensils, arms, turquoises, gold and silver braid, horses and "yabus," or baggage ponies. Whenever the railway is completed to this place Kandahar must become the great

KANDAHAR.

emporium of British and Indian produce for Central Asia. Its chief manufactures are silks, felts, and rosaries of a soft crystallised silicate of magnesia. The melons, grapes, and other fruits of the district are abundant, and of excellent quality.

The old citadel, of which many ruins are still standing, is situated a few miles outside the walls of the present city. This citadel has been the scene of remarkable sieges and defences.

The strategical value of Kandahar is increased by the fact that it is the first place where an army advancing from Herat towards the Indus would naturally rest to recruit its strength. It also gives access to the Ghazni and Kabul road through the Tarnak valley. Its proximity to the desert on the south renders one at least of its flanks safe from being turned. As it is further accessible from Persia and India west and east, it has incessantly changed hands during the period of its history—Persians, Usbegs, Afghans, and in recent times the English, having more than once occupied and relinquished it.

On the great military and trade route between Kandahar and Kabul the chief stations are Kalat-i-Ghilzai and Ghazni, the former of which is a strong fort standing on a commanding plateau on the right bank of the Tarnak. The fortified town of Ghazni, about as far south-west of Kabul as Kalat-i-Ghilzai is north-east of Kandahar, lies on the left bank of the Ghazni river near a spur of the Gilkoh range, and 7730 feet above the sea. It is surrounded by a lofty stone and brick wall on the top of a mound, and contains a citadel erected on an abrupt knoll at its north end. Besides Afghans it is inhabited by Hazaras and a few Hindki traders, dealing chiefly in corn, fruits, madder, sheep's wool, and camel's hair cloth from the Hazara country. Ghazni is memorable for its brilliant capture by Sir John Keane in the first Afghan war. Three miles to the north-east are the ruins of the old city, destroyed in the twelfth century by the Prince of Ghor, who, however, spared the tomb of the renowned Mahmud of Ghazni. The entrance to this mausoleum, which is still preserved with careful veneration, was formerly closed by the famous sandal-wood gates brought by Mahmud from Somnath in Guzerat, but sent back to India in 1842.

Crossing the watershed north of Ghazni, between the Helmand and Indus basins, we enter the territory of Kabulistan, which takes its name from the ancient city of Kabul, the present capital of the State. Kabul lies 7 miles above the confluence of the Logar and Kabul vers, at the western extremity of a spacious plain in an igle formed by two converging ridges. It is now an pen town, though formerly encircled by brick and mud alls. There are no noteworthy public buildings, and its interest, apart from its being the seat of government for nearly a hundred years, arises chiefly from its position at the junction of routes from Turkestan, Herat, Ghazni, the Kuram valley, and the Panjab by way of Jelalabad. This happy situation has made it an emporium of Central Asiatic trade, notwithstanding the difficulty of the passes connecting it with the Oxus valley. To the south-east stands the Bala Hissar, or citadel, on a commanding knoll at the extremity of the spur overlooking the city. Kabul imports from India calicoes, indigo, spices, drugs, and all kinds of British goods; from Russia broadcloth, silks, velvets, gold and silver lace, paper, and hardware, mostly by the long and toilsome route through Bokhara and Bamian. The province yields wheat, barley, and fruits in considerable quantity and of good quality.

Of the alternative routes between Kabul and Jelalabad, the northern and more frequented leads over the Lattaband Pass, while the southern follows the narrow

Khurd Kabul defile, where about 3000 men, women, and children perished on the occasion of Elphinstone's ill-fated retreat in January 1842. East of the Karkacha hills lies the equally ill-omened Jagdalak Pass, where the massacre of the retreating troops was continued, a few officers and men alone escaping to Gandamak. At this place, where the last treaty with the British was signed in 1880, a rapid descent leads down to the well-watered plain of Jelalabad, about midway between Kabul and Peshawar. This town forms an irregular quadrilateral surrounded by walls at the junction of roads from India, Kabul, and over the Hindu-Kush from Yarkand. It is noted for the brave and successful defence by Sir Robert Sale in 1841-42 against overwhelming numbers of Afghans. Hither it was that Dr. Bryden, sole survivor of the Kabul disaster, found his way in January 1842.

Between Jelalabad and Peshawar are the towns of Lalpura and the fort of Ali-Musjid. The latter was stormed by the British at the beginning of the war with the late Amir Shir Ali.

Kelat, the capital of Baluchistan, is almost the only town in the country. It is a small fortified place in the centre of the province of Kelat, 6000 feet above the sea, badly built, and presenting an appearance of extreme squalor and decay. Its 12,000 inhabitants include representatives of nearly all the surrounding races—Brahuis, Baluchis, Afghans, Tajiks (here called Dehwars), Jats, and Hindus. Here resides the Khan, surrounded by a bodyguard of troopers, described by Bellew as tatterdemalions.

A far more important place strategically is Quetta (Shal), the capital of a district near the head of the Bolan Pass and close to the Pishin valley, from which it is separated by Mount Takatu, 10,504 feet high. By treaty with the Khan, Quetta has become an advanced

British military station at a vital point on the southernmost route from India to Afghanistan, and about midway

KELAT.

between Shikarpur and Kandahar. Its occupation secures the Pishin valley, holds all the unruly Mari, Bugti, and other border tribes in check, keeps open the roads of the Khojak and Gwaja passes over the Khoja Amran range, and thus facilitates a rapid advance on Kandahar. The valley of Quetta lying 5500 feet above the sea, and enclosed by mountain ranges which rise from 5000 to 6000 feet higher, is an extremely romantic spot, surrounded by rocky mountains.

In Makran, Kej and the other so-called "towns" are mere clusters of hamlets, or insignificant fishing villages on the coast.

9. *Highways of Communication: Passes.*

One of the chief results of the recent hostilities in Afghanistan was the revelation that not two or three, as had been supposed, but at least a score of practicable routes give access from the plains of India to the Iranian plateau. From above Peshawar nearly to Karachi the intervening highlands are almost everywhere pierced by rivers and mountain torrents flowing down to the Indus, many of which run through gorges and ravines affording good passes to the interior. "What we have learnt chiefly in connection with them is this—that most excellent roads are easily constructed along even the worst of them" (Captain Holdich).

Still more surprising was it to find that the Hindu-Kush itself is crossed throughout its whole length by mountain tracks more or less practicable during the summer months from the Khawak westwards to the Irak Pass leading to Bamian. The Paghman range also, parallel and equal to it in height, is crossed by "durras" or paths leading from nearly all the large villages north of Kabul over the intervening Ghorband valley and Hindu-Kush down to Afghan Turkestan.

At the western extremity of the Paropamisus the Tajand and Murgh-ab valleys also afford ready access from what is now Russian Turkmenia to the Herat district. Two comparatively easy routes lead southwards to this city, one from Sarakhs through Zulfikar and over the Borkhut hills (900 feet), the other from Merv through Penjdeh and up the Kushk valley.

From Peshawar, north-western terminus of the Indian railway system, the great historic route to Kabul enters the Afridi hills near Ali-Musjid, thence following the

THE KHAIBAR PASS AT ALI-MUSJID.

Khaibar river over the Khurd Khaibar Pass (3370 feet) north-westwards and south of the Kabul river to Jelalabad. Here, crossing the Nangnahar plains, it ascends through the narrow Jagdalak defile to the Karkacha hills and the dangerous Khurd Kabul Pass, with an alternative northern route over the Lataband Pass and the hills near Butkhak down to Kabul.

Farther down, the scarcely less important Kuram route to the capital runs by Thal and the Kuram river to near Fort Kuram, north-westwards, over the Paiwar range and Pass to Ali-Khel at junction of Rivers Karaia and Hazardarakht. Thence it follows the latter river over the Surkai Kotal between the Kuram and Kabul basins to the Shutargardan Pass (10,800 feet), and so on by Dobandi, Khushi, and the Logar valley, north to Kabul.

South of these two the ancient Gomul route ascends from the Derajat plains over the Kotal-i-Sarwandi water-parting to the Gomul or Gwalari Pass and thence to Ghazni.

The southernmost route to Afghanistan follows the new line of railway, now completed, from Sakkar on the Indus across the plains and Kachi desert to Quetta, near Pishin, at the head of the Bolan Pass. There are also alternative routes through the Bolan to Quetta, and through the Nari river valley to the Lora river and Pishin valley, and thence over the Khoja Amran range, by the Khojak and Gwaja Passes, and across the Dori and other streams to Kandahar. The latter, though the longer, is the easier of the two, and is followed by the railway through Quetta and over the Gwaja Pass towards Kandahar.

The usual routes from Herat to Kabul are the northern, round by Maimana and Bamian, and the southern round by Kandahar and Ghazni. But the direct route across the Aimak and Hazara highlands up the Hari-rud and east of Obeh is also occasionally used by the natives, and has been frequently traversed in eight days on horseback. The southern crosses the Zamindawar Durani domain by Farah, Girishk, and Khushk-i-Nakhud, to Kandahar. Here it follows the Tarnak valley to Kalat-i-Ghilzai, Ghazni, and over the Sher-i-Dahar Pass between the Helmand and Kabul basins, and down the Shiniz river valley to its junction with the Logar, where it bifurcates through the Wardak and Logar valleys to Kabul.

Beyond those from the Indus through the Bolan and Mula Passes to Quetta and Kelat, no regular routes are yet laid down in Baluchistan; but the longitudinal valleys running east and west parallel with the coast are often traversed, and give easy access from the eastern highlands to Persia. In 1891 an expedition was made to Baluchistan by Sir R. Sandeman for the purpose of settling the Panjgur district and opening the old trade routes between Makran and Persian Sistan. The chief difficulties appear to have been the tribal jealousies, which have been so far allayed that caravans now pass from Quetta to Sistan for Khorasan and other parts of Persia. The British Consul-General at Mashhad speaks favourably of this route, by which a camel-load of goods may reach Mashhad from the Chaman terminus of the Quetta railway in about forty days, whereas from Bandar-Abbas it would take from seventy-five to ninety days. When all arrangements are complete, it is expected that this Sistan route will be preferred to all others, not even excepting that by Kandahar and Herat. The Mashhad traders hope that by this means a great impetus may be given to British trade between Khorasan and India, which would thus be able to compete on more equal terms with Russia.

10. *Afghan and Baluchi Administration.*

During the four decades following the death of the Amir, Dost Muhammad, and the restoration of Shah Shuja by the British forces in 1838, Afghanistan was so torn by tribal and dynastic feuds, that the civil administration of the country was mainly reduced to the enforced collection of tribute and revenue. But since the accession of the present Amir, Abdur-Rahman, a great improvement has taken place in almost every branch of the public service. In theory the government is monarchical under a hereditary Amir ("Prince"), whose authority, however, largely depends on his personal character and power to enforce his mandates. His dominions comprise the four administrative provinces of Kabul, Kandahar, Herat, and Afghan Turkestan, which with the district of Badakhshan and dependencies are each under a *hakim* (formerly *naib*), responsible to the Amir for the revenue and maintenance of public order. Under the hakim ("governor") the great tribal chiefs retain almost sovereign jurisdiction over the clansmen, dispensing justice somewhat after a feudal fashion, and this loose system of administration has hitherto given free scope to spoliation, embezzlement, and exactions of all kinds, especially in the districts far removed from the seat of government.

But during the last decade the Amir's authority has been greatly strengthened, and his vigorous and enlightened rule has been marked by the introduction of many salutary reforms and improvements carried out by the aid of several experienced Englishmen invited to Kabul for the purpose. Under the superintendence of Mr. Salter Payne, extensive workshops have been erected at Kabul, at the cost of several hundred thousand pounds, and these workshops are provided with plant and machinery of the most varied description. Here are appliances for sawing and planing timber, for minting silver and copper coins, for producing guns, rifles, cartridges, gunpowder, and other war material, soap, candles, furniture, jewelry—in fact, almost everything required to meet the daily wants of the Amir's subjects. "The effect of all this upon the future of Afghanistan cannot be overrated. The Amir never expects any pecuniary profit from these works. His one object is to civilise and refine his fanatical masses. The Afghanistan of thirteen years ago, when Abdur-Rahman succeeded to the throne, is as totally different from the Afghanistan of to-day as the Afghanistan of to-day will be different from the Afghanistan of ten years hence, if the Amir is spared to rule over it" (Salter Payne).

Some improvement has also taken place in the general administration of Baluchistan, especially since the accession of Mir Muhammad, Khan of Khelat, in 1893. His father Khudadad, who had succeeded Nasir Khan II., of the Brahui dynasty, in 1857, had been guilty of so many excesses, culminating with the murder of his Minister and others, that the British authorities were compelled to intervene, and call upon the discredited ruler to abdicate, measures being at the same time taken to prevent a recurrence of such abuses. At present the political power of the Khan of Khelat is little more than nominal. On the one hand he is subsidised, like the Amir of Kabul, by the Indian Government, whose supremacy is acknowledged by several treaties; on the other his jurisdiction is limited by the great tribal chiefs, so that the greater part of Baluchistan forms a confederacy of chiefs under the suzerainty of the Khan of Khelat. The rest of the country is now practically British territory, the districts of Quetta and the Bolan being administered since 1876 by British officials on behalf of the Khan, while the so-called assigned districts (Pishin, Shorarud, Kachk, Kawas, Harnai, Sibi, and Thal Chotiali), the territories of the Mari and Bugti tribes, the district of Khetran, and the tract between the Zhob valley and the Gomul Pass, are now under direct British rule. There is no standing army, beyond a small corps maintained by the Khan; all the recent fortifications erected in connection with the general scheme of imperial defence lie within the British territory; the forts scattered over the Baluch confederacy are of an obsolete type; in a word, Baluchistan as an independent political factor has ceased to exist.

11. *Statistics.*

AREAS AND POPULATIONS OF EAST IRANIA AND AFGHAN TURKESTAN.

	Countries.	Area in sq. miles.	Population.
Aralo-Caspian Basin	Wakhan	3,000	3,000
	Badakhshan	8,500	158,000
	Kunduz	11,000	400,000
	Balkh	15,000	64,000
	Andkhui	6,000	60,000
	Shabirkhan	2,500	270,000 (Shabirkhan, Akcha, Sar-i-pul, Maimana)
	Akcha	3,300	
	Sar-i-pul	2,000	
	Maimana	15,000	
	Gurzivan	2,500	5,000 (Gurzivan, Darzab)
	Darzab	1,500	
	Total Afghan Turkestan	70,300	960,000[1]
Hindu-Kush (Southern Slopes)	Kafiristan	7,000	5,000,000 (Kafiristan, Afghanistan Proper)
	Afghanistan Proper	217,000	
	British Baluchistan / Quetta District	176,000	145,000 / 27,000
	Grand total	470,300	6,132,000

APPROXIMATE CLASSIFICATION BY RACES AND RELIGIONS.

Religion	Race	Stock	Number
Fire-Worshippers	Wakhis	Galcha Stock.	3,000
Pagans	Siah-Posh Kafirs	Galcha Stock.	250,000
Muhammadans	Safis and Chagnans	Galcha Stock.	100,000 (Safis and Chagnans, Kohistanis)
Muhammadans	Kohistanis	Galcha Stock.	
Muhammadans	Badakhshis	Galcha Stock.	160,000
Muhammadans	Afghans	Iranic Stock.	3,000,000
Muhammadans	Tajiks (Dehwars)	Iranic Stock.	800,000
Muhammadans	Baluchis	Iranic Stock.	200,000
Muhammadans	Sistanis	Iranic Stock.	50,000 (Sistanis, Kurds)
Muhammadans	Kurds	Iranic Stock.	
Muhammadans	Hindkis	Indic Stock.	500,000
Pagans	Lassis	Indic Stock.	50,000 (Lassis, Luris, Jats)
Pagans	Luris	Indic Stock.	
Pagans	Jats	Indic Stock.	
Muhammadans	Hazarahs	Mongol Stock.	500,000 (Hazarahs, Aimaks)
Muhammadans	Aimaks	Mongol Stock.	
Muhammadans	Usbegs	Tatar Stock.	400,000
Muhammadans	Turkomans	Tatar Stock.	50,000
Muhammadans	Kizil-Bashis	Tatar Stock.	150,000
Muhammadans	Brahuis	Mongoloids?	350,000

Pagans and Fire-Worshippers	303,000
Muhammadans, mostly Sunnis	6,260,000
Total	6,563,000

[1] Grodekov's estimate.

Chief Towns.

	Pop.		Pop.
Kabul	75,000	Khulm	10,000
Kandahar	60,000	Ghazni	8,000
Herat	50,000	Kelat	6,000
Mazar-i-Sherif	25,000	Sar-i-pul	3,000
Andkhui	15,000	Kunduz	3,000
Shabirkhan	12,000	Faizabad	2,500
Maimana	?	Jelalabad	2,000

Trade Returns (Rough Estimates) for Afghanistan.

Imports from India (1893), £68,000; from Russian Central Asia and Bokhara, £394,000; total, £462,000.

Exports to India, £34,000; to Russian Central Asia and Bokhara, £398,000; total, £432,000.

Revenue, including Subvention from Indian Government, about £1,000,000.

Army (1891): About 50,000 available troops of all arms. The artillery comprises one elephant, six mule, and two field batteries.

War materials are now manufactured at the Kabul arsenal under the superintendence of Englishmen in the Amir's service.

Distances.

	Miles.		Miles.
Kabul to Herat	600	Kabul to Peshawar	165
Herat to Mashhad	200	Sukkur to Sibi, by rail	140
Kandahar to Sukkur	410	Sibi by Chaman railway to Kandahar	265
Kunduz to Balkh	105	Quetta to Shikarpur, by rail	182
Kandahar to Herat	335	Kohat to Ghazni	264
Kandahar to Kabul	290	Dera Ismail Khan to Ghazni, *via* Gomul	250
Balkh to Bokhara	260	Dera Ismail Khan to Kandahar, *via* Sakhi Sarwar	424
Balkh to Andkhui	100		
Kabul to Ghazni	60		
Kabul to Jelalabad	75		

CHAPTER II

THE INDIAN EMPIRE

1. *Boundaries—Extent—Area.*

WITH a few comparatively unimportant exceptions, British India forms a vast geographical and political system, which, if it cannot everywhere yet boast of strictly "scientific frontiers," enjoys none the less at many points the advantage of the grandest natural boundaries of any region on the globe. For it consists mainly of a vast peninsular mass shut off from the Asiatic mainland by the lofty Brahui and Suliman ranges towards the north-west, and on the north by the still loftier Hindu-Kush and Himalaya, while it is elsewhere washed by the Arabian Sea and Bay of Bengal. In outline it presents the form of a somewhat irregular equilateral triangle with its base rooted in the Himalayas, whence it tapers across 28 degrees of latitude southwards to its apex in the Indian Ocean. Of this triangle the three sides fall about 100 miles short of 2000 miles each, the distance between the extreme frontiers of the Panjab and Assam west and east, and from these points to Cape Comorin, at the apex, being as nearly as possible 1900 miles. The coast-line, although broken on the east side only by the small Chilka lagoon near Jaganath, and on the west by the more important inlets of the Katch and Cambay Gulf, has a total length of nearly 9000 miles. The land frontier is conterminous for nearly 6000 miles with the surrounding States of Persia, Afghanistan, Russia, China, and Siam. The north-west frontier, as provisionally laid down by the Anglo-Russian Commission of 1895, coincides west of Lake Victoria on the Great Pamir with the Pamir River to its junction with the Wakhan-su at Kalah Panjah, converging point of British India, Afghanistan, and Asiatic Russia. Eastwards the line follows the crest of the mountain range a little south of Lake Victoria as far as the Benderky and Orta-Bell Passes, and thence northwards in the direction of Kizil-Rabat on the Ak-su River, and eastwards to the Chinese frontier. But if Kizil-Rabat is found to be north of the latitude of the lake, then the line is to run to the nearest point on the Ak-su south of that latitude, and to be prolonged thence eastwards to the Chinese frontier. It was agreed that the territory between the Hindu-Kush and a line running to the Chinese frontier shall belong to Afghanistan, while all the passes over the Hindu-Kush itself remain definitely under complete British control. Within these borders there is an area estimated at 1,560,000 square miles, with a population, according to the census of 1891, of 287,000,000, and including Ceylon of over 290,000,000, or about one-fifth of mankind.

2. *Relief of the Land—Geological Retrospect.*

A good idea of the general relief of the land will be had by supposing it to subside about 500 or 600 feet below its present level. Such a slight subsidence, altogether imperceptible in the northern highlands, would have the effect of flooding all the plains at their base and converting the rest of the triangular mass into an island, shorn of a narrow strip along the east coast, but elsewhere almost intact. In other words, the Himalayas in the north would continue to present much the same outlines that they now do. The southern region of the Deccan also, that is, the peninsula proper, forming an elevated plateau 2000 to 3000 feet above the sea, fringed on the north by the Vindhya range, and on the west by the Western Ghats, would be materially affected only on the east side, where a strip of low-lying and partly alluvial coast-lands intervenes between the low and interrupted scarp of the east coast. But the space occupied by the Indus and Ganges valleys, known emphatically as the "Plains of India," and lying mainly between the Himalayas and the northern scarp of the Deccan, would disappear altogether, their place being occupied by a broad strait or channel connecting the Arabian Sea with the Bay of Bengal.

That such was the actual condition of things, even in comparatively recent times, has till lately been the generally accepted conclusion of geologists, who held that the Indus and Ganges valleys are old marine beds filled up by the alluvia brought down by those great rivers and their numerous tributaries from the Himalayas and Vindhyas. And although this view is now shown to be absolutely erroneous by Blandford, Oldham, and others, its mere expression serves to give us a clear conception of the physical geography of India. For we thus see that this region consists of three distinct geological areas: the Deccan, the Indo-Gangetic alluvial plains, and the extra-peninsular Himalayan uplands.

Pre-Tertiary Times: Indo-Africa.

Thanks to the labours of the *Indian Geological Survey*, the results of which have been lucidly summarised by

Mr. R. D. Oldham,[1] it is now possible to trace with some confidence the physical history of this region back to the close of the palæozoic era. At that remote epoch the peninsular area proper, including the Aravalli Hills, the Vindhya range, and the extension of the so-called "Eastern Ghats" across the present Ganges-Brahmaputra delta to the Assam and East Himalayan Mountains, was already, and has since for the most part remained, dry land. But the extra-peninsular area, that is to say, most of the Himalayan region, together with the western (Suleiman) and eastern (Arakan-Yoma) mountains, was still a marine bed, which appears to have been alternately upheaved and submerged several times during the secondary era, and to have assumed its present outlines not earlier than middle or even late tertiary times. The connection between the peninsula (Deccan) and the Asiatic mainland was in fact very slight in pre-tertiary ages, whereas its connection with the African mainland by continuous land across the present Indian Ocean down to the lower chalk period is now placed beyond doubt. Thus the geological record shows that what may be called the present *Indo-Asiatic Continent* was preceded by an *Indo-African Continent, Gondwanaland*, as it has been called by Suess in contradistinction to the supposed continent of *Lemuria*, postulated by Sclater on other grounds, and for a later period.

As the recently upheaved Himalayas form the great orographic feature of Indo-Asia, so the far older but now degraded Aravalli range, with its eastern extension, the red sandstones of the Vindhyan system, was the most conspicuous feature of Gondwanaland. From their colour, and the remarkably uniform character maintained over a vast area passing under the later traps of the Deccan,

[1] "The Evolution of Indian Geography," in *Geographical Journal*, March 1894.

these sandstones appear to have been deposited by rivers in freshwater lakes and lagoons. They are the waste of

ROCK PILLARS, SPITI VALLEY.

the Aravalli Mountains, just as the later alluvium of the Indo-Gangetic plain are the waste of the Himalayas. "As the deposits of the Gangetic plain can be shown to be derived from the Himalayas, and to have been formed during the elevation of those mountains, it is difficult to avoid the conclusion that the upper Vindhyan sandstones were similarly formed of the *débris* washed down from the Aravalli range, and that the period during which they were being deposited represents that of the elevation of the Aravalli Mountains and of their maximum development" (Oldham, *loc cit.*).

To the same Indo-African continent, and most probably to the same age as the Aravalli system, belongs the series of hills, such as the Nallamalai and Yellakonda groups, extending north from Madras roughly parallel with the east coast, with which they are also geographically connected. The present coast-line thus appears to have undergone relatively little modification of form and trend since Permian or late palæozoic times. Hence the remarkable absence of secondary or tertiary marine deposits in the interior of the Deccan, which shows no indication of subsidence during the upper palæozoic epoch.

The great earth movement to which were due the Aravalli and Vindhyan heights, was followed by a long period of quiescence, during which was accumulated the vast Gondwana system of fluviatile deposits, showing evidence of glaciation in Rajputana and elsewhere. Throughout this period the connection was still maintained north-eastwards with the Assam and East Himalayan uplands, and also south-westwards through the Laccadives, the Mascarenhas, and South Madagascar with Africa, as plainly shown by the allied floras and faunas of the lower Gondwana (Indian) and Karoo (Cape) series. The plants of the Indian and South African coal-measures are all absolutely identical, and the remarkable *Dicynodon* and other allied forms of fossil reptiles are equally characteristic of both regions.

The late Jurassic and early cretaceous (neocomian)

marine fossils on the South African seaboard, on the east coast of India, and in the Khasi Hills (Assam) are also either identical or closely allied. But those of the same age in Western India are quite distinct, and belong to a different marine zoological province, which extended along Western Asia and North Africa into Europe, and down the east coast of Africa as far south as North Madagascar. The present Indian Ocean was thus evidently decomposed into two distinct basins by the Indo-African continent of the secondary era.

In the northern basin the marine waters still flowed over the whole of the present Himalayan region, except perhaps the eastern section, which, so far as known, shows no trace of marine deposits, and which probably formed continuous land with Indo-Africa throughout the secondary and late palæozoic eras. On the other hand, the extensive series of marine sediments in Burma, the north-western Himalayas, and the Iranian plateau beyond the Indus, show that in pre-tertiary times all these regions were still mostly, though not continuously, under water. "Unconformable breaks show that there were alternations of land and sea; but there are no extensive sub-aërial formations, and there is no evidence of any considerable disturbance or compression of the rocks. There were periods of placid accumulation of sediments, interrupted by times when they were quietly raised above sea-level and exposed to denudation; but there are not at present sufficient data to allow of our attempting any detailed restoration of the geography until the close of the Jurassic period" (Oldham, *ib.*).

Tertiary and Later Times: Indo-Asia.

The long period of quiescence, which had prevailed throughout most of the secondary era, was followed in the peninsular area by the greatest outburst of igneous energy of which there is any distinct record on the surface of the globe. The eruptions, which have swallowed up much of the old red sandstones of the Vindhyan system, and covered a space of some 200,000 square miles in the Deccan with tuffs and lava sheets several thousand feet thick, were probably associated, or at least coincident, with the great earth-movement in the extra-peninsular area which began before the close of the secondary era, and which brought about the transition from the Indo-African to the Indo-Asiatic continent. This movement thus witnessed the three great events which, so to say, complete the physiographical history of the Indian region taken as a whole, that is to say, the submergence of the land stretching south-westwards from the present west coast, the union of the northern and southern marine basins in the present Indian Ocean, and the upheaval of the western and central Himalayas, of the Suleiman range in the extreme west, and of the Arakan Yoma in the extreme east, with its southern extension through the Andeman and Nicobar groups towards Malaysia. But so far from being in the nature of a cataclysm or sudden convulsion, the movement would appear to have been mainly an extremely slow process extending over a great part of the tertiary era, and in fact still continuing. Thus the Himalayas, rising almost imperceptibly during eocene and miocene times, had attained an elevation of probably not more than 20,000 feet in the pliocene period, since when they have increased their altitude by only about 9000 feet. A clear proof of their comparatively recent origin is afforded by the presence at heights of 18,000 or 20,000 feet of nummulitic limestones, which were deposited on the marine bed not later than the middle and upper eocene age, consequently well within the tertiary era. In the eastern and western mountains also, rocks that had no existence before that era are now found upheaved thousands of feet above sea-level.

Still more modern are the Siwalik foot-hills of the Himalayas, whose pliocene fossils show them to be upper tertiary formations. Originally fluviatile deposits similar to those now being formed on the neighbouring plains, they became cut off from the lowlands and raised to their present height by quite recent disturbances.

The lowlands themselves, that is to say, the great Indo-Gangetic plains, are in the nature of a depression caused by the same earth-movement to which the Himalayas owe their existence. As the land rose in the north above the marine waters, it sank in the south from the peninsular area down nearly to sea-level, above which it has since been raised several hundred feet by the waste of the Himalayas themselves. Hence it is that this depression shows no trace of marine deposits from the meridian of Delhi eastwards to Assam, and in fact may well have originally formed part of the old peninsular area,—its northern seaboard, as it were,—with escarpments towards the great ocean, which in palæozoic and mesozoic times nearly separated the Asiatic from the Indo-African world.

So long as the connection was maintained between the East Coast range and the Assam Hills, the Indo-Gangetic plain could have but one seaward outlet for its running waters, that of the Indus, which has maintained itself to the present day. But when a breach was made, either by erosion or subsidence, in the eastern barrier between the Rajmahal and Assam Hills, much of the drainage was diverted from the Indus and carried through the Ganges to the Bay of Bengal. This diversion "must have been a gradual process, whose final stage, the permanent diversion of the Jumna into the Ganges, may even have taken place within the historic period. Before this the waters of the Jumna must have flowed westwards; then it may have wandered and flowed alternately into the Ganges and Indus, or into that dry river channel which can still be traced through the desert of Western Rajputana. In its latest stage it probably, like the Cassiquiare in South America at the present day, divided its waters between the eastern and the western drainage; but now no further change can take place, for the river has cut its channel deep below the general level of the plain, and must perforce remain a tributary of the Ganges" (*ib.*).

Possibly owing to their recent origin and extremely slow upheaval, the Himalayas, despite their great elevation, do not form a divide between the Indian and Tibetan hydrographic systems. The Indo-Gangetic plain receives all the drainage, not only of the southern but also of the northern slopes, through the Indus, Ganges, and San-po (Brahmaputra), whose head-waters all rise on the north side of the range and pierce it at various points on their seaward course. The explanation in this, as in other cases where similar phenomena are presented, may be that as the land rose the rivers were able to keep open their channels by erosion, the two processes going on simultaneously at about the same rate of progress. Thus the equilibrium was maintained throughout tertiary times and down to the present day without any great disturbance of the local hydrography.

In the peninsular area the subsidence of the land connection with Africa had the result of giving the Deccan a continuous western seaboard, perhaps originally coincident, or nearly so, with the line of the Western Ghats. This chain, now little more than the seaward scarp of a plateau, may thus be regarded as forming in secondary times, not a coast range, but a low inland water-parting between two great hydrographic systems, one of which disappeared in the Indian Ocean during the tertiary era. The other still persists, and comprises all the great rivers of the Deccan except the Narbada and Tapti, which alone have a westerly course to the Indian Ocean, and apparently to that section of it which in Jurassic and cretaceous times lay to the north of Indo-Africa. All the other large peninsular streams flow across the Deccan eastwards to the Bay of Bengal, having their rise, as might be expected from the geological history of the land, on the slopes of the Western Ghats, near the present Indian Ocean, that is, on the east side of the old Indo-African water-parting, upheaved in eocene or miocene times to its present elevation. Towards the south of the peninsula the Palghat depression between the Western Ghats and the Travancore Hills, if not originally a marine inlet, may perhaps represent one of these ancient water-courses. But the old physical features of the district are so obliterated that it is no longer possible to determine its original trend, which may have been either westwards or eastwards, at a time when the land extended to and far beyond the neighbouring Laccadive Islands.

With regard to the "Eastern Ghats," their non-existence has been made so evident by the geological survey, that the expression should henceforth be banished from the map of India. "They are a figment of the imagination, the name belonging to the roads that lead to the Mysore plateau having been loosely applied to sundry groups of hills that have no connection with the Mysore plateau, nor with each other. One of these groups, it has been suggested by Mr. Oldham, may be connected with the history of the eastern coast from Madras to Masulipatam. In this case, however, it will be seen, by looking at the geological map, that the parallelism is not very well marked, and there is no range at all having the same relation to the south coast of Madras."[1]

The Himalayan Orographic System.

From the Great Pamir focus of the continental highland systems, the Himalayas seem to break away south-eastwards in three main parallel lines—the Karakorum and Kailas or Gangri ranges, enclosing between them the valley of the Shayok, and the Himalayas proper, enclosing with the Gangri the Upper Indus valley. The Karakorum or northernmost range is known as the Tsungling, or Mustagh ("Ice Mountains"), to the natives, who reserve the term Karakorum to the pass of that name. Beginning at the knot of Pusht-Khar in 74° 30′ E. long., it forms an eastern continuation of the Hindu-Kush, sweeping round the northern frontier of Kashmir, and stretching thence in a south-easterly direction to the neighbourhood of the sources of the Indus in Tibet. Of its eastern continuation beyond the Chang-Chenmo Pass nothing definite is known, and it is still uncertain whether it forms a connection with the Kailas range about the sources of the Indus and San-po, or merges gradually with the Tibetan plateau. The highest elevations occur in the section between the Karakorum Pass and the Gilgit valley, where the Dapsang (28,000 feet), and the peak long marked K[2] on the Indian Survey maps, and now named Godwin-Austen from its first explorer (28,278), are, next to Mount Everest, the highest peaks on the globe. The northern extremity is broken by long transverse valleys, while the southern presents much more abrupt escarpments towards the Indus valley.

The general direction of the Mustagh from north-west to south-east is maintained at a mean elevation of

[1] K. T. Blandford, *Geograph. Jour.*, March 1894, p. 193.

18,000 to 19,000 feet for some distance beyond the Karakorum Pass, after which it trends southwards, and again rises to imposing heights along the southern edge of the Tibetan plateau. The snow-line seems to rise on the north side to 18,000, on the south to 18,600 feet, and the Karakorum Pass leading from the Shayok valley to Yarkand is no less than 18,200 feet above sea-level. The Karakorum, rather than the more northern Kuen-lun, forms the true water-parting between the inland Asiatic and southern drainage. All the streams flowing from its southern slopes make their way through the Indus to the Indian Ocean, while those rising on its north side belong to the closed basins of Tibet or Eastern Turkestan.

In 1887 Captain Younghusband wound up his long ramblings of 7000 miles around and across Central Asia by passing from Kashgaria to India by the shortest but far most difficult route, which leads over the Mustagh Range by two passes of that name (the Old and New), neither of which had previously been crossed either from the north or from the south by any European traveller. Striking south from Yarkand, this explorer reached Dora on the Upper Yarkand-darya by the Tisnaf river valley and the comparatively easy Tupa Dawan (10,400 feet) and Chiragh Saldi (15,000 or 16,000) Passes. Beyond Dora the track led over the Aghil Dawan (16,000 to 17,000) Pass down to the Shaksgam, a hitherto unknown head-stream of the Yarkand, which is fed by the snows of Ghusherbrum and Godwin-Austen, and which, after its junction with the Sarpo Leggo from the Mustagh glaciers, joins the main stream below Dora. From the heights above the Shaksgam valley a superb view was obtained of the Mustagh-Karakorum Mountains, whose aspect, seen from this point, "is extremely bold and rugged, as they rise in a succession of needle peaks like hundreds of Matterhorns collected together; but the Matterhorn, Mont Blanc, and all the Swiss mountains would have been several hundred feet below me, while these mountains rose up in solemn grandeur thousands of feet above me. Not a living thing was seen, and not a sound was heard; all was snow and ice and rocky precipices" (*Geo. Proc.*, 1888, p. 506).

Beyond the Shaksgam the route ascends the Sarpo Leggo torrent right up to the foot of the Mustagh glacier, where it branches to the east over the old and to the west over the new Mustagh Pass, both attaining an extreme altitude of more than 19,000 feet, and both so difficult that they may be regarded as impracticable except by a *tour de force.* Owing to the accumulations of ice the old had not been used by any native for thirty or forty years, and even the new had been abandoned for the last ten years. Captain Younghusband selected the former, by which, after strenuous efforts and extreme risk, he reached the Baltoro glacier on the southern slope of the range, and made his way thence through the village of Askoli in Baltistan to Skardo on the Upper Indus above the Vale of Kashmir. As he looked back on the pass from the Baltoro snowfields, "it seemed utterly impossible that any man could have got down such a place."

Noteworthy in this alpine region are the numerous glaciers, the largest of which is the Baltoro, 33 miles long, and flanked on either side by two giant peaks over 27,000 feet high. Yet, vast as they are, these glaciers are mere remnants of the enormous ice and snow fields, which formerly covered the whole region of the Western Himalayas. These highlands are also exposed to sudden floodings, avalanches, and landslips, often causing widespread ruin in the upland valleys.

Much light was thrown on the general aspect of the

Mustagh highlands by the memorable expedition of Mr. W. M. Conway, who in 1892 explored some of the highest crests, passes, and glaciers of this alpine region, crossing the greatest glacier pass that exists anywhere in the temperate zone, and reaching the greatest height yet attained by any one on the surface of the globe. Starting from Srinagar, capital of Kashmir, on 15th April, Mr. Conway first followed the new Gilgit road over the Tragbal and Burzil passes, across the main Himalayan range down to the Indus, which river was crossed at Bunji, and then skirted along its right bank to the Gilgit valley. This valley is taken as a typical example of the action of mud avalanches, whose influence in modifying the surface of the land Mr. Conway thinks has not yet received due attention by geologists and geographers. The lower parts are buried beneath accumulated masses of *débris*, round and angular stones embedded in mud to depths of probably from 500 to 1000 feet and more, the Gilgit river flowing in a gorge or cañon, which has not been so much cut through as built up by these accumulations. "If the valley were filled up in this fashion to a depth of 2000 or 3000 feet, it would resemble the Pamirs, and all the deeply filled valleys which are characteristic of the Central Asian plateau from the middle of Tibet on the east to the upper region of the Oxus on the west. Mud avalanches, I maintain, have done all this work of filling up the valleys, and done it too with great rapidity" (*Geographical Journal*, Oct. 1893, p. 291)

After surveying the wooded Bagrot valley sloping southwards from the Rakipushi heights, the party struck north into the rock-bounded Hunza-Nagyr valley, whose turbulent chiefs were brought under direct British control in December 1891. The little Hunza and Nagyr principalities, described in Mr. E. F. Knight's *Where*

Three Empires Meet (1893), lie on the north and south sides respectively of the Hunza river, as far as the parting of the streams, where Nagyr has a river of its own fed by the Hispar and other great glaciers of the region surveyed by Mr. Conway. The scenery of these alpine valleys, all draining to the Indus, is grand beyond the powers of description. "The mountains fling themselves aloft on either hand with astonishing precipitancy, as it were, into the uttermost heights of heaven; so steeply, in fact, that a spring avalanche falling from the summit of Rakipushi on the south must almost reach the bottom of the valley, whilst I myself saw within a short distance of the houses of Hunza town (Baltit) the snowy dust of a great avalanche, which descended grandly from near the top of the noble peak that rises close behind the place. Rakipushi is 25,500, the Hunza peak 24,000 feet high, their summits being separated by a distance of 19 miles. Both are visible from base to summit at one and the same time from the level floor of the valley between them, which is not more than 7000 feet above the sea. No mountain view that I saw in the Mustagh surpasses this for grim wonder of colossal scale, combined with savage grandeur of forms and contrast of smiling foreground" (*ib.*, p. 294).

Mr. Conway was unable to determine the exact area of the Hindu-Kush covered by the now familiar names, Mustagh and Karakorum, which on some maps have a north-west trend, though no range is disposed in this direction; all run parallel to each other, mainly from west by north to south by east. One extends from Rakipushi along the south side of the Hispar valley to the Hispar pass; a second to the north forms the north bank of the Hispar valley, and includes the Hunza Peak; a third still farther north, but not yet surveyed, separates Gujal, that is, the Upper Hunza valley from the

Taghdumbash Pamir, close to the proposed new British frontier towards Chinese Turkestan. But both Mustagh and Karakorum appear to be originally the native names, not of ranges, but of passes, as has been pointed out by Colonel Godwin-Austen, who proposes to apply the term Mustagh to that section of the range running west

FOOT OF THE HISPAR GLACIER.

of K^2 as far as the Hunza-Nagyr valley, and Karakorum to the eastern section up to the Chang Chenmo plain.

Passing from Nagyr to the Hopar basin, the explorer surveyed the Hopar, Hispar, and Barpu glaciers, all of which formerly converged in a single glacier stream; but the Hispar has now retreated some 20 miles into the mountains, while the Hopar, of which the Barpu is a branch, has greatly shrunk in width, leaving exposed the Hopar plain on its west bank. Above Barpu was unexpectedly discovered a vast series of glacier basins, a view of which was obtained from the crest of the ridge between the Barpu and Hispar valleys. "As we mounted the view developed; the great glacier basins below revealed their distant recesses, and the cirque of giant peaks behind, all white with *nevé,* reared themselves aloft against the blue sky and showed the smallness of their outlying satellites, which had seemed to rise so high above us from our camps. The view on the other side was of peculiar interest to us, for we looked for the first time into the Hispar valley and beheld the long avenue of peaks that lined the way up the Hispar glacier towards the unknown snowy regions through which lay our intended route into Baltistan."

Crossing the Hispar Pass into the Askoli district of North Baltistan, the explorer surveyed the great Baltoro glacier and for the first time ascended the Crystal and Pioneer Peaks. Later, the Skoro Pass was crossed in frightful weather at the enormous altitude of 17,400 feet, and farther on the exploring party entered the pleasant and fertile Shigar valley, which to their eyes, accustomed for months to the wilderness of barren rocks and blinding snows, "seemed beyond measure luxuriant. The air, too, was full of colour, and bathed all nature in its tender glow. Busy peasants, driving oxen to tread out the corn, and singing as they drove, made the fields animated and musical. Birds twittered among the trees, butterflies flitted about in countless numbers, and we walked along as in a dream. The picturesque architecture of the group of mosques in the principal village of Shigar showed that we had returned to regions where men have leisure for art." From this point the return route lay through Skardo, capital of Baltistan, on the Upper Indus, to Srinagar and Abbotabad, which was reached on 28th October 1892, after an absence of seven months in the Mustagh highlands.

The great Hispar Pass, as determined by this expedition, extends from the end of the Hispar glacier to the end of the Biafo glacier for a distance of over 80 miles, and is consequently the longest glacier pass in the world outside the Arctic regions. Before Mr. Conway's expedition it appears never to have been crossed in the memory of any living person.

During his first expedition to the Pamir in 1889 Captain Younghusband, starting from Shahidula above the Karakorum Pass (18,550 feet), and following the course of the Raskam (Upper Yarkand-daria), crossed from that basin to the valley of the tributary Oprang River, by the hitherto unvisited Aghil Pass, leading southwards to the K^2 peak, now named Mount Godwin-Austen (28,278 feet). This remarkable depression, however, was found to cross, not the Mustagh range proper, but an unknown chain, which is separated from it by the Oprang valley, and to which its discoverer has given the name of the Aghil range. "It runs in a general north-west direction, parallel to and intermediate between the Mustagh range and the western Kuen-lun Mountains. It is about 120 miles in length, and is broken up into a series of bold upstanding peaks, the highest of which must be close on 23,000 feet."

From the Aghil Pass and the Oprang valley superb views were obtained of the loftiest and most extensive glacial region in the temperate zone. Standing on the pass, the observer looks south-westwards up the Oprang valley, "which is bounded on either side by ranges of magnificent snowy mountains, rising abruptly from either bank, and far away in the distance could be seen the end of an immense glacier flowing down from the main range of the Mustagh Mountains. This scene was even more wild and bold than I had remembered it on my former journey [1887, when the explorer crossed the Mustagh Pass on his expedition from Manchuria across the continent to India], the mountains rising up tier upon tier in a succession of sharp needle-like peaks, bewildering the eye by their number; and then in the background lie the great ice mountains, white, cold, and relentless, defying the hardiest traveller to enter their frozen clutches." And from a northern gorge in the Oprang valley a splendid prospect was commanded, both of the Godwin-Austen and of the Ghusherbrum group, four of whose peaks rise to a height of over 26,000 feet. A glacier descending from this group terminates on the edge of the stream in a great wall of ice from 150 to 200 feet high. "Another glacier could be seen to the south, and yet a third coming in a south-east direction, and rising apparently not very far from the Karakorum Pass. We were therefore now in an ice-bound region, with glaciers in front of us, glaciers behind us, and glaciers all around us" (*Geo. Proc.*, 1892, p. 210).

Lower down the Oprang is joined on its left bank by another huge glacial stream descending from near the Shimshal Pass (25,460 feet). This glacier, which was explored for some distance southwards, also afforded a glorious view of the Mustagh Mountains, whose appearance "was truly magnificent as they rose up in solemn grandeur for thousands of feet above me, sublime and solitary in their glory, their sides covered with the accumulated snow of countless ages, and their valleys filled with glistening glaciers. With infinite toil and difficulty I had insinuated my way through the chinks in their seemingly impregnable armour of rock and ice, and my feelings now as I looked on the wonderful scene before me can only be appreciated by one who has himself penetrated the great mountain solitudes of the

Himalayas, and stood alone, as I was then, deep in the inmost recesses of the mightiest range of mountains in the world; separated from the haunts of civilisation by chain after chain of inhospitable mountains, and far from the abodes of even the wild and hardy hillmen of the Himalayas—alone, where no white man had ever yet set foot, where all was snow and ice, pure, white, and unblemished, and where not even the rustle of a single leaf, the faintest murmur of a stream, or the hum of the smallest insect, rose to break the spell of calm repose which reigned around" (*ib.*).

The Himalayas proper—that is, the "Abode of Snow,"[1] as they have been named by the Aryan inhabitants of the plains—constitute, if not the largest, by far the most elevated highland system on the globe. With a breadth varying from 180 to 220 miles, they stretch in a continuous curve of about 1500 miles along the Indo-Tibetan frontier between 72° to 96° E. long. from the western limits of Kashmir to the eastern extremity of Assam. The main direction for nearly two-thirds of the distance to Mount Everest (29,002 feet), culminating point of the globe, is north-west and south-east, and thence nearly due east to the Indo-Chinese frontier. Throughout this vast distance a mean elevation is maintained of from 17,000 to 19,000 feet, while as many as forty peaks are known to exceed 24,000 feet—that is, a height greater than the loftiest summits of the Andes, or probably any other range beyond the Asiatic continent. The Himalayas, which do not form a single chain, but a number of more or less parallel ridges, with spurs often projecting in various directions, may be regarded as forming the southern scarp of the great Central Asiatic tableland, towards which they slope gently, while falling abruptly

[1] From the primitive Aryan root *hi, hu,* preserved in the Greek χεῖμα = Latin *hiem-s* = *winter, storm.*

down to the Indian lowlands. Far inland lie the inmost ridges, which from the coast cannot be distinguished from the more advanced chains and transverse sections, often projecting far into the plains, above which they rise in a succession of steep rocky barriers to the Tibetan tableland. The southern foot of the main ridge is skirted by the marshy "Tarai," forming a watery hollow trough of great depth, extremely favourable to the growth of a luxuriant and even rank vegetation, but also perpetually shrouded in noxious exhalations rising from the dank ground. The Tarai, which traverses the British and Nepal frontier for nearly 500 miles east and west, lies at a lower depth than the plains from which it is separated by the outer and lowest ridges of the system.

A prominent feature of the Himalayas consists of the narrow gorge-like valleys of the advanced spurs, entirely destitute of waterfalls, and seldom presenting favourable sites for human abodes. But a few of the more gently sloping valleys, at elevations of from 6000 to 7000 feet, have been chosen for the summer retreats and sanitaria of the English officials, and even these are occasionally subject to sudden and destructive landslips.[1]

The Himalayas may be divided into a western, a central, and an eastern section. The first begins at Mount Nanga-Parbat (26,629 feet), where the Indus suddenly trends southwards between Kashmir and Gilgit. Although there are here no well-defined ridges, there are several longitudinal valleys between which the Indus and other rivers flow for hundreds of miles before they can find an outlet southwards. Here also several peaks, besides the Nanga-Parbat, rise above 23,000 feet, the Nanda-Devi attaining an elevation of 25,661 feet.

[1] In the year 1880 the station of Naini Tal was partly destroyed by one of these landslips, which partly filled in a lake at the foot of the hills.

NANGA-PARBAT.

The central section, forming the so-called Nepal highlands, and stretching from the source of the Indus to the Tista, a tributary of the Brahmaputra, is intersected by numerous transverse valleys running north and south, and contains the highest summits on the globe. The most conspicuous peaks are the Dhawalagiri (26,826 feet) in the west, Gauri-sankar or Mount Everest (29,002) in the east, and Kanchinjinga (28,156) north of Darjiling on the Sikkim frontier. In the extreme north tower the glittering summits of the main chain, forming, as it were, the topmost foamy crests of these billow-like formations, which, after sinking twice to a depth of 10,000 feet, again suddenly fall to little over 1000 feet above the level of the plains. But before reaching the lowlands there is another abrupt rise to from 3000 and 4000 feet, formed by a long sandstone ridge rolling away towards the so-called "Bhaver," a dry wooded tract, which in its turn sinks through a succession of long undulations down to the Tarai.

In 1892 a scientific expedition was undertaken to the Central Himalayan regions of Kumaon, Garhwal, and Hundes for the purpose of studying the extensive trias deposits on the Tibetan frontier west of the Manasarowar Lakes. Starting from Naini Tal, the party, including Dr. Diener and Messrs. L. Griesbach and C. S. Middlemiss, first struck north-east to Milam (11,250 feet), the highest inhabited place in the Bhot Mahals, or Tibetan border region. The Milam glacier, terminating at a height of 11,340 feet two miles north-west of the village, is 12 miles long, being the largest of the ice-streams discharged by the Nanda Devi group. Beyond Milam the route lay over the Utadhura Pass (17,590 feet) to the upper Girthi valley, where some rich fossiliferous trias deposits were discovered on the southern slopes of the Bambanag range (19,000 feet). From this point the journey was continued over the Kiangur Pass (17,000 feet) and the Kisgarh-Chalda Pass (17,400) to the pastures of Lachambelkichak and Chitichun in Hundes.

After scaling the Chitichun Peak No. 1 (17,740), the two Chanambaniali summits (18,320 and 18,360), and Kangribingri (19,170), the party returned by the Yandi Pass to Milam, and made a fresh start by the Utadhura and Kiangur Passes for the Rimkin Paiar territory, which is claimed by the Tibetans. Here the route was continued along the southern slopes of the watershed as far as the Niti Pass (16,628 feet), the most frequented of the Central Himalayan divide, which led down to the Dhauli Ganga valley, the farthest point reached by the expedition. Scarcely anywhere in the world, not even in the Grand Cañon of the Colorado, could more magnificent geological sections be found than those presented by the limestone zone of the Central Himalayas near Rimkin Paiar and on the Silakank, where in some places the whole series of marine deposits from the lower silurian to the chalk was exposed to view. In sheltered places along the watercourse occurred dense thickets of salix and birches, and isolated birches 16 to 23 feet high ranged up to 14,000 feet, though in the Central Himalayas the tree limit lies generally between 11,000 and 12,000 feet. In the Dhauli Ganga valley the forests, extending to 11,000 feet, consist chiefly of the magnificent deodar cedar, which at Juma Gwar forms beautiful woodlands (*Proceedings Berlin Geograph. Soc.*, June 1893).

The eastern section of the Himalayas, running west and east through Sikkim, Bhutan, and north Assam, while maintaining a mean elevation of 16,000 feet, presents no peaks comparable to the giants of the central and western sections. The highest known summit is Chumalarhi (23,933 feet). But much of this region still remains unexplored, and the eastern uplands, where the San-po suddenly disappears in a profound abyss, have never yet been visited by European or native surveyors.

South of the northern plains rises the triangular

plateau of the Deccan, which has a mean elevation of from 2000 to 3000 feet, with a general incline eastwards to the Bay of Bengal. The northern scarp of this extensive tableland is formed by the Amarkantale Vindhya range, whose secondary sandstone formations are continued north-eastwards beyond Panna and Rewah nearly to the Ganges below Benares. Here is the water-parting between the streams flowing north to the Ganges basin and west to the Arabian Sea. The left bank of the Sone, which joins the Ganges above Patna, is skirted by the Khaimur range, separated by a broken plateau from the Panna ridge, which traverses Bundelkhand, and is noted for the deep gorges and isolated crags on its north-western slopes.

The steep southern slopes of the Vindhyas present the aspect of a weather-beaten coast-line, as if the valley of the Narbada now flowing at their base had once formed a deep inlet of the sea. This valley is separated southwards from that of the Tapti River by the parallel Satpura range, which runs from the classic Amarkantak, the source of the Narbada, westwards for nearly 600 miles at a mean elevation of 3000 feet, and culminates with the Pachmarhi hills (4500 feet), rising abruptly from the Narbada valley at Dhupgarh, east of Betul.

This culminating point of Central India is one of the most hallowed regions in the Hindu world. Here is the renowned shrine of Siva, the *Mahadeo*, or "Great God," a term sometimes applied to the whole range. The road from Jilpa, the last village on the plains, lies through a romantic region that has been vividly described by J. Forsyth. After crossing the jungle it surmounts the scarp of the Pachmarhi plateau, which presents the aspect of a beautiful English landscape; and here, through breaks in the dense woodlands, a first glimpse is had of three isolated peaks all aglow in the fiery sunset, and standing out from the purple clouds banked up in the background. East of the plateau the rocky heights descend from an altitude of 2000 feet down to the vast level forest of Sal, while the scarps of the plateau are furrowed with mysterious abysses, one of which, the sacred and almost inaccessible Jambo-Dwip, forms an awe-inspiring natural marvel on the path of the pious pilgrim. These woodlands are the home of the bison and "sanbar," prince of red deer.

East of Asirgarh the Bombay-Allahabad Railway, and the main highway to Central India, cross this chain at a depression 1240 feet above the sea. But west of this point the system is continued to the Western Ghats by a highland tract 40 to 50 miles broad, with a mean height of 2000 feet, and several peaks from 3000 to 4000 feet.

The Western Ghats begin immediately south of the Kandeish valley, which separates them from the Satpura Mountains. From this point the Ghats—that is "Passes"—run close to the coast along the western edge of the Deccan southwards to the Nilgiri hills, where they meet the eastern coast ranges. The prevailing formation is trap, and indurated lava in the northern and central parts, culminating with the Mahabaleshwar Peak (4800 feet), and succeeded by sandstones and granites in the southern part. Like most coast ranges, the Ghats slope gently inland towards the central tableland, but fall abruptly down to the narrow strip of lowlands separating them from the sea. Here they are scored by the beds of deep watercourses, which in the rainy season are flooded by foaming torrents rushing over precipices and romantic waterfalls down to the coast. From the Tapti valley to the Nilgiris the Ghats maintain a mean elevation of about 4000 feet at a uniform distance of 30 to 40 miles from the sea.

The Nilgiris, or "Blue Hills," which culminate with the Dodabetta (8760 feet), form the converging point of the western and eastern ranges, by which the plateau of the Deccan is here enclosed. They cover an area of 700 square miles, and are noted especially for their genial and healthy climate, rendering them a favourite resort of Europeans enfeebled by the enervating heats of the plains.

The eastern coast ranges differ from the Western Ghats chiefly in three respects. They are much less elevated, with a mean height of scarcely more than 1500 feet; they do not form a continuous chain, being broken up into distinct sections by the valleys of the Godavari, Kistna, and other streams flowing to the Bay of Bengal; lastly, they run at a much greater distance from the coast, the intervening lowlands averaging from 50 to 80 miles. They stretch from the Mahnadi River valley near Kattak for about 500 miles south-eastwards to the nucleus of the Nilgiris, beyond which they fall abruptly southwards to the so-called "Gap," a narrow, deep, transverse fissure, scarcely 400 feet above sea-level. North of the Godavari the system attains an elevation of over 5000 feet.

South of the Nilgiris the Palni hills to the west of Madura are crowned by peaks 6500 and 7100 feet. These hills, like the Western Ghats, are extremely salubrious, and form a sanitarium for Europeans.

Beyond the above-mentioned "Gap," the extremity of the peninsula is occupied by the independent system of the Cardamum Mountains from about the 8th parallel to Cape Comorin. In these highlands, which culminate in the lofty Anamalli hills (9700 feet), are found the highest elevations south of the Himalayas. They seem to be connected with the mountain system of Ceylon by "Adam's Bridge," a chain of rocky islets stretching between the Gulf of Manar and Palk Strait from the mainland to the northern extremity of the island.

There remains to be noticed the somewhat isolated Aravalli range, running north-east and south-west across the Rajputana country, which they separate into two natural divisions—desert plains in the north-west, fertile and well-watered rolling lands in the south-east. At their southern extremity is the somewhat detached Mount Abu (5653 feet), highest point of the system, which has a mean elevation of about 2000 feet. Between Meywar and Marwar, where they rise to 4330 feet near the village of Jargo, the hills are crossed by the Dasuri Pass, which is alone practicable for wheeled traffic. The isolated character of the Aravalli range would be made evident by the already suggested subsidence of 600 feet, when they would appear as a long narrow rocky island about midway between the Baluch and Vindhya hills at the western entrance of the strait connecting the Arabian Sea and Bay of Bengal.

3. *Hydrography: The Indus, Ganges, Brahmaputra, Godavari, Kistna, Narbada, and Tapti Rivers.*

In its water system, as in many other respects, India presents a most striking contrast to the Iranian tableland. While this arid upland region is characterised chiefly by an inland drainage, and by a deficiency of large rivers, the Indian peninsula has absolutely no inland drainage at all, and possesses, in proportion to its size, a greater number of streams, all flowing seawards, than perhaps any other country in Asia. In the north nearly all these streams are collected into three vast systems, flowing either through the Indus to the Arabian Sea, or through the Ganges to the Bay of Bengal, or through the Brahmaputra and its affluent the Megna to the same bay.

Even the Brahmaputra forms no exception to this general disposition, for its numerous channels are mingled with those of the Ganges delta before reaching the coast. But in the southern plateau of the Deccan there are almost as many river mouths as there are rivers, most of the large streams here forming separate systems, and finding their way in independent channels to the sea. This is true not only of the Mahanadi, Godavari, Kistna, Pennar, Kavari, and others, draining eastwards to the Bay of Bengal, but also of the Narbada, the Tapti, and the innumerable little mountain torrents rushing from the Western Ghats to the Arabian Sea. Thus it happens that, whereas the coast north of the Vindhya hills is broken only by the Indus and Ganges-Brahmaputra deltas, the southern seaboard is scored by at least fifty watercourses from the mouth of the Narbada to that of the Mahanadi. At the same time the volume of water sent seawards through the two great northern deltas is vastly greater than that of all the southern estuaries combined.

The Indus, like nearly all the great Asiatic rivers, has its farthest sources, not on the seaward slopes of the outer range, but behind the Himalayan escarpment of the Tibetan tableland itself. It rises on the north side of the Kailas range in 31° 20′ N., 82° E., near the sources of the Satlaj and San-po, and within 60 miles of the Karnali, farthest head-stream of the Ganges. The Indus flows first north-west through Ladak between the Kailas and main Himalayan range nearly to Gilgit, in 36° N., 75° E. Here it trends sharply southwards, maintaining this direction for the rest of its course through the Panjab and Sind to its delta in the Arabian Sea between Katch and Karachi. In its upper course it receives no important tributary except the Shayok, joining its right bank from the Karakorum range. But on emerging from the

THE INDUS AT TORBELA, NEAR GILGIT.

Himalayas it collects all the southern drainage of the Hindu-Kush through the Kabul River, which joins its right bank at Attock, almost on the frontier of British India. Lower down it receives the waters of the Suliman uplands mainly through the Kuram and Gomul Rivers. But the chief accession to its volume is from the united waters of the Jhelum, Chenab, Ravi, and Satlaj, all flowing from the western Himalayas and through the Panjnad joining the left bank of the main stream at Mithun-Kot in the Derajat, towards the Sind frontier. These four great tributaries, with the Indus itself, give their name to the Panjab—that is, the "Five Waters"—beyond which province the united stream receives no further affluents.

It is remarkable that throughout its entire course of 1800 miles the Indus flows by no important towns, the only places of any consequence on its banks being Sakkar on its right bank, with the opposite town of Rori on the left bank, and Hyderabad near the head of its delta. Multan, Lahore, Amritsar, Wazirabad, and all the other large cities in its basin, which has a total area of 373,000 square miles, lie not on the main stream, but on or near the Chenab or other great tributaries. This fact may seem remarkable, inasmuch as the Indus was the first great stream occupied by the Aryans during their migrations from the north-west into the peninsula. The real cause is the shifting character of the banks below Kalabagh. During the rainy season the Indus is subject to sudden inundations which spread for miles along both banks, often causing great devastation, and preventing the foundation even of villages in its immediate vicinity.

From its source to the sea the Indus has a total fall of about 18,000 feet—that is, 8000 to Leh in Ladak, 9000 between that place and Attock, 1000 thence to

LEH IN LADAK.

the coast, a distance of nearly 950 miles. Hence the current in the upper reaches is extremely rapid, and even below Attock it runs at the rate of 6 miles an hour, mostly between high cliffs as far as Kalabagh. Here it enters the plains, suddenly widening to an average breadth of from a half to over one mile, with a mean velocity of rather less than 3 miles an hour. At low-water the tides are felt for nearly 80 miles from the mouth, and the Indus, like most of its great tributaries, is navigable to the foot of the hills for light craft. The delta is very extensive, reaching inland to Hyderabad, and from Karachi to the Rann of Katch, or about 130 miles both ways. The mean annual discharge through the mouths or through irrigating canals is estimated at over 150 billion tons, being about 41,000 cubic feet per second in December, and fully ten times that quantity during the August floods.

Although taking their name from the Indus, the Hindus still regard the Ganges as pre-eminently the great river of India. And in this they are so far justified that, although of shorter length than the Indus, it has a larger area of drainage, comprised entirely within the limits of the peninsula. For the Ganges differs in this respect from the other great Asiatic streams, that it rises, not behind the scarp of the plateau, but on the seaward face of the higher Himalayan range. Its two chief head-streams, the Bhagirati and Alaknanda, flow from an immense mass of snow 14,000 feet above the sea in the native Garhwal district, 31° N., 79° E. After a southerly course of about 80 miles the two streams unite a little above Hardwar, 30 N., where they burst through the outer barrier of the Himalayas, thenceforth flowing in a south-easterly direction through the rich alluvial plain of Northern India to the head of the Bay of Bengal.

In the Upper Ganges basin small lakes, usually of a

temporary character, are occasionally formed by landslips falling across the fluvial beds and blocking the running waters. Thus was formed some years ago the Gudyar Tal lake in the Bireh-ganga valley, British Garhwal, about 16 miles above its confluence with the Alaknanda. In 1869 this lake was entirely filled up by a later landslip, the water being forced over the natural barrier and causing disastrous flooding lower down the valley. At Gohna in the same valley, 8 miles above the confluence, a similar lake was formed by a series of landslips in 1892-93. In August 1894 this basin, 5 miles long, 700 yards broad, and 775 feet deep, burst its dam and tore with a velocity of 22 miles an hour down the Alaknanda valley, destroying all bridges and buildings between Gohna and Hardwar, and rising in some gorges to a height of 160 feet. Gohna, which stands 150 miles above Hardwar at an elevation of 3500 feet, is enclosed by lofty ridges from 12,000 to 15,000 feet high, culminating in Trisul and other snowy peaks over 20,000 feet above sea-level.

In the same district of Garhwal rises the Jamna, chief tributary of the Ganges, which pursues a nearly parallel course south of the main stream to their junction at Allahabad. The Jamna carries to the common artery the drainage of Rajputana, Sindhia, and Bandelkand, collected by the Chambal, Betwa, and Ken, all of which join its right bank below Agra. Below the junction the united stream still continues to receive several large affluents, of which the chief are the Son from the south, the Gumti, Gogra, Gandak, and Kusi from the Himalayas. At Hardwar the Ganges has a discharge of 7000 cubic feet per second, in the cold season, when the water is at its lowest; at Benares its volume has increased to 19,000, with a breadth in the rainy season of 3000 feet and a rise of 43.

For about 500 miles from its mouth it maintains a nearly uniform depth of about 30 feet, and a width of over one mile, while the fall from Hardwar to the sea scarcely exceeds 1000 feet. Hence the Ganges would afford one of the finest water highways to be found in any country but for the troublesome and even sometimes dangerous navigation of its shallow tortuous channel and numerous mouths. Of these the southernmost and most frequented is the Hugli, which gives access to large vessels for 100 miles as far as Calcutta. Beyond this point large boats ascend for upwards of 1000 miles along the main stream, and for perhaps five times that distance along its numerous tributaries, northwards to the Himalayas, southwards to the Vindhyas. "The navigation of the Brahmaputra and its affluents, of the Lower Ganges and its many branches, is quite magnificent, and offers probably one of the finest spectacles of its kind to be seen in the world. Not only every trader and landholder keeps many vessels, but every cultivator or peasant has his boats, and almost every labourer his canoe; thus the craft may be reckoned by hundreds of thousands. At several points on the great rivers the vessels congregate for several months consecutively, and form floating cities and marts, where many thousands temporarily dwell, where much barter takes place, where monetary transactions are arranged, and banking business is done" (Sir R. Temple).[1]

The united Ganges-Brahmaputra delta is of vast extent, probably the largest in the world, and of most complicate character, constantly shifting its channels with the annual inundations, and continually advancing towards the sea, which it discolours for a distance of 60 miles with over 235,000,000 cubic yards of matter yearly brought down to the coast. The delta extends

[1] *India in* 1880, p. 319.

for over eighty miles along the Bay of Bengal, and stretches 200 miles inland, discharging through its innumerable channels 100,000 cubic feet per second during the dry and 500,000 during the wet season from April to August. At this time there is a rise of 32 feet above its ordinary level, which is sufficient to flood the whole country for 100 miles about the junction of the Ganges and Brahmaputra, leaving nothing visible except the tree-tops and the villages built on mounds raised above the highest level of the floods. Besides these inundations the delta is exposed to cyclones and to the phenomenon known as the "Bore," when a tidal wave five to ten feet high rushes up the Hugli with a roar at the rate of 18 miles an hour, often causing a rise of several feet as far up as Calcutta, or even 20 miles above it.

The upper course of the Brahmaputra had long been one of the most interesting geographical problems awaiting solution. But the explorations of the natives employed by the Indian Survey Office have at last practically confirmed the generally accepted view that the San-po of Tibet, the Dihong of Assam, and the Brahmaputra form a continuous water highway, which has been traced throughout its whole course with the exception of a small gap, where the San-po plunges into a ravine and traverses a still unexplored region of the Himalayas.

Although belonging properly to Tibetan geography it will be convenient here to deal with the San-po[1] as forming undoubtedly the true upper course of the Brahmaputra. Its source has not been visited, but from information obtained by the Pundit's journey in 1865, it

[1] The word, which in Tibetan means "holy water," occurs in a great variety of forms, such as Tsangbo, Tsanrbo, Tsanpu, Dzangbo, Sampo, Sambo, Sampu, Sanpu, Sanpo, etc.

may be fixed with tolerable certainty in 82° E., 30° 35′ N., at a height of nearly 16,000 feet above sea-level, a little east of Lake Manasarowar, source of the Satlaj. This lake with the Rakus-tal partly fills the depression between Mounts Gurla and Kailas, sources of the Ganges and Indus. Between the lake and the San-po the water-parting is very low; yet it suffices to send the Satlaj on a journey of 1000 miles to join the Indus on its way to the Arabian Sea, and the San-po for 1800 miles in the opposite direction to the Ganges delta.

Flowing first eastwards along the northern base of the inner Himalayan range, the San-po receives several tributaries on both banks, and east of Shigatze it trends north-east with a huge bend, the apex of which lies above the intersection of the 94th meridian with the 30th parallel. It then turns south-east, passes through the above-described gorge in the Himalayas, and apparently reappears about 100 miles lower down as the Dihong of Assam. The unexplored gap is occupied by fierce and lawless tribes, whose hostile spirit, combined with the rugged character of the land, has hitherto defeated every effort to penetrate into this region, and clear up the mystery by actual observation.

Near the Buddhist monastery of Tadum (13,000 feet above the sea), where the Mariam-la route enters its valley, the San-po is already navigable for light craft; but lower down the navigation is obstructed at several points by shoals and rapids. It is also crossed by ferries and by light suspension bridges at many places where the stream is narrowed by projecting bluffs. At Chetang, a little below the junction of the Kichu from Lassa and 600 miles from its source, it is as large as the Rhine, and in the dry season 1400 feet wide, with a discharge of about 30,000 cubic feet per second, which during the summer rains is probably increased to over 700,000.

Yet it was long uncertain what became of this vast body of water. Now, however, the explorations, especially of Mr. Needham, place beyond doubt the connection of the San-po with the Brahmaputra, not through the Subansiri, the Dibong, or the Lohit (Brahmakanda), but through the Dihong, which is joined near Sadiya by the Dibong from the north and by the Lohit from the east. In 1888 Mr. Needham ascended the Lohit through the Mishmi hills to the Zayul district near its source a little west of the Lu-Kiang, which is almost certainly the Upper Salwin river. Two points were thus settled, first, that the Irawadi, unless it be the Lu, cannot possibly rise on the Tibetan plateau; second, that the Lohit cannot possibly be the Upper Brahmaputra, that is, the San-po. Mr. Needham also ascertained that the Dibong, still identified by some authorities with the Kenpu (Gakbo) rising far to the north between the San-po and Lu-Kiang, has in fact a relatively short course of not more than 180 miles altogether. The Subansiri also, which joins the right bank of the Brahmaputra some distance below the triple confluence near Sadiya, is now known to have its rise on the southern slopes of the eastern Himalayas. The identity of the San-po with the Dihong is thus established by a simple process of elimination, and this inference is independent of the somewhat untrustworthy statements of the Pandit K. P. regarding his surveys of the San-po during his long detention in the Pemakoi district below Gia-la-Sindong, where the great river disappears in the Himalayan gorges above the plains of Assam.

But even independently of the San-po, the Brahmaputra proper represents a vast river system, filling the whole of the Assam valley, where it collects the waters of the Eastern Himâlayas from the north, and those of the Naga, Khassia, and Garo hills from the south and east. Here its chief affluents are the Dubong, Subansiri,

Lohit, or Brahmakunda, the latter of which was long regarded as its true upper source. The main stream flows through the centre of Assam, nearly due west, to the 90th meridian, where it turns sharply to the south to the already described Ganges delta. Including the San-po, it has a total course of 1800 miles, an estimated area of drainage of rather over 360,000 square miles, and a mean discharge of 146,000 cubic feet per second in the dry season. But this discharge is vastly increased during the summer months, when the incessant rains of this watery region convert the river into a great inland sea, flooding the whole of the Assam lowlands,. and cutting off all communication except by boats and causeways raised 10 or 12 feet above the level of the roads.

In the Deccan six considerable rivers find their way in independent channels to the coast—the Narbada and Tapti westwards to the Arabian Sea; the Mahanadi, Godavari, Kistna (Krishna), and Kavari eastwards to the Bay of Bengal. Of these by far the largest are the Godavari and Kistna, which are 900 and 800 miles respectively, and jointly drain the greater part of the region between the 14th and 22nd parallels, representing a total area of over 206,000 square miles. They both have their farthest head-streams on the slopes of the Western Ghats, whence they follow nearly parallel winding courses through the Bombay Presidency and the Nizam's Dominions. As they approach the eastern sea-board, the main streams gradually converge, and after traversing the narrowest part of the Madras Presidency, they enter the sea through two large deltas, which, during the floods, overflow into the intervening Lake Colair. This lake, or lagoon, which is the largest in India, is 47 miles long by 14 broad, and is entirely formed by the overflow of the Godavari and Kistna, whose lower courses between the hills and the coast are dammed by enormous dykes and connected by an extensive system of canalisation available both for irrigation and navigation purposes.

A somewhat similar parallelism is observed by the Narbada and Tapti, whose valleys are separated by the Satpura range. The two streams rise from the Satpura range in the very heart of the peninsula, and gradually converge towards their respective estuaries near Baroch and at Surat in the Gulf of Cambay. The Narbada presents scenes of enchanting beauty, especially in its upper reaches about Jabalpur. Here it winds for a short distance with a narrow transparent stream of greenish-blue waters, between two glittering walls of snow-white marble, with here and there a vein of dark-green or black basaltic rock considerably heightening the effect of the marble. Near its mouth a fine prospect is also commanded from the noble railway bridge which crosses the estuary at Baroch. Although much obstructed by rapids, the Narbada is navigable by boats for 250 miles to the falls of Dari. It has a length of 800 miles, or about double that of the Tapti, and the united drainage of these two rivers is somewhat over 63,000 square miles.

4. *Main Natural and Political Divisions.*

To the three main physical divisions of the peninsula correspond, on the whole, its three great political administrations. Thus the northern highlands and lowland plains are mostly comprised in the Presidency of Bengal with its dependent feudatory States, while the southern plateau of the Deccan is divided between the Presidencies of Bombay and Madras, the former embracing the Western Ghats and Malabar coast-lands, the latter including the eastern or Coromandel coast-lands. Through the universal acceptance of British rule the whole region has also in recent times acquired a general political unity, answering to such general physical unity as is derived from its peninsular form and tropical climate. But beneath this broad uniformity we are everywhere confronted with a dualism, betrayed especially in the social, religious, and political worlds. Thus we find a society based on caste intermingled with communities which ignore all class distinctions—Brahmanism invaded even in its most hallowed precincts by Islam; territory administered directly by the paramount power everywhere in contact with tributary and even with semi-independent States.

In the northern highlands, besides the independent States of Bhutan and Nepal, we have the semi-independent feudatory States of Sikkim, Garwhal (Tirhi), and Kashmir, collectively occupying most of the southern slopes of the inner Himalayan ranges. These States are attached to the Presidency of Bengal, an expression the original meaning of which has become considerably modified. Bengal is still understood to comprise all the British territory not included in either of the other two Presidencies of Bombay and Madras. But as a matter of fact the Bengal Presidency is itself now subdivided into nine distinct administrations (including the three Lieutenant-Governorships), which, with those of the two southern Presidencies and attached Native States, will be found fully tabulated at the end of this chapter.

Kashmir—Jammu—Swat—Chitral—Chilas.

In the extreme north-west the basin of the Upper Indus, probably the grandest alpine region in the world, is almost entirely comprised in the territory of the feudatory prince Golab Singh, Maharaja of Jammu and Kashmir. This State, which by the treaty of Amritsar,

March 16, 1846, accepted the paramount sovereignty of England, embraces within its borders a great variety of climatic, physical, and ethnical conditions, stretching as it does from the hot plains of the Panjab for 280 miles to the eternal snows and glaciers of the Western Himalayan and frontier Karakorum ranges, and from the Hindu-Kush for 400 miles east to Tibet. It is essentially a highland region, almost everywhere mountainous, but having one splendid valley (Kashmir), broad, long, and populous. Moreover, there are many broad upland valleys, extremely fertile, well sheltered by the towering Himalayan crests from the northern blasts, and watered by copious streams all draining to the Indus or to its tributaries.

Physically speaking, the whole country may be divided into three zones, rising in successive terraces from the Panjab lowlands to the Karakorum range. The lowest and southernmost of these zones comprises the more advanced hilly districts with a mean elevation of 2250 feet above sea-level. This is succeeded by the central zone between the Himalayas proper and the Kailas range, from 7000 to 9000 feet high, beyond which follows the truly alpine region of Baltistan, or "Little Tibet," between the Kailas and Karakorum, with a mean elevation of 11,000 feet, and culminating with the Godwin-Austen peak, 28,278 feet, next to Everest the highest point on the globe. Here also are the Baltoro and many other glaciers, which, vast as they are, seem to be but the poor remains of the prodigious icefields which must have formerly covered the whole region of the Himalayas. The melting of the snows in the fierce summer sun, combined with the precipitous slopes and the silent action of underground waters, exposes all these upland valleys to sudden floodings, avalanches, and landslips, often causing widespread ruin.

In the central zone lies the lovely vale of Kashmir, 4500 square miles in extent, hemmed in on all sides by snow-clad peaks and watered by the Jhelum, which in its placid winding course flows through the Wular and several other beautiful lakes. Thus pent up, and with an elevation of over 5000 feet above the sea, this romantic valley presents somewhat the appearance of a

THE KANJAT VALLEY NEAR CHALT.

vast cirque with a narrow southern outlet, through which the Jhelum escapes towards the Indus. Kashmir has ever been the theme of Eastern song, an earthly Eden, where prevails a perennial spring, and of which the Mogul emperor, Shah Jahan, was wont to say that he would prefer to sacrifice all his vast Indian dominions rather than be deprived of this delightful retreat. Here the picturesque elements are the snowy peaks, the romantic gorges, the numerous lakes, streams, and waterfalls, the magnificent woodlands and rich flowery meads, —a combination of natural beauties scarcely to be found elsewhere concentrated in an equal area.

Of the numerous passes leading into the Kashmir valley, and practicable for pack animals, the chief are the Banihal (9700 feet) from the south, the Punch (8500) from the west, and the Pir Panjal (11,500) from Gujarat in the Panjab. After the annexation of the Panjab, Kashmir, which was formerly under Sikh rule, became a vassal state of the British Raj under Gulab Sing, grandfather of the present Maharaja. In recent years the relations with India have become more frequent and more intimate, and the administration has been much improved, thanks to the influence of the British Resident, Walter R. Lawrence, by whom many improvements have been introduced, though not yet (1896) everywhere enforced. Unfortunately the shawl industry, for which Kashmir was formerly renowned, has been ruined, partly by change of fashion and partly by the Franco-German war, by which the export trade was arrested. The finest goats' hair used in the manufacture of these shawls came, not from the country itself, but from Turfan in Yarkand.

Nearly one-third of the whole population, or 2,544,000 souls, are concentrated in the vale of Kashmir, which might easily support a much larger number. Female infanticide probably exists in the sub-Himalayan tracts, and till lately "suttee," or widow-burning, was still practised in Jammu, where it was more fanatically enforced than elsewhere in India.

East of the Panjkora valley lies the Swat country, unexplored till 1878, when one of the Indian native surveyors mapped it out from the source of the Swat River in the great transverse range between Bunj on the Indus and the Chitral valley to its junction with the

1 *The Valley of Kashmir*, by Walter R. Lawrence, 1895.

Panjkora. Swat formed at one time a powerful state under a venerable chief of great repute for sanctity, called the Akhund, who exercised considerable influence over the unruly tribes of the district.

Little was known of the highland state of Chitral before the year 1893, when the political confusion and disorders following on the death of Aman-ul-Mulk, its former *Mehtar*, or ruler, led to the intervention of the Indian authorities, and eventually to a more direct British control over the administration of the country. Chitral lies on the southern slopes of the Hindu-Kush, about 200 miles by the nearest route from Peshawar, and at a mean elevation of 5200 feet above sea-level. Northwards it reaches the crest of the Hindu-Kush range, and is conterminous southwards with the petty states of Dir and Asmar. The bulk of the inhabitants, estimated at from 150,000 to 200,000, are concentrated in six large villages extending 3 or 4 miles along both banks of the fertile Kunar (Kashkar) River, the fortified residence of the Mehtar lying on the right bank. This potentate has at different times been tributary to Badakhshan, Afghanistan, and Kashmir, but since December 1893, his territory forms an integral part of British India.

Fresh troubles arose in January 1895, when the Mehtar, Nizam-ul-Mulk, was murdered by his younger brother, Amir-ul-Mulk, who usurped the throne, bid defiance to the British authorities, and besieged the Agent in the capital. During the ensuing campaign, which was distinguished even in the annals of Indian border warfare for brilliant execution and many acts of great daring and endurance on the part both of the English and native forces, the country was rapidly reduced, the usurper put to flight, and order secured by the permanent military occupation of various strategical points along the main routes leading through Swat

CHITRAL VALLEY.

southwards to Peshawar, and through the Dir district eastwards to Kashmir. The new Mehtar, Shuja-ul-Mulk, became a British pensioner, receiving an annual subsidy of 12,000 rupees, besides 8000 per annum in return for the loss of revenue derived from the Khushwaktis district, which was now detached from Chitral. At the durbar held in the capital by Sir George Robertson (September 1895) it was announced that State affairs would be left to the Mehtar, assisted by three headmen and the advice of the Political Agent. Traffic in slaves and murderous outrages were strictly forbidden, while the principality was promised security from foreign aggression by the British garrison. At the same time the Baraul Valley towards Kafiristan was transferred to the Khan of Dir, and Bajaur farther south left under tribal rule, arrangements however being made for the improvement of the roads and bridges along all the main lines of communication.

Captain Younghusband, who in 1893 accompanied a political mission to the district, speaks favourably of Chitral: "It is a delightful country to be stationed in, if it were only not so much cut off from civilisation. The people are a hardy, cheery group of mountaineers, who delight in polo, sport, and shows of any kind. The climate is delightful, and the mountains, though not wooded like those of Kashmir, have patches of forest on many of their slopes, and the villages in the valley bottoms are crowded with fruit trees." The culminating point is the twin-peaked Tirich Mir (25,000 feet), which fills up the head of the Kuñar valley at a distance of about 30 miles from the settled parts.

East of Upper Chitral lie Yasin and Gilgit, the latter of which during the last few years has been the site of a British residency under Major Biddulph. This officer was here stationed on the Kashmir frontier with a view to control the tribes occupying a district of some strategic and political importance. The River Gilgit drains eastwards to the Indus near the Nanga-Parbat peak, which marks the north-western extremity of the central Himalayan chain. This region, where the Dard and Afghan races meet about half-way up the valley, has at all times proved most inaccessible to external influences.

The Chilas district lies on the north-west frontier of Kashmir, between the northern slopes of Nanga-Parbat and the left bank of the Indus, which here flows at an altitude of scarcely more than 3400 feet above the sea. But as Nanga-Parbat towers to a height of nearly 27,000 feet, it presents towards the Indus valley one of the grandest mountain views in the world. Before the British occupation in 1893 Chilas had never been visited by any European traveller. It is described as an arid, treeless region, which is approached from Astor by the high and difficult Mazeno Pass. The village of Chilas on the Indus was occupied in consequence of an attack made by the natives on a British detachment at Taliche (Talech) on the road to Gilgit below Bunji on the Indus. The Chilasi, who have always been fierce marauders, raiding even into Kashmir, belong to the ethnical group to which Leitner has given the name of Dards. They are consequently a branch of the Galcha Aryans, who form the bulk of the population in Nagyr, Hunza, Yasin, Chitral, Kafiristan, and generally of the whole region between Baltistan and Afghanistan proper. A few in the eastern districts are Buddhists, but all the rest Muhammadans, who show the same abhorrence of the cow that other Muhammadans do of the pig and dog.

Nepal.

South-east of Kashmir follow the Native States of Garhwal (Tihri), Nepal, Sikkim, and Bhutan, occupying most of the southern slopes of the Himalayas nearly to the great southern bend of the San-po, where it disappears in the North Assam highlands. The central section is almost exclusively comprised in the nominally independent kingdom of Nepal bordering east on Sikkim and Darjiling, west and south on British territory. Nepal thus consists of a comparatively narrow strip 550 miles long and 160 broad, limited northwards by the crests of the inner Himalayan range (here culminating with Gaurisankar, the highest peak on the globe), and falling in a series of five continuously diminishing terraces and deep intervening troughs down to the Indian plains. One of these troughs is the already described Tarai, which sinks even to a lower level than the open plains, and forms the chief physical feature along its southern border. In some places it is overgrown with a low jungle, very sparsely inhabited, while it consists elsewhere of uninhabitable wastes covered with a coarse growth of grass. Some parts of the Tarai frontier, however, are very fertile. The heart of the country comprises a delightfully well-watered and productive caldron-shaped valley, in which is situated Katmandu, capital of the State. The total area is about 54,000 square miles, with an estimated population (1894) of 2,000,000.

Nepal, which is despotically ruled by a hereditary minister of the warlike Ghurka tribe, under a titular Maharaja, abounds in mineral wealth, including copper, iron, lead, arsenic, and sulphur. The principal valley of Katmandu is well known; it is a sort of second Kashmir. Beyond that, the interior is very little known, and such is the jealousy of the Government that no Englishman is allowed to pass beyond the Katmandu valley under any pretext. Hence no surveys have here been yet carried out, and the British Government is still without exact information regarding the relations between Nepal and

China. A mission bearing presents of a prescribed value is said to proceed every five years from Katmandu to Peking, although Nepal, while independent in internal administration, is in the position of a feudatory State to India.

Sikkim.

A somewhat similar position was till lately held by the adjoining petty State of Sikkim, which stretches for 52 miles between Nepal and Bhutan, and for 66 between Tibet and India, with an area of about 2600 square miles, and a population (1891) of 30,500 Rongs, or Lepchas, as their Gurkha neighbours call them. Physically, Sikkim forms an eastern continuation of Nepal, from which, however, it is separated by a lofty transverse ridge of the Himalayas, 11,000 to 12,000 feet high, culminating northwards with Mount Kubru (24,015 feet). The crest of this ridge is so sharply defined that it may be traversed for 40 or 50 miles at a uniform level almost without a break. Till recently Sikkim owed allegiance both to Tibet and India. The Maharaja, who resides at Tamlang (Tumlong), on the Indian side of the Chola range, in winter, and in summer in the Chumbi valley on the Tibetan side, accepted allowances from both countries, about £200 from Lhassa, and £1200 from Calcutta, the latter grant being made on the condition of his affording every facility for the trade between the two countries. Now, however, he is absolutely a feudatory of the British Empire. By the treaty of March 1889, China recognises the British protectorate, involving direct and exclusive control over the internal administration of Sikkim. A British officer has also been appointed to advise the Maharaja and his council, and to reorganise the administration. The result has been a great improvement in the financial and commercial relations, the revenue being now about double the expenditure, while the imports and exports advanced from 15,000 Rx. in 1891 to over 24,000 Rx. in 1893.

The Chumbi valley, traversed by the Am-mo-chu, a head-stream of the Brahmaputra, is a south-eastern corner of Tibet, wedged in between Sikkim and Bhutan, near a lovely lacustrine district in east Sikkim. The lakelets, which lie at elevations of 10,000 to 16,000 feet at both sides of the border range, are mostly tarns or closed basins, evidently due to glacier action.

The existence of glaciers in the Kanchinjinga highlands having often been denied, the question was set at rest in 1891, when Mr. C. White, British Resident at the Court of Sikkim, made an expedition to this alpine district from Talung. Crossing the Yeumtso-la Pass (15,800 feet), the explorers entered the valley of the Zemu at a point where it was seen to flow from a very large glacier descending from the flanks of Kanchinjinga eastwards down to the level of 13,800 feet. This Zemu glacier, probably the largest in Sikkim, terminates at the head of the stream in a huge wall of ice 400 to 500 feet deep, and is fed from the north by many minor glaciers, of which as many as a dozen were counted; eight others were seen coming down from the slopes in the north-east, some reaching the main glacier, others ending abruptly in jagged walls of ice, or in more gradual slopes. From the altitude of 17,500 feet, the highest point reached on the Zemu glacier, Kanchinjinga was seen towering in an almost vertical wall of rock and ice to a further height of nearly 10,000 feet, while to the south of Simiolchum (Si-imvovonchim, 22,300 feet), a gap in the range falling to 17,450 feet, revealed a magnificent group of lofty peaks not yet entered on the maps. Two other peaks, both over 22,000 feet, were also discovered, one in front, the other to the right of Kanchinjinga. It is stated that this hitherto unvisited ice-bound district may now be reached in a fortnight from Darjiling.

KANCHINJINGA.

Bhutan.

Still less known than Nepal is the State of Bhutan, lying mainly between Tibet and Assam, east, north, and south, and stretching from Sikkim for 400 miles eastwards to the unexplored region separating the San-po from the Dihong. The surface is intersected by two parallel ranges intervening between the inner Himalayas and Assam, the first enclosing a bleak and almost uninhabitable tableland; the second skirting the "Duars," a fertile tract ceded in 1866 to the British in return for a yearly subvention of Rs. 50,000. The subsidy, however, is dependent on the good behaviour of the people, and this, combined with the occupation of two strong positions (Baxa and Diwangiri) near the frontier, has sufficed to maintain peace along the borders since the troubles which led to the war of 1865.

In the north the country is extremely wild, but elsewhere Bhutan affords some of the grandest and most romantic scenery in the world. In the more sheltered districts Bhutan produces millet, wheat, and rice in abundance. But it is mostly uninhabited, with a population of scarcely more than 300,000 (?) altogether. It is nominally ruled by the Dharm Raja, a sort of incarnate Buddha. But the real head of the State is the Deb Raja, elected every three years by the Penlows or chiefs from their own body. Commercial relations are confined mainly to Tibet and Assam, to which countries musk, madder, coarse cloth, and horses are exported in exchange for cottons, woollens, tea, gold, silver, and embroidered work. The capital, Punakha or Dosen, occupies a position of great natural strength on the Bugni river, 96 miles north-east of Darjiling. But a better-known place is Tasichozong (Tassisudon), on the Gudada River, centre of the peculiar form of Lamaism prevalent in the country.

India Proper—The Panjab.

From the foregoing account of the northern highland States, it appears that the Himalayas still belong to some extent politically, as they mostly do physically and ethnically, rather to Central Asia than to India proper. They form, in fact, the outer and southernmost barrier of the great Central Asiatic tableland, and although Indian influences and elements of race are here everywhere more or less perceptible, the bulk of the Himalayan aborigines belong not to the Aryan, but to the Tibetan Mongoloid stock. Hence the Indian peninsula may be said properly to begin with the plains, which stretch along the base of the great northern barrier from the eastern scarp of the Iranian plateau to the Bay of Bengal.

These plains are wholly comprised in the Indus and Ganges basins, and lie nearly altogether in the Presidency of Bengal. In the extreme west, however, the lower portion of the Indus valley, embracing the province of Sind, is exceptionally included in the Bombay Presidency. But higher up the whole of the Indus valley, as far as the Hindu-Kush and the Kashmir uplands, is included in the Panjab Province, one of the separate administrations of the Bengal Presidency. In the north this province is very hilly, comprising several ranges separating the upper courses of the large rivers between the Jamna and the Indus. In the extreme north-west the hills beyond the Indus form a sort of connecting link, broken by the Kabul River valley, between the Himalayas and the Suliman range, which skirts the province along its western border for nearly 400 miles. Farther south the open plains fall very gradually from an elevation of about 1600 to 200 feet above the sea at the confluence of the Indus with its united tributaries. Here the riverine tracts are generally extremely productive, whereas the so-called "doabs," or "two-waters"—that is, the spaces enclosed between the great rivers, which give their name to the province—often consist of mere wildernesses of scrub and jungle, but generally afford extensive pasture for cattle, camels, sheep, and goats. The country between the Indus and Jhelum is generally rugged, cultivated in parts only, and thinly inhabited, intersected by the Salt range, with a mean elevation of 3000 feet, and culminating with the Sakesar peak, 5010 feet. On these heights the salt frequently crops out, and there are several large salt deposits. West of the Indus the range is continued under the name of the Kalabagh hills as far as the Suliman highlands, and the long narrow alluvial strip between these highlands and the river is known as the Upper and Lower Derajat.

Besides Kashmir there are over thirty Native States attached to the Panjab. Among the most important of these is Bahawalpur, occupying a strip about 17,000 square miles in extent between Rajputana on the south-east and the Satlaj, Panjnad, and Indus on the north-west. Practically the most important are the so-called Phulkhian States (Patiala, Jhind, and Nabha), with a joint area of 7500 square miles in the hilly and plain tracts between Delhi and Lahore. North of these are the "Simla Hill States," traversed by the Upper Satlaj, and reaching eastwards to the Tibetan frontier. None of these are of any size except Basahar (Rampur), which has an area of over 3000 square miles.

Sind.

The southern province of Sind, although, as stated, included administratively in the Bombay Presidency, cannot be separated physically from the Panjab. They both merge eastwards in the Rajputana wastes, and the western parts of the Sind lowlands still consist of waterless steppes yielding little beyond a scanty pasture for herds of buffaloes, asses, and camels. Even in the more productive Indus delta barren and swampy tracts are everywhere intermingled with the cultivated fields.

With but few exceptions, the cultivation in this province depends on a large series of canals drawn from the Indus, which river is to Sind what the Nile is to Egypt, and has caused the province to be called "the lesser Egypt." These channels are filled when the river rises in the summer and are dry when it subsides in the winter.

The whole of this low-lying delta region is for a long way up stream exposed to the inundations of the Indus, which reach east to the "Thar" or great Rajputana desert, and west to the foot of the Baluchistan hills. East of the delta the Gulf of Katch penetrates far inland, skirting the north side of the Gujarat or Kathiawar peninsula, and gradually merging in the so-called "Ranns" of Katch. These remarkable formations, consisting of two portions, the Great and Little Rann, with a total area of 9000 square miles, become sandy, saline swamps, and inland lagoons or arms of the sea, according to the season of the year. When flooded they connect the Gulfs of Katch and Cambay, thus converting the Native States of Gujarat into an island. Northwards they are confined by the Allahband—that is, the dam or mound of Allah, and both dam and lake owe their existence to an earthquake which occurred in the year 1819.

Rajputana.

In most coloured maps of the peninsula a large space towards the north-west will be noticed marked off from British territory proper, as belonging to Native States. In the very centre of this region is the small British enclave of Ajmir-Merawa, about 2700 square miles in extent, which is administered by a British officer. It consists partly of an elevated plateau, partly of a picturesque hilly district at the northern extremity of the Aravalli range, and is politically distinct from the surrounding plains; for with the exception of this enclave the whole of the region in question is divided amongst a large number of feudatory Native States attached to the Bengal Presidency. These States are disposed in two distinct geographical and political groups, under the control of the Rajputana and Central India Agencies respectively.

The Rajputana Agency has a total area of no less than 129,000 square miles, and is bounded north-west by the Panjab and Sind, north-east by the North-West Provinces, south-west by Sind and Gujarat, south-east by the Central India Agency. Rajputana is divided by the Aravalli range into two unequal parts, of which the north-western or larger consists to a great extent of sandy, arid, and unproductive wastes, with some arable and even fertile tracts towards the north and north-east. Here is the *Thar*, or great sandy desert of Northern India, intersected everywhere by long parallel dunes 50 to 100 feet high, with few streams or wells, and a scant vegetation of tufty grass and scrub.

THE PALACE OF AKBAR, AJMIR.

Considerably more elevated and fertile is the south-eastern division of Rajputana, which is diversified by wooded rocky hills, and watered by the Chambal, Banas, and some other large rivers flowing north to the Ganges basin. The country between the Chambal and Patar consists of a rich black loam, highly productive and well cultivated. But even in this division most of the surface is stony, rugged, under jungle, and unfertile, except close to the river banks.

Of the twenty Rajputana States the largest are Udaipur (Maiwar), Jaipur, Jodhpur (Marwar), and Jaisulmir. But all except Shahpura and Lawa belong to the first rank in the empire, being under treaty with the Imperial Government. Of the Bhil tracts between Sirohi and Dungarpur some are directly administered by British Commissioners, while others are either tributary to Udaipur or under the control of Rajput princes.

The Marwar and Bikanir tracts are essentially pastoral, abounding in cattle, sheep, and a superior breed of camels. But elsewhere Rajputana is mainly agricultural, yielding grain, cotton, and opium in considerable quantities.

Central India Political Agency.

The Central India Agency comprises all the region lying between Rajputana and the British Central Provinces, and stretches from Gujarat eastwards to Chota-Nagpur and the North-West Provinces. It is divided by British territory into two sections—native Bandelkhand and Baghekhand lying to the east, and the Central India portion to the west of the Jhansi and Lalatpur districts of the North-West Provinces. The eastern section forms a part of the Deccan plateau, here intersected by the Khaimur and Panna ranges, and watered by the Ken, Betwa, and Son, flowing to the Jamna and Ganges.

The western section consists of a broken upland tract stretching from the Vindhyas northwards to the Jamna, mostly fertile, and well watered by the Chambal, Parbatti, Sind, and Betwa. Wheat, rice, cotton, sugar, and especially opium, are the staple products, and iron, copper, and coal, besides diamonds, exist in many places. The diamantiferous district, yielding several thousand pounds' worth yearly, lies some 14 miles north-east of Panna, and the stones here found are of four different tints, all, however, inferior to those of the Kistna valley.

The Central India Agency has a total area of 86,000 square miles, and comprises nearly sixty feudatory States, of which the largest and most important are Gwalior (Sindhia's), Indore (Holkar's), Bhopal, and Rewah. Most of the numerous petty States have feudal relations with one or other of the larger ones, while their autonomy is guaranteed by the paramount power.

North-West Provinces—Behar—Bengal.

The Ganges valley forms a well-defined natural region, occupying the whole space between the Himalayas and the Deccan, and comprising the separate governments of the North-West Provinces, Oudh, and Bengal, including Behar, in the Bengal Presidency. These vast alluvial plains, watered by the Ganges, Jamna, and their numerous tributaries, form the very heart of British India, in which nearly one-half of its entire population is concentrated. The fertility of the soil and the thrifty habits of the people have here produced the same results as in the rich alluvial valleys of the Yang-tze and Hoang-ho. In both of these regions the density of the population is estimated by several hundreds to the square mile, amounting in the North-West Provinces to 380, in Oudh to 470, and in Bengal to 484. These proportions are considerably more than treble those of France and other European States, which are regarded as fairly well peopled, and the census returns for 1891 showed a total of no less than 118,000,000 for the three above-mentioned provinces alone. Yet many parts are covered with dense jungle, tenanted only by wild animals, while others are either barren wastes, uninhabitable swampy tracts, or rugged uplands, necessarily but very thinly peopled. Hence the density in the more populous alluvial and cultivated districts is much higher than might be supposed even from these astonishing figures. Such teeming multitudes, which the official returns show to be steadily increasing, especially in Bengal (including Behar), could not possibly be supported even in these exuberant lands except on the most frugal diet, and as a matter of fact rice, the cheapest of all grains, forms the staple, in many cases almost the exclusive, article of food for the great majority throughout the Ganges basin.

In this basin are comprised, besides the lowland plains, extensive highland tracts, consisting in the north chiefly of the outer Himalayan ridges, and in the south of the more advanced spurs and offshoots of the Vindhyas, which near Benares and other points approach to within a few miles of the Ganges. About Darjiling, in British Sikkim, the hills rise to elevations of from 6000 to 8000 feet, commanding a view of Kanchinjinga (28,156 feet), lying 45 miles due north on the frontier of Tibet and independent Sikkim. These Darjiling hills are extremely interesting, not only for their magnificent scenery and glorious vegetation, but also as forming the "divide" between the Ganges and Brahmaputra basins and between the Hindu and Buddhist religious worlds.

It would be difficult to conceive a greater contrast than that which exists between these cool or cold mountains and the watery plains of the Lower Ganges, whence we pass by a natural transition to the still more watery region of the Lower Brahmaputra valley. All this low-lying tract, as far east as the Garo hills and British Burma, is included in the Bengal Provinces, which thus

consist altogether of four different sections—Behar, enclosed west and north by the North-West Provinces and Nepal; Chota-Nagpur, stretching from Behar southwards to the Central Provinces; Orissa, lying between Chota-Nagpur and the coast; and Bengal proper, comprising the united Ganges-Brahmaputra delta, and stretching north to Sikkim, west to Behar, east to Assam, the Lushai hill tribes, Burma, and Arrakan, but overlapping or enclosing the independent Hill Tipperah country on the south-west frontier of Assam.

Below the Himalayas Behar is mainly alluvial and level, but rises westwards to the Rajmahal hills in the Santhal Parganas. Chota-Nagpur, on the contrary, is an upland rugged country, embracing the eastern spurs of the Vindhyas, with elevations of from 2000 to 4400 feet; while Orissa consists of a flat diluvial tract between these hills and the coast, and an extensive hilly district in the interior occupied by petty tributary States. These hills also form a continuation of the Vindhyan system, which here culminates in the Parasnath (4480 feet), close to the East Indian Railway, about midway between Benares and Calcutta. Far higher are the elevations on the east frontier of Lower Bengal, where the Tipperah highlands, forming a continuation of the Lushai and Manipur ranges, attain an altitude of from 11,000 to 12,000 feet.

West of these hills stretches the great delta with its "thousand mouths," its intricate network of countless channels and backwaters, its highly-cultivated and densely-peopled inland Backergange tract, its almost impenetrable coast region of the Sundarban,[1] covered

[1] This name, which wrongly takes a plural form, *Sundarbans*, in English writings, appears to be derived, not from *sundar-ban*, "fine forest," or *sundar-band*, "fine embankment," but from *sundri* (*Heretiera littoralis*), a tree still abounding in the district, and *ban*, "forest."

with dense jungle, and a prey to wild beasts, terrific cyclones, and deadly exhalations. Here land and water still struggle for the mastery, while unbridled nature laughs at the feeble efforts of man to tame the jarring elements. The work of the day is swept away by the raging night-storm, and the patient labour of years often suddenly disappears in a chaos of widespread ruin. Nevertheless, the Saugor Light, firmly established at the entrance of the Hugli, shines like a beacon of future promise that one day even this wild region will be brought under the control of man.

The *Bhati*, "Flooded Lowland," as the Sundarban was formerly called, would thus be restored to its original condition; for Mr. Rainey has shown that within a comparatively recent period this alluvial tract stood at a considerably higher level than at present, and was extensively cultivated and inhabited. All over the delta are scattered numerous ruins of brick houses, large tanks, and even temples and palaces, such as the *Satgumbaz*, a curious structure with seventy-seven domes and "not less than twenty-six arched doors on the four sides, which open into a vast hall 140 feet in length by 96 in width, or an area of 13,440 square feet." In the same locality are the remains of many mosques, reputed to number 360, one of which dates from the year 1459. But a great change was caused some centuries ago by the shifting of the Ganges eastwards to its junction with the Brahmaputra. "The rivers in this tract since then, having no considerable bulk of fresh water poured into their upper channels, have been mainly fed by the tide coming up from the sea, which renders these streams more or less impregnated with brine, and quite unfit for human beings to drink" (J. R. Rainey, *Geo. Proc.*, 1891, p. 280).

Assam.

Beyond the Brahmaputra section of the delta lies the province of Assam, occupying the north-eastern extremity of the empire between Bhutan and Tibet on the north, Burma on the east and south-east, the Manipur Lushai and Manipur hill States on the south. The administrative province embraces the Brahmaputra and Surma (Barak) river valleys, with the intervening Naga, Jaintia Khasi, and Garo hill tracts.

But Assam proper is confined to the Brahmaputra valley, an extensive alluvial plain 450 miles long by about 50 broad, everywhere enclosed by lofty ranges, except towards the west, where the Brahmaputra escapes towards the Ganges. This plain, however, is diversified by innumerable rivers, boundless woodlands, extensive prairies, and even by isolated ridges at some points approaching close to the Brahmaputra. The number of watercourses is probably greater than in any other country of equal extent, no less than sixty considerable streams having been enumerated in this narrow tract, all connected together by a labyrinth of channels and branches. Thanks to this superabundance of water, Assam is one of the most fertile regions in India. But owing to the dense forests and lofty enclosing ranges impeding the free circulation of the air, the moist climate is oppressively hot, and all the more unhealthy because the rainy season continues here longer than in any other part of the empire except Lower Burma. It lasts from March to November, when the low-lying riverine tracts are often completely flooded. The slow evaporation of these liquid masses charges the atmosphere with dank vapours, generating ague, dysentery, and other malarious disorders. Even during the cold season dense fogs usually prevail in the plains from midnight to noon, enveloping everything in an impenetrable misty veil. Assam, however, is beyond the reach of the hot winds, which in May and June convert many parts of the peninsula into a glowing furnace. It abounds in coal, iron of excellent quality, sulphur, salt, and petroleum, and after the rainy season the natives search the streams for the gold-dust brought down with the alluvia from the hills.

The Central Provinces.

The Deccan, in the largest sense of the term geographically, is distributed politically between the Central Provinces, forming a portion of the Bengal Presidency, and the two southern Presidencies of Madras and Bombay, together with the Native States, of which by far the most important are the Nizam's Dominions and Mysore.

The Central Provinces form a British enclave almost everywhere cut off from British territory proper by intervening feudatory States. But the broad political boundaries are the Chota-Nagpore States of Bengal on the north, the tributary Native States (of which some belong to the Orissa province and others to the Madras Presidency) on the east, the Nizam's territory on the south, the Central India Agency on the west and north-west. They form an irregular square (almost a triangle) about 600 miles long east and west, by 500 north and south, with a total area, including the Berars, of over 130,000 square miles. They constitute the northern portion of the Deccan plateau, here divided into two sections by the Satpura range, with a mean elevation of 1500 to 2000 feet, and a general eastward tilt. But the surface is everywhere diversified with hilly plains and river valleys, and on the south it is enclosed by the upland Bastar tract reaching from the coast to the Godavari, and stretching thence, under different names, westwards to the Khandeish plateau.

This extensive region is traversed by the Narbada, Mahanadi, Wainganga, and Wardha, flowing generally in deep beds, and navigable for long distances during the rainy season. Large tracts are still covered with dense virgin forests, but others are well adapted for the cultivation of cereals, and cotton of the best quality is grown, especially along the right bank of the Wardha.

The Nizam's Dominions.

In the Nizam's Dominions are included East and West Berar, forming the Hyderabad Assigned Districts. These were, under the treaties of 1853 and 1861, assigned by the Nizam to the British Government, which, on its part, undertook to maintain a body of troops, to be styled the Haidarabad Contingent; but the sovereignty is still retained by the Nizam.

The Nizam's Dominions still form by far the largest and most important of all the Native States in the empire. They comprise the very heart of the Deccan, lying mainly between the two great rivers Godavari and Kistna north and south, and between the Presidencies of Bombay and Madras west and east, and stretching about 475 miles both ways, with an area of close on 100,000 square miles. It consists of an elevated plateau, sloping from 2500 down to about 1000 feet towards the coast ranges which skirt its south-eastern frontier. Much of the surface is still waste and covered with low brushwood, but the soil is naturally fertile, and where irrigated produces heavy crops of cotton, cereals, oleaginous plants, and even dates.

In the Kistna valley within the Nizam's territory are the famous Partial and Kollur diamond-fields, where the "Great Mogul," the "Koh-i-núr," the "Pitt" or "Regent," the "Orloff," and many other magnificent gems were found. The rough stones yielded by these mines were formerly cut and polished in the town of Golconda, about 100 miles farther north, and from this circumstance the diamonds were popularly supposed to be produced at or near Golconda, which is not a diamantiferous district.

The Madras Presidency.

This Presidency comprises roughly the whole of the so-called "Eastern" and a considerable section of the Western Ghats, the coast-land stretching thence to the Bay of Bengal, and all the southern portion of the peninsula from the Nizam's territory to Cape Comorin. It is thus bounded by the sea on two sides, and landwards by Orissa, the Central Provinces, the Nizam's Dominions, and the Presidency of Bombay. Besides Mysore, Kurg, Travancore, and some other smaller Native States, the chief historic divisions are the Northern Sircars and the Carnatic on the Coromandel coast, South Kanara and Malabar on the west coast, the Balaghat or uplands near the junction of the Eastern and Western coast ranges, and the Nilgiri highlands connecting these ranges south of Mysore. The Presidency stretches across 12 degrees of latitude (8° to 20° N.) on the east side, and across 6 (8° to 14° N.) on the west side, with an extreme length of 1000 miles, breadth 380 miles at the parallel of the capital, and total area 148,000 square miles. It is traversed on both sides by the Ghats, and is generally mountainous towards the south, where the Nilgiris, Palni, Shevaroy, and other hills occupy most of the apex.

The Malabar or West Coast is a narrow but very fertile and highly-cultivated tract intervening between the shore and the Ghat Mountains. Several shallow inlets, called "backwaters," run sometimes for 150 or

280 miles parallel with the coast. For 170 miles from Cape Comorin the east coast is also low, but rocky and fringed with reefs. Navigation is here further obstructed by the so-called "Adam's Bridge," a sandbank stretching from the mainland to the northern extremity of Ceylon, with two open channels only. But even these are too shallow for large vessels, so that all the deep-sea navigation is diverted round the island from the Gulf of Manar and Palk Strait. The Coromandel coast, running from Point Calimir nearly due north to the Kistna delta, retains the character of a low sandy seaboard, the beach shoaling very gently, and preventing large vessels from approaching the land. Beyond the Kistna delta the Golconda coast, as it used to be called, trends north-eastwards for nearly 300 miles to Vizagapatam, and although the hills here approach somewhat nearer to the sea the intervening strip is deltaic, being so low and flat that it is sometimes subject to inundation. But on the northern coast, stretching for 230 miles from Vizagapatam to the Chilka Lake or Lagoon on the frontier of the Bengal Presidency, the hills approach nearer to the shore, which is unbroken by any inlets or large river mouths.

Mysore and Kurg.

A large portion of the interior is occupied by the Native State of Mysore and the province of Kurg, which are enclosed by the Madras Presidency on all sides except towards the north-west, where Mysore impinges on the Bombay Presidency. It consists of an extensive tableland 290 miles by 230, with an area of over 27,000 square miles, filling the angle where the Western and Eastern coast ranges merge in the Nilgiris. The surface is very undulating, and diversified in many places by the remarkable rocky formations known as *Drugs*, huge piles rising either isolatedly or in clusters from 1000 to 1500 feet above the plateau. Many of these have perennial springs on their summits, which have often been converted into almost impregnable strongholds.

The plateau culminates at Bangalore, which lies due west of Madras, over 3000 feet above sea-level. But the highest peaks are found more to the west, where the Kuduremukha, 6215 feet, forms a striking landmark visible from the sea. Here also two peaks in the Baba-Budan Mountains rise to elevations of 6214 and 6317 feet, and on the former is the tomb of Baba-Budan, a Muhammadan saint, from whom the range takes its name.

The chief rivers of Mysore are the Kavari, Penner, Paler, and Pennair. None of the streams are here navigable, and many are utilised to form artificial reservoirs, of which there are no less than 38,000 in the State. Some of these are of considerable size, the Sulikere tank, which is the largest, having a circumference of 40 miles.

The Native State of Mysore forms no part of the Madras Presidency, but for some years before 1881, when it was restored to the native ruler,[1] it was separately administrated on the model of the Panjab by a Chief Commissioner, directly responsible to the Supreme Government.

A similar arrangement has been permanently adopted for the adjoining territory of Kurg, a hilly tract 2000 square miles in extent, occupying the crests and eastern

[1] This ruler descends through a collateral line from Raj Vodejar, founder of the present State of Mysore in 1610. For thirty-six years the throne was usurped by the Muhammadan soldier of fortune Hyder Ali and his son Tipu Sahib (1763-99). The native dynasty was restored by the English, but in 1831 the Raja was deposed for gross maladministration, and from that time till 1881 Mysore was practically administered by British Commissioners. The present ruler bears the title of Chama Rajendra Vodeyar.

slopes of the Western Ghats between 12° and 13° N. latitude. The term Kurg, generally written Coorg, is a corrupt form of *Kudagu* or *Kodumale*, meaning "steep mountains," which is a sufficiently accurate description of the land. It consists of a series of steep ridges and deep gorges, densely wooded in the east on the Mysore frontier, more open towards the west. Nearly the whole surface is drained by the head-streams of the Kavari, which here rises in the Brahmagiri range at Tale Kavari, a place of great repute among the Hindus. Here are some temples yearly visited by thousands of pilgrims, who regard the Kavari as a holier river than even the Ganges itself. In their course to the main stream all the mountain torrents form romantic waterfalls, conspicuous among which is that of the Jessy near Merkara, the capital of the territory.

Respecting sanctity, however, opinions differ; some Hindus claim it for one river, others for another. But the Ganges still maintains its supremacy among the majority of Hindus.

The cardamom plant is indigenous in Kurg, where it is extensively cultivated at elevations of from 2500 to 5000 feet above the sea.

Travancore.

The two Native States of Cochin and Travancore, occupying the south-western extremity of the peninsula, are under the direct control of the Madras Government. They have a joint area of 8000 square miles, of which about five-sixths are comprised in Travancore, which embraces the western slopes of the hills and the low-lying coast-lands from Cochin to Cape Comorin. The lowlands are very fertile, and watered by numerous small streams flowing from the hills, which have here a mean elevation of 4000 to 5000 feet, culminating with the

Augustier Peak 7000 feet high. Here the sovereignty as well as the inheritance of property passes in the female line, a custom probably due to the practice of polyandria, formerly universal along the Malabar coast, and still surviving among many of the low-caste hill-tribes in this region.

Presidency of Bombay.

Including the already described northern province of Sind, the Bombay Presidency, occupying the north-western section of the peninsula, stretches from the Panjab and Baluchistan for 1100 miles southwards to Mysore, with an average breadth of about 200 miles between the Arabian Sea (Indian Ocean) and Central India. But towards the interior the frontier line is extremely irregular, being determined in this direction by the limits of the Panjab, Rajputana, the Central India Agency, Berar, the Nizam's Dominions, the Madras Presidency, and Mysore. It has a total area of over 125,000 square miles, and, exclusive of Sind, comprises three distinct natural divisions—Gujarat, the western portion of the Deccan, including Khandeish, and the Konkans.

Gujarat.

Politically, Gujarat comprises mainly the feudatory States of Kattiawar and Katch, and Baroda, besides a portion of British territory proper about the mouth of the Narbada—some of the richest lands of the empire—and the Gulf of Cambay. Physically, it includes the Katch and Kattiawar peninsulas, which consist mostly of rich and highly-cultivated alluvial plains, varied by a few low ridges and isolated eminences. Towards Central India it is skirted by a chain of hills running from

Mount Abu at the southern extremity of the Aravallis southwards to the western extremity of the Vindhyas.

The Deccan and Konkan.

The Deccan, including the plains of Khandeish, stretches thence over the vast upland between the Eastern and the Western coast ranges, southwards to Mysore and Madras, including the Nizam's Dominions, and a part of the Bombay Presidency. Geographically, too, it includes the Balaghat districts of the Madras Presidency. This region is watered in the north by the Tapti, in the south by the head-streams of the Godavari and Kistna, which drain the whole of the plateau eastwards to the Bay of Bengal. The hilly district of the *Dangs* in the north and the Kanara district in the south are mostly covered with dense forests, and the Western Ghat mountains are mostly wooded; but elsewhere the country is open, generally well cultivated, and very fertile along the banks of the rivers.

The Konkan comprises the narrow strip of coast-lands extending from Bombay between the Western Ghats and the sea southwards to the Portuguese territory of Goa. These coast-lands are everywhere intersected by creeks and short rapid streams or torrents, flowing from the Ghats in separate channels to the sea, and in some places form tolerably sheltered harbours. Hence this rockbound coast is mostly of difficult access, and along the whole seaboard of the Presidency the ports of Karachi, Bombay, and Karwar alone afford a complete refuge to shipping during the prevalence of the south-west monsoon. The lesser harbours are, however, being improved, and at several seasons are already useful.

French and Portuguese Possessions.

The only remaining political divisions on the mainland are the fragments of territory still remaining to France and Portugal. Here the French possessions consist of five isolated portions, with a joint area of 178 square miles, all subordinate to a Governor, residing at Pondicherry. The several settlements are: in Bengal—Chandernagore, on the right bank of the Hughli, 17 miles north of Calcutta; in Madras—Pondicherry, considerably larger than all the rest put together, on the Coromandel coast, 12° N. lat.; Karakal, next in size, in the Kavari delta on the same coast; Yanaon, or Yanan, on the old Golconda coast at the northern extremity of the Godavari delta; lastly, Mahe, on the Malabar coast, nearly under the same parallel as Pondicherry.

The Portuguese settlements consist of Goa, Daman, and Diu, all on the west coast, and within the Bombay Presidency, with a total area of 1096 square miles. Goa comprises a small territory, 64 miles long north and south, by 20 broad, on the Malabar coast near the southern limits of the Presidency. It is a fertile, well-watered, and cultivated tract, divided into the two districts of Salsette and Bardes, with a small sheltered harbour, five miles from the now deserted town of Old Goa. The new town of Panjim, or Villa Nova de Goa, lies at the entrance of the harbour, which is defended by several forts. Daman is situated on the Gujarat coast, at the entrance of the Gulf of Cambay, over against Diu, which is a small island on the Kattiawar coast, 170 miles north-west of Bombay.

Ceylon.

With the exception of Diu, all the islands in the Indian waters are either British territory or subject to the

Supreme Government. They consist of one large island, Ceylon, at the apex of the peninsula; the four groups of the Andamans, Nicobars, Mergui, and Moscos, in the Bay of Bengal; and the two groups of the Làccadives and Maldives in the Indian Ocean.

Ceylon, "Pearl of the Eastern Seas," is almost connected with the mainland by Adam's Bridge, a chain of low coral reefs and sandbanks, 62 miles long, running between the Gulf of Manar and Palk Strait. But from the northernmost extremity to Point Calimere on the Coromandel coast the distance is only about 40 miles. The island has the form of a pear, tapering northwards with a total length of 270 miles, an extreme width of 146, and an area of over 25,000 square miles. The surface rises gradually from the northern plains to the central highlands, which consist of a series of ridges and intervening upland valleys, culminating with the Pedrotallagalla Peak, 8260 feet, which overlooks the elevated plateau of Nuwara Eliya, itself 6000 feet above the sea. The other chief summits are Tolapella (7720 feet), Kirrigalpota (7810), and Adam's Peak (7420), an isolated mass on the south-western edge of the central highlands, which was long supposed to be the highest point in the island.

The central highlands form a complete water-parting, whence a large number of rivers flow in every direction seawards, thus rendering Ceylon one of the best-watered countries in the world. The largest of these streams are the Mahavila - Ganga, running from the Nuwara Eliya plateau northwards to the east coast near Trincomali; the Kalani-Ganga, Kala-Ganga, and Maha Oya, draining to the west coast.

The soil is extremely fertile, even in the upland districts, and is almost everywhere clothed with a rich and varied vegetation. The chief sources of wealth are the coco-nut, cinnamon, coffee, and especially tea plantations, besides tobacco in the northern lowlands. The cinnamon groves are restricted chiefly to the south-western districts about Colombo, and the tea and coffee plantations to the upland valleys and mountain slopes. Coffee, at one time the staple product, has in recent years suffered terribly from disease, the yield having fallen from over 825,000 cwt. in 1879 to 32,000 cwt. in 1894. On the other hand, tea, by which coffee has been largely replaced, thrives well, the exports having risen from 2,392,000 lbs. in 1884 to 85,376,000 lbs. in 1894. Some of the Ceylon tea is of exquisite flavour, and samples have been sold in the London market at quite fabulous prices.

In the forests, satin-wood, ebony, calamander, and other valuable trees arrive at great perfection. Ceylon also abounds in minerals, such as plumbago, iron, manganese, nitre, alum, and salt, besides a great variety of precious stones—rubies, sapphires, amethysts, garnets, and the cat's-eye.

Ceylon is a Crown Colony, entirely separated from India and administered by a Governor appointed by the Queen, an Executive Council of five, and a Legislative Council of fifteen. It is divided into six provinces under Government agents, with a supreme civil and criminal court in the capital, Colombo. This city is connected with the old highland capital, Kandy, by a railway, which has recently been extended to the coffee plantations of the central province. On the south-west coast is the important harbour of Point de Galle, a port of call for steamers plying in the Eastern waters.

Maldives and Laccadives.

Nearly 500 miles due west of Ceylon is the group of the Maldives—that is, Malediva, or "Thousand Islands."

KANDY.

It forms a chain of coral islets, comprising 17 atolls, each enclosing deep lagoons fringed with reefs, and richly clothed with coco-nut palms. They also yield millet, fruits, and edible roots. The group is governed by a hereditary Sultan, who resides in the island of Male (Mol), and pays a yearly tribute to the Ceylon Government.

Some 200 miles north of the Maldives are the Laccadives, also of coral formation, comprising 20 atolls, besides numerous islets and reefs, mostly barren, or producing nothing but coco-nuts. They form five separate groups, which, with Minicoy, midway between the Laccadives and Maldives, are now attached to the South Kanara district of the Madras Presidency on the opposite Malabar coast.

The inhabitants of the Maldives are Muhammadans of Malay stock; those of the Laccadives, a half-caste Indo-Arab race, also Muhammadans. All alike are extremely inoffensive, hospitable, and friendly to Europeans. The Maldivians especially are noted for their kind treatment of shipwrecked sailors, seldom accepting any pecuniary return for the care bestowed on them.

The Andamans and Nicobars.

The Andamans, with the little Cocos group at their northern extremity, and the Nicobars farther south, form the scattered links of a chain suggesting a former connection of Pegu with Sumatra. The Andamans consist of four large and several smaller volcanic islands, some 200 miles west of the Tenasserim coast, with a total length of 200 miles, and an area of 2700 square miles. They are surrounded by dangerous coral reefs, generally mountainous, culminating with Saddle Peak, 2400 feet, in North Andaman, and mostly clothed down to the water's edge with a dense tropical vegetation. In South Andaman is the well-sheltered harbour of Port Blair, chosen as a penal settlement for all India in 1868, when this archipelago was annexed and placed under a "Chief Commissioner of the Andaman and Nicobar Islands," responsible to the head Government. At Port Blair, just south of Mount Harriett (1296 feet), Lord Mayo, Governor-General of India, was assassinated, while on a tour of inspection, by one of the convicts in 1872. East of Middle Andaman, the largest of the group, is Barren Island, a remarkable active volcano, 7 miles in circumference and 1700 feet high.

The Andamans are notoriously unhealthy, owing to the hot and damp climate, the thermometer ranging from 72° to 96° F., with a rainfall averaging about 150 inches, and malarious exhalations rising from the numerous mangrove swamps and creeks intersecting the islands like fiords in all directions. But thanks to their position and excellent harbours, such as Port Cornwallis and Stewart's Sound on the east side of North Andaman, Portman Harbour, Macpherson's Straits, Port Mouat, Port Blair, and other inlets in South Andaman, these islands afford numerous places of refuge to vessels overtaken by cyclones in the Bay of Bengal. Little Andaman, southernmost of the group, was first surveyed in 1887 by Mr. Maurice Portman. It is 27 miles by 15, is encircled by a coral-fringing reef, and its limestone caves contain edible birds' nests.

The Andamans are identified by Mr. Portman with the Bazacatas or Aginates of Ptolemy, and their present name may possibly be a corruption of Hanuman, the monkey race of the Ramayana. The Malays look on the natives as descendants of this race, and still call them "Orang Handuman." They have visited the islands for hundreds of years in quest of edible birds' nests, trepang, and *slaves.* Hence the suspicious and somewhat treacherous character of the natives, who, however, are far from being ferocious savages and cannibals, as they have been described by superficial observers. They are naturally of a gentle, kindly disposition, and "they always treat well those who behave well to them, and as they have been strictly looked after by the local government they look on all white men as friends. They are charming companions in the jungle—full of life and fun, quite ready to see the comic side of everything, and always cheerful and good-tempered. They pass their lives in hunting or dancing, their wants being easily satisfied" (Mr. Portman, *Geo. Proc.*, 1888, p. 575).

Although the women are ill-favoured, the men are often good-looking, with none of the thick lips, high cheek-bones, and flat noses of the Negro type. "They are probably the most intensely black race in the world" (*ib.*); but although usually grouped with the Aetas, Semangs, and other Negritoes of the Philippines and Malay Peninsula, the Andamanese cannot be regarded as a dwarfish people, the average height being rather over 5 feet, whereas the true Negritoes fall below 4 feet 6 inches.

In the Andaman Sea, east of the Andaman group, are the two volcanic islets of Narcondam and Barren, visited in 1891 by the botanist, Mr. D. Prain. Narcondam, which rises to a height of 2330 feet above the sea, is clothed with a dense vegetation, from the "beach forest" of coco-nuts, plaintains, *Pisonia, Terminalia,* etc., up to the summit of the cone, which shows no trace of a crater, and which in any case has long been extinct. The crater on the inner cone of Barren Island is merely a cinder-heap, almost destitute of vegetation; but elsewhere the surface is covered with *Terminalia catappa,* various species of *Ficus*, and other forest growths. From Mr. Prain's observations it seems highly probable that these islands, with Flat Rock, nearly in a straight line with them farther south, form a northern continuation of the volcanic system which traverses Java and Sumatra. But the further extension of this system would appear to be indicated rather by the extinct volcanoes of Popah and Han-shuen-shan in Burma and Yun-nan, than by the mud volcanoes of Ramri on the Arakan coast, as has hitherto been generally assumed. The igneous line would thus run everywhere parallel to the eastern base of the tertiary ridge passing through Arakan and the Andamans (*Jour. Asiatic Soc. of Bengal*, lxii. part ii.).

The Nicobars, lying nearer to Sumatra than to the Andamans, form two groups, separated by Sombrero Channel—Great and Little Nicobar in the south; Nancowry, Kachal, Camorta, Car Nicobar, and a few others, in the north. Nancowry, about the centre of the archipelago, is 225 miles from Port Blair and 550 from Rangoon. The hills in the south are generally covered with forests to their summit, and in the north with grass. The whole group is under the Commissioner resident in Port Blair, while the Mergui Archipelago, as already stated, is attached to Burma.

5. *Climate of India.*

The general features of the climate of India are mainly determined by five conditions—latitude, the northern highlands, the elevation of the Deccan plateau, the neighbourhood of the western desert, and proximity to the Indian Ocean. The latitude produces tropical heats, tempered on the southern plateau by the general elevation of the land, intensified on the northern plains by the Himalayas, which refract the vertical summer solar rays, while in winter intercepting the cold atmospheric currents from the bleak Central Asiatic tablelands. The great desert intervening between the upper basin of the Ganges and the lower basin of the Indus helps to cause the hot blasts to blow over the North-West Provinces. The Indian Ocean, surrounding the peninsula on two sides, supplies a superabundance of moisture during the prevalence of the southern monsoons. None of the Ghats or southern highlands are sufficiently elevated to arrest any large portion of the rain-bearing clouds, which at this time roll up continuously from the seething surface of the surrounding seas, sweeping over the Deccan plateau, penetrating far northwards through the head of the Bay of Bengal, and precipitating all their remaining humidity on the southern slopes of the Himalayas. From these conditions it results that, while great heats prevail everywhere, the provinces south of the Satpura range are, on the whole, cooler than the Indus and Ganges basins, and that an unusual quantity of moisture is pretty evenly distributed throughout the peninsula. At certain points the amount of this moisture surpasses the records taken on any other part of the earth's surface, varying on the Malabar coast from thirty to forty feet, and in the caldron-like Assam valley exceeding fifty.

But in such a vast region stretching across thirty degrees of latitude, in Ceylon approaching to within six degrees of the equator, in Kashmir impinging upon the Pamir, in Nepal rising to the highest summits on the globe, there is necessarily much diversity amidst this general uniformity.

6. *Flora and Fauna.*

Although less than half of the peninsula lies within the tropics, the average temperature of the land is everywhere so high that the organic world of the torrid zone naturally predominates greatly over that of the temperate. Owing to their low elevation the eastern parts of the great northern plains are in this respect quite as tropical as the southern plateau of the Deccan. In the north-western part of these great plains and in the upper part of the Deccan, wheat, barley, millets, pulses, European vegetables, and other plants characteristic of the temperate zone are cultivated successfully. The development of the wheat culture within the last few years has been remarkable. But the great staples of food and commerce are rice, jute, indigo, oil-seeds, poppy, betel, all distinctly tropical growths. It is noteworthy that, with a few signal exceptions, the vegetable products of India are on the whole inferior in quality to those of other countries. Thus the cotton, tobacco, sugar, and rice here grown are all surpassed by those of America, while the maize, wheat, wine, fruits, and vegetables cannot be compared with those of Europe, and the betel-nut, cinnamon, spices, and dates are excelled by the corresponding products in the Eastern Archipelago and other parts of Asia. The most notable exceptions are the Malabar coco-nuts, the Bengal indigo, jute, and opium, the coffee and tea of Ceylon, the Nilgiris, Western Ghats, Assam, and Himalayas, all of which are unsurpassed, in some cases unapproached, in flavour. The indigenous uncultivated plants also, such as the cedars, pines, teak, ebony, india-rubber, rhododendrons of the Himalayas are fully equal, if not superior, to those of other regions.

Amongst the useful plants whose cultivation has been more recently developed, coffee, tea, cinchona, and the Australian *Eucalyptus globulosa* take a conspicuous part. The eucalyptus has already been naturalised in the Nilgiris, and according to the official report for 1881 there are now over 4,500,000 cinchona trees in Southern India, yielding a sufficient supply of bark for the medical depots of all the Presidencies, with a surplus of 3000 lbs. for sale to the public.[1] The cultivation of coffee has become one of the staple industries of Southern India. The plantations now extend almost continuously along the slopes and crests of the Western Ghats from North Mysore to Cape Comorin, and occupy in the Nilgiris alone no less than 12,000 acres, yielding an annual crop valued at £200,000.

But the development of tea culture in recent years is still more remarkable, and may be regarded as one of the chief factors in the future economic relations of India and Ceylon with the rest of the world. In the hard-fought battle with the long-established Chinese trade the Anglo-Indian growers have already won the day, having in little over a decade (1881-93) displaced China teas in the British market to the extent of 76 million pounds. In that period the imports from China fell from 112 to 36 million pounds, while the consumption of Indian and Ceylon teas rose from 48 to 172 million pounds. "This great industrial revolution has been accomplished by an international rivalry almost without a parallel in the history of the world. The Chinese and the British growers have fought with all the characteristic qualities of the two races. The economies in production effected by British capital and co-operation have been pitted against the parsimony of the Chinese peasant; British energy and dash against the inertia with which the Celestial clings to an established livelihood, however slender the subsistence which it yields.

"As regards quality China has not a chance against India and Ceylon. Her rule-of-thumb methods produce

[1] In connection with this industry it may be mentioned that Dr. King, head of the Government cinchona factory in Darjiling, succeeded in 1881 in producing sulphate of quinine from cinchona bark.

an article inferior in flavour and in high-class strength to that which the scientific appliances, the costly machinery, and the chemistry of averted fermentation enable the British tea-planter to send to the market. The Indian and Ceylon teas have now no rival as regards quality in any sea-borne teas in the world."[1]

Having captured the British market, the Anglo-Indian growers have extended the struggle with their Chinese rivals to other tea-consuming countries, especially Australia, and since the Chicago Exhibition (1893) the United States. They have invested altogether about £26,000,000 of capital in this industry, and their plantations are now widely scattered over the peninsula, from Assam in the extreme east to Kangra in the far west, and to Travancore and Ceylon in the extreme south.

Notwithstanding the reckless destruction of timber that has been going on for ages, large well-wooded tracts are still found, especially on the slopes of the hills in every part of the country. Measures have of late years been taken for the preservation and increase of the forests, which cover an area of about 70,000 square miles altogether. The chief species are the conifers (cedar, pine, fir), the oak, elm, maple, plane, ash, ebony, teak, banyan, sandal-wood, mango, bamboo, sâl, and palms, including the date, palmyra, coco- and betel-nut, and other varieties.

These forests, with the jungle of the plains, are still tenanted by vast numbers of wild animals, birds, and especially reptiles. So destructive are many of these, that about 20,000 human beings and 50,000 head of cattle are yearly destroyed by wild beasts and venomous snakes. Man suffers mostly from the cobra and other reptiles, while the herds are ravaged chiefly by the tiger,

[1] *The Times*, August 9, 1894.

panther, and other large beasts of prey. India is probably the indigenous home of the tiger, which is found in every part of the country, and which, in the Royal Bengal species, attains his highest development. He preys chiefly on deer, flocks, and herds, but will sometimes turn upon man, and once he has tasted human flesh prefers it to any other. The "man-eater," as he is

BENGAL TIGER.

then called, is one of the greatest scourges of the villages lying on the skirt of the jungle. At present special measures are taken by the authorities to secure the speedy destruction of these animals.

Scarcely less formidable is the gray panther, or rather leopard, which also occasionally becomes a man-eater. But the cheeta, a somewhat smaller tawny-coloured species, is kept by native princes and trained for hunting. He is conveyed blindfolded in a cart to within a short distance of a herd of deer, when the hood is suddenly removed. In a few wonderful bounds he has seized the quarry, or, missing it, abandons the pursuit, having spent all his energy on a single effort.

Other large wild animals are the bear and wild boar, very generally the rhinoceros, chiefly in the woods at the foot of the Eastern Himalayas, the bison, gayal (*Bibos frontalis*) in the Ghats and North Assam hills; the elephant, still met in large herds in Nepal, the hilly districts of Eastern Bengal, the Nilgiris, and some other parts; two species of the alligator, the harmless "sharp-nosed," and dangerous "snub-nosed," frequenting not only most of the large rivers, but many of the numerous tanks scattered over the country. Deer and antelopes abound in immense variety, while the ibex, ovis ammon, and fine-fleece-bearing goat and sheep are numerous, especially in the Western Himalayas.

As a rule, the domestic animals, like the cultivated plants, are inferior to those of most other countries. The sheep, oxen, camels, and especially the horses, are generally of indifferent stock, although some hardy breeds of ponies occur in the Himalayas, and the camel of Bikanir in Rajputana is noted for its great size, strength, and swiftness. Large herds of oxen, camels, sheep, and goats occur chiefly in the Panjab and Rajputana, but India is on the whole more an agricultural than a grazing land. Hence, although there are a vast number of wild tribes in the more inaccessible hilly districts, there are, strictly speaking, no pastoral nomads, except, perhaps, the Ladakhi Champas and a few others of Mongoloid stock, who according to the seasons migrate between the southern and Tibetan slopes of the inner Himalayas in search of a scanty pasturage for their flocks.

7. *Inhabitants: Hindus, Dravidians, Kolarians, and Tibeto-Burmese.*

It is not to be supposed that the inhabitants of India belong to one homogeneous type. There is scarcely a country in the world containing a greater diversity of tribes and races in every stage of civilisation, from the cultured European and philosophic Hindu down to the most degraded savages. A certain outward physical uniformity, noticeable especially in the prevailing brown, olive-brown, and dark-brown complexions, has no doubt been brought about during the course of ages by the climatic conditions. It is also true that the great bulk of the population is ultimately reducible to two distinct stocks—the mixed Aryan Hindus,[1] chiefly in the northern plains; and the Dravidians, in the Deccan. But besides these at least two others are also largely represented—the Kolarian, chiefly in the Vindhyan and Satpura ranges between the Aryans and Dravidians, north and south; and the Mongoloid, inhabitants of the Himalayas, the Assam highlands, and British Burma.

Whether the absolute aborigines or not, the Kolarians are at all events the first arrivals in the peninsula, where they have scarcely anywhere risen above the lowest grades of human culture. Next came the Dravidians, some of whom, if true Dravidians, still remain at the same low level as the lowest Kolarians, while the great

[1] To avoid confusion it is necessary carefully to note the twofold meaning which this term Hindu has acquired. In its original ethnical sense it means, as here, the Aryan as opposed to the non-Aryan peoples of India. Hence the word *Hindustan*, or "Country of the Hindus," is properly restricted by native writers to the Indus and Ganges basins, the true home of the Indian Aryans. But in a religious sense Hindu is synonymous with the Brahmanical cult, and is opposed, not to the non-Aryans, but to the Muhammadan and other forms of belief prevalent in the peninsula.

majority became in course of time susceptible to the civilising influences of the Hindus, who were the last arrivals from the north-west. The land was now full except on the remote northern and north-eastern frontiers, which were gradually occupied by Mongoloid Tibeto-Burman tribes from Central and South-Eastern Asia.

The subdivisions of the Kolarians and Tibeto-Burmans are chiefly of a tribal—that is, social—character, while those of most Dravidians and all the Hindus are based essentially on linguistic considerations. The Kolarians and Tibeto-Burmans themselves speak a great variety of different dialects, but their classification depends even more on the tribal organisation than on the diversity of those dialects. This is also true of many low-caste Dravidian tribes, especially in the Nilgiris and Malabar highlands. But the vast majority of the Dravidians and all the Hindus are grouped in different branches, bearing much the same relation to each other that, for instance, the great branches of the Latin family bear to each other in Southern Europe. All have long been fused together in one common ethnical, social, and religious system, while still separated one from the other mainly by their different languages, all derived in Europe from the common Latin stock, in India either from a common Sanskritic or from a common but now extinct Dravidian mother-tongue. These points should be borne in mind in estimating the value of the subjoined general grouping of all the Indian races. It is also to be noted that in a comprehensive classification of the human family the Hindus and Tibeto-Burmans would appear as mixed branches of the Caucasic and Mongol stocks respectively, whereas both the Dravidians and Kolarians would stand quite apart, their possible affinities to the other great families of mankind being still undetermined:—

I. Hindus.

(*Aryan Stock.*)

Language	Dialects / Tribes	Region	Number
Kashmiri	Rambani; Bhadarwahi; Padari; Doda; Kishtwari	Kashmir; Jummu	2,543,000
Panjabi	Sikhs; Jats; Changars	Panjab	17,800,000
Sindi		Sind	2,000,000
Gujarati and Kachi		Gujarat	10,600,000
Marathi and Konkani		Bombay; Central Provinces; Berar	18,900,000
Hindi and Urdu	Urdu; Marwari; Gwalior, etc.	N.W. Provinces; Rajputana; Upper Bengal	100,000,000
Bengali		Lower Bengal	41,300,000
Uriya		Orissa; Ganjam; Chota-Nagpore	9,000,000
Assamese		Assam Lowlands	1,500,000
Nepali (Parbhatia)		Nepal	2,000,000

II. Dravidians.

Language	Dialects / Tribes	Region	Number
Telugu		Sirkars, Nizam's Berar, Mysore	20,000,000
Tamil		Karnatic, Travancore Mysore, N. Ceylon	16,000,000
Kanarese		S. Kanara, Mysore, Kurg	9,500,000
Malayalim		Malabar Coast	5,500,000
Tulu		Kanara, Malabar Hills	500,000
Kodagu		Kurg	150,000
Oraon	Tirki; Ekhar; Barar; Minjar	Chota-Nagpore	370,000
Rajmahal		Rajmahal Hills, N. of Chota-Nagpore	42,000
Khondi	Betiah; Beniah; Maliah	Ganjam; Orissa	320,000
Gondi	Dher; Gottur; Koi; Badiya; Madi; Wardha, etc.	Chota-Nagpore; Central Provinces	1,380,000
Tuda		Nilgiris	750
Kota		Nilgiris	1,100
Sinhalese ?		S. and W. Ceylon	1,700,000
Veddhas ?		Travancore and E. Ceylon	? 3,000

III. Kolarians.

Tribe	Subdivisions	Locality	Number
Santhal	Saran; Murmu; Marli; Kisku	Baghalpur	1,710,000
Munda	Bumij; Ho; Larka	South of the Santhals	850,000
Kharia		Singbhum district; Chota-Nagpore	? 1,000,000
Mal-Paharia; Juang		Orissa, N. of Kattak	
Gadaba		Bustar Hills, left bank lower Godavari	
Korwa		Barwah	
Kurku		About source of the Nerbadda	
Mehto		Chota-Nagpore	
Savara		N. Sirkars	
Bhil	Kala	Vindhyas	900,000
	Ujvala	Gujarat	
	Mina	Malva, Bundi	

IV. Tibeto-Burmans.

(*Mongol Stock.*)

Tribe	Subdivisions	Locality	Number
Ladakhi; Gaddi; Champa; Bunan; Khamba; Balti		Ladakh, Baltistan	100,000
Garhwali	Rongbo; Kohli; Kakka	Garwhal	? 50,000
Kanawari			
Magar; Sarpa; Gurung; Pahri; Sunwar; Kachari; Chepang; Kusunda; Newar; Bhramu; Kiranti		Nepal	200,000
Lepcha; Yayu; Murmi; Limbu		Sikkim	? 80,000
Lhopa; Dimal; Towang; Mechi; Char Duar; Thebengea		Bhutan	? 750,000
Miri; Lutukotia; Angka; Mishmi; Dafla; Abor		North Assam Highlands	? 50,000
Kachari	Hojai; Garo; Mech; Koch; Rabha; Chutia; Tipperah	Goalpara and Garo districts, S.W. Assam	230,000
Singpo; Kuki		S. Assam Hills	? 200,000
Mikir		Nowgong district, Central Assam	62,000
Khasi		Khasia Hills, S. Assam	140,000

			Locality	Population
Nagas.	Central.	Hatigorria; Sema; Lhota; Banpar; Rengma; Primi, etc.	Naga Hills, S. and S.E. Assam	? 200,000
	Eastern.	Tablung; Sangloi; Tengsa; Banfera; Mutonia; Mohongia; Namsang.		
	Western.	Angami; Liyang; Arung; Mao; Muram; Luhupa; Maring		250,000

V. Sundries.

Shans	Ahom; Khampti	Assam	? 100,000
Malays		Maldives; Nicobars ?	250,000
Negritoes		Andaman Islands	10,000
Indo-Arabs		Laccadives	15,000
"Moormen" (Arabs)		Malabar, Ceylon, etc.	200,000
Baluchis	Mari; Bugti, etc.	Sind and Derajat	500,000
Afghans	Afridis; Waziri; Yusafzais, etc.	Derajat and Peshawar	
Swatis, Chitrali, Hunzas		Peshawar, Hindu-Kush	80,000
Persians		Sind	50,000
Parsis		Gujarat, Surat, Bombay	
Eurasians		India generally	536,000
Europeans			

Caste.

The religious and social system of the Hindus is everywhere in India based on the institution of caste, which was originally introduced to uphold the political supremacy of the fair Aryan intruders over the dark aborigines. But before its introduction a considerable intermixture had already taken place, except perhaps amongst the very highest classes of the Aryan conquerors.[1] The indigenous

[1] Dr. Gustav Oppert's investigations, spread over many years, tend to show that these Aryan invaders never were numerous, and that their influence on the aborigines was more social and religious than ethnical. Thanks to their higher culture, they imposed their religion on the masses everywhere throughout the peninsula, and their Aryan speech (Sanskrit) on most of the populations in the Indo-Gangetic regions (*On the Original Inhabitants of Bharatavarsa or India*, 1894).

elements being by far the most numerous, the Aryans were thus threatened with ultimate absorption, and in fact had in many places become largely assimilated to the natives. They could be saved from extinction only by checking further alliances. Marriage with the dark races was accordingly forbidden by the laws of Manu, and a definite rank was assigned to each shade of colour which had already been developed, while the prohibition itself was referred to divine prescription. Hence caste originally meant colour, and had therefore an ethnical value. But once established, the institution gradually acquired an indefinite development, and the four original castes of *Brahmans* (priestly order), *Kshatryas* (warriors), *Vaïshyas* (citizens, traders, agriculturists), and *Sudras* (the menial classes) have in course of ages expanded into minute subdivisions almost past counting. The process seems to be even still going on, and the last census returns give 2500 main divisions, and in Madras alone nearly 4000 minor distinctions.

These, however, probably include the *Pariahs*, or outcastes, a term which originally simply meant "hillmen," and which thus throws considerable light on the institution. It shows that while the three highest orders were reserved for the ruling Aryans, the Sudra mainly comprised the aborigines who had been reduced to a state of thraldom or Helotism; whereas the Pariah embraced the independent highlanders who were excluded from all the social privileges of the Hindu system. Refusing to submit to the conquering race, and successfully maintaining their independence in the inaccessible mountain fastnesses of the Vindhyas, the Satpuras, and the Ghats, where so many of them are still found, they were declared to be outlaws; and the term pariah, or highlander, thus came to be synonymous with outcaste. Hence the outcastes must, to some extent at least, be regarded as the last remnants of the aboriginal elements, and the surviving representatives of a pre-Aryan or prehistoric culture.

Although still flourishing, the institution of caste has been somewhat though not largely affected, first by the settlement and spread of Muhammadans in the land, and then by the establishment of British rule. Hinduism, as a religious system, has always met with the utmost possible toleration both from the Moslem and Christian governments. Hence the Brahmanical or Sacerdotal caste has survived all the political changes by which the land has been convulsed during the past twelve hundred years; but the Kshatrya, or military caste, naturally lost its vitality under Muhammadan princes, and under the present political system, except in the feudatory Hindu States of Rajputana and the Deccan.[1] On the other hand, sacerdotalism and secular tradition have been strong enough to maintain the class distinctions of the Vaishya, and especially of the Sudra order, which last, with the Pariah, comprises most of the Dravidians and Kolarians of the Deccan. Its main subdivisions at present are:—1. Husbandmen; 2. Graziers; 3. Artisans; 4. Writers; 5. Weavers; 6. Field labourers; 7. Potters; 8. Mixed, or broken, mainly sects who have discarded caste and attend to the service of the temples; 9. Fishers and hunters; 10. Barbers; 11. Washermen; 12. Low castes of various degrees, merging in the outcastes.

Redemption from this social yoke will ultimately be found in the spread of education, in such internal upheavals as are foreshadowed by the Brahmo Samáj and other monotheistic movements, in the silent influences of the higher European culture, quickened by the de-

[1] Some of the Kshatryas, such as the Khatri of the Panjab, have even turned to trade. Many of these hold the same social rank in the north that the Baniyas (Banians) do in the Central Provinces and Southern Presidencies, while others still possess an important status civilly and politically.

velopment of the railway system and other levelling institutions.

The Brahmo sect is described by Professor Monier Williams in a paper read before the Royal Asiatic

ARYAN BRAHMIN FROM KASHMIR.

Society, in November 1880, regarding Hindu religious reformers.

Many of the castes still preserve clear indications of the physical distinctions on which they were originally based. This is especially true of the Brahmans, who are everywhere in the peninsula conspicuous for their intelligence, retaining much of the common Aryan inheritance, and displaying the noble cast of countenance which is characteristic of that race.

Religious Sects.

The religious system of the Hindus retains little of the primitive belief of the Aryan race, a few Vedic hymns and formulas recited without being understood by the priests being nearly all that survives of the old cult. In modern Brahmanism there are many sects, some of which have sunk to the lowest depths of the grossest superstition. Such are the Aghoris (Aghor-Pants), many of whom belong to the Brahman order. But the two most widespread sects are the worshippers of Siva and Vishnu, who typify the opposite poles of religious thought, the Vaishnava appealing to the deity he worships as the author of all good, while the followers of Siva seek in man alone and his efforts the attainment of supreme happiness. But apart from this fundamental difference, the two sects often meet on common ground. By Hindus generally Brahm is regarded as the creator, Vishnu the preserver, and Siva the destroyer.

All the civilised Dravidian races—Telugus, Tamils, Kanarese, Malayalims—have long conformed to the Hindu religious system. But nature-worship of a very crude type still prevails among the wild tribes—Tudas, Kolas, Kudagus, Gonds, Khonds—as well as among most of the Kolarians. All these rude hillmen still retain their primitive usages, practise sorcery, and believe in evil spirits. English and German missionaries, however, have been for some time at work amongst them, and have already succeeded in forming a large number of Christian communities, especially in the Santhal and other Kolarian districts.

Within the last twenty years, some judicial trials have disclosed practices of the worst social tendency among a sect in Western India called the Maharajas.

The Kashmirians and Nepalese.

On the other hand, many of the Hindus, especially in the north, have accepted Islám. The Kashmirians, among the finest of the Hindu races, became Muhammadans some centuries ago, and are mostly Sunnis. They are described as almost European in appearance, and in Kashmir we miss the slender frames, prominent

KASHMIRIAN.

cheek-bones, and other unpleasant features so prevalent in other parts of India. The men are of a square, herculean build, well proportioned, and with a frank expression, while the women are fresh-looking and often decidedly beautiful, with an almost Jewish cast of countenance.

Those of the better classes are scarcely darker than the average natives of Italy.

The Kashmirians are a shrewd, witty, and cheerful people, but superstitious and somewhat sensuous. They are skilful artisans and traders, but over-shrewd perhaps in bargains; and although crime in the ordinary sense is almost unknown in the country, Wilson, in his *Abode of Snow*, draws a far from flattering picture of their present social state.

Still there are some Kashmiri Brahmans remarkable for their intellectual ability.

The non-Aryan or Tibetan inhabitants of Kashmir are found chiefly in Ladakh and Baltistan. Frederick Drew, who has carefully studied this region, attributes to the prevalence of polyandry the sparse population of these upland tracts. Here the partly Hinduised Gaddi tribe produces a startling effect by probably the most astonishing of all head-gears, while the Khampas are noted chiefly for their remarkable lung-power. Living in alpine valleys towards the Tibetan frontier, they find respiration difficult at any lower elevation than 11,000 feet above sea-level. Their favourite camping-grounds are the shores of an extensive salt lake in a secluded valley over 13,000 feet above the sea, where they live in tents and practise polyandry. But the kindred Balti race, having adopted the Muhammadan faith, have become polygamists, and are consequently now so numerous that many are compelled yearly to migrate southwards.

In Nepal there is a mixture of races, languages, and religions. The ruling people are the warlike Ghurkas, of mixed Tibetan stock, but assimilated in speech and religion to the Hindus. The rest of the population are partly a mixed Indo-Tibetan race, like the Buddhistic Newaris, partly Bhutiyas—that is, pure Tibetans—who are mostly nomad shepherds, speaking ten or twelve distinct languages.

The Panjabi Hindus and Sikhs.

In the Central Panjab the chief ethnical element consists of the Jats, a tall, hardy, and robust race, with genuine Caucasic features. These Jats of the Indus valley have never adopted the institution of caste in its integrity, and are regarded by the rest of the Hindus with a feeling which is embodied in the expression "Baheka"[1] or "aliens." The Hindus themselves are of a superstitious type, addicted to many peculiar observances. At the birth of a son the priest is summoned to cast his horoscope (yaman-putri), and after forty days to give him a name. During the five first years his hair remains uncut, after which he is usually taken to Ivalamuki, where flames are often seen bursting from the ground, and here his hair is cut with much ceremony by a Brahman. Before his twelfth year his head is shaven, and he is instructed by the family priest in the "sandhya" and "gayatri," or sacred texts from the Vedas, and then receives the sacred thread. He is now considered to have reached his majority, and has to observe the six duties incumbent on all Hindus. He wears a solitary tuft of hair on the crown of his head and assumes the "dhoti," or loin-cloth, with the holy marks in red or white on his forehead.

On attaining his fourteenth year his parents cast about to find him a suitable wife of equal rank with the family. The father of the girl sends the family barber with six dates and a rupee to the boy's house in token of his willingness to accept the alliance. The inmates welcome the

[1] Baheka corresponds exactly to the term "Overn," applied in a like sense by the natives of the Isle of Wight to intruders from the English mainland.

messenger by smearing the entrance with oil, after which the friends meet, the barber throws the dates and rupee into the bridegroom's lap and makes the sacred marks on his forehead.

The usages of the Sikhs differ greatly from those of the orthodox Hindus. They never employ the services of Brahmans, nor do they pay any attention to the Vedas or other sacred Hindu writings, replacing them by a so-called "granth" or "book" of their own, which contains their religious code. They marry somewhat later in life than the Hindus, and are a far more vigorous race. The wedding is conducted by the "granthi," who simply reads some appropriate text from the granth. The Sikhs, who never cut the hair or beard, wear close-fitting trousers and a high turban, invariably containing a bit of steel which must never be laid aside.

Amongst the Muhammadans, Hindus, and Sikhs there are a great number of men called by the name of fakir, and many other names, who lead a life of pious indolence and contemplation at the expense of the poorer classes. These ascetics wander about over long distances or pass their days under the trees, amid the tombs, or at the burning-places, or else herd together like monks in a monastery, under a "mahant" or abbot. Most of them prefer begging under a religious cloak to honest work. In times of political danger or excitement they are mischievous in carrying news, false or exaggerated, from place to place.

The Assamese.

Assam presents even a greater variety of races, religions, and languages than Nepal. The bulk of the inhabitants consists of Hindus, Muhammadans, immigrants from Bengal, and numerous Tibeto-Burman tribes on the highlands enclosing the Brahmaputra basin on three sides. The Muhammadans generally understand Hindustani, which serves as the common medium of intercourse throughout most of the peninsula; and since the Government schools have been opened the educated classes have become familiar with English. But the language of the great majority is the Assamese, a Prakrit dialect closely allied to Bengali. Assam takes its name from the Ahoms,[1] the former rulers of the country, who were originally of Shan (Siamese) stock, but who have become nearly everywhere assimilated in speech and religion to the Hindus. They are a very fine, strong-built race, of rather fair complexion, extremely intelligent, and capable of a high degree of culture. The Ahom dynasty was overthrown in 1810 by the Burmese, who were in their turn ejected by the English in 1827. Since then the Ahoms have become some of the most loyal subjects of the Queen.

The surrounding hills are still peopled by numerous semi-independent wild tribes, such as the Garo, Khasi, Naga, Mishmi, Abor (properly Padam), Kuki, and others, mostly, if not altogether, of Tibeto-Burman stock, whose habits and customs are still but little known. Much valuable information, however, has been supplied regarding the Nagas by G. H. Damant and some of the officers engaged in suppressing the unruly Angami tribes in 1879-80. Mr. S. E. Peal, however, points out that the true form of this word is not *Naga*, but *Noga*, from a root *nog*, *nok*, meaning "people." They are so named in the Borunjis, or "History of the Kings of Assam," dating from the thirteenth century; they are still always called *Noja* (for Noga) by the Assamese, and *Naga* only by the Bengali Babús, probably through a popular etymology and confusion with the naga ("snake") worshippers of

[1] *Ahom* is the same word as *Assam* or *Assom*, *h* interchanging with *s* in Burmese phonetics.

India (*Geo. Proc.*, 1889, p. 90). But it is to be feared that the form Naga is now too firmly established to be set aside, more especially as it has been extended to the land itself as well as to its people, as in the geographical expression "Naga Hills."

The Gonds and Bhils.

Few races present matter of greater interest to the student of human culture than the uncivilised Dravidian and Kolarian tribes of Central India. Many of the Gonds, whose domain in the highlands north of the Deccan is from them called Gondwana, were formerly employed in the coal-pits of the Narbada valley and its tributaries. From their infancy they are accustomed to look on every rock, every river, gorge, and cavern, as the abode of a special spirit, who may be propitiated and rendered harmless by some simple rite.

Amongst the Bhils of the Vindhyas there are many superstitions showing a striking analogy to those of the West. When a Bhil goes out to fight or rob, if the byru bird is on his right hand he will prosper; if on his left, nothing will induce him to go. The belief is very strong in witchcraft, and in the powers of the Burwa, or witch-finder, who is consulted in all important cases. "Should any person die without apparent cause, the friends inquire of the burwa, who selects the ugliest old woman in the village, and oracularly attributes the death to her spells. She is thereupon seized and tried, much in the same way as in Europe two centuries ago" (Col. Kincaid).

The Nilgiri Hill Tribes.

Many of the dark aborigines of the Nilgiris and other parts of the south, although classed with the Dravidians, seem to bear a much greater resemblance to the Kolarians. Such are the *Kallar*, or "Robbers," on the Tanjore frontier, who "by no means disown their profession or consider it discreditable. Indeed, the caste ranks high among the Sudras, and they have a king, the Tondiman Raja, who has always been a faithful ally of the British. The present well-educated and enlightened Raja receives a salute of twelve guns when visiting Madras. Unscrupulous as thieves, they are men of their word, and to this day are employed by the English residents of Trichinopoly to watch their houses — a trust they faithfully fulfil, and keep off all other thieves. Their skill in tracking equalled that of any savages. Their ordeals and marriage customs agree generally with those of the Bhils, and like them they live in continual dread of witchcraft, being often driven to cruel deeds in revenge for supposed injuries. They are now fast becoming peaceable cultivators" (M. J. Walhouse).

In the Nilgiris dwell the Tudas, Kotas, and one or two other remarkable tribes, Dravidian in speech, but otherwise quite distinct both from the Dravidians of the plains and in some respects even from each other. The practice of polyandry would seem to point at the Bhutyas of Tibet as their remote ancestry, but for the fact that this custom is not confined to that race. Some of them, although not leading a nomad life, present in many particulars a great resemblance to the European gypsies. They betray no trace of a religion beyond what may be implied in a firm belief in witchcraft. They are all very peaceful and inoffensive, occupied either with agriculture or stock-breeding, but these hill tribes seem to be dying out. The Kotas number little over a thousand, while the Tudas have been reduced to about 700 or 800.

The Parsis.

Amongst the minor heterogeneous ethnical elements of the peninsula, one of the most interesting are the Parsis of Bombay, the direct descendants of the Ghebrs, or Old Persian fire-worshippers, who fled to India when the Muhammadan invasion burst with all its fury over the Iranian tableland in the seventh century. Since then they have kept entirely aloof from the surrounding peoples, preserving their race and religion alike intact from all extraneous influences. They are remarkable for their general intelligence, business habits, and commercial ability; and they sympathise with the English far more than with any of the native races. In proportion to their numbers, they are the wealthiest and most influential section of society in Bombay, as well as the most loyal and devoted subjects of the Queen in India.

PARSI OF BOMBAY.

The Indian Muhammadans.

In this respect they present a striking contrast to the Muhammadans, who must always be regarded as an element, if not of danger, at least of anxiety, to the central government. The Indian Muhammadans, who are chiefly Sunnis, with an influential Shiah minority, are concentrated chiefly in Bengal, the North-West Provinces, and the Panjab, and number altogether (1891) 57,000,000, or 19 per cent of the whole population, so that the Empress of India rules over far more Mussulman subjects than any other sovereign in the East. The Lieutenant-Governor of Bengal alone "has in his jurisdiction as many millions of Moslems as the Sultan of Turkey, and thrice as many as the Shah of Persia. The Indian Muhammadans are met with on all the coasts, and are emphatically the sailors of the Indian seas. In the interior they are urban rather than rural, employed in some branches of commerce, in retail dealing, in skilled and refined industries, in the army, in public and private service, but seldom connected with agriculture, save in the capacity of landlords. In Sind, however, the agricultural population is Muhammadan, both landlords and cultivators. In eastern and northern Bengal, in the region comprising the Brahmaputra basin, and in the united delta of the Ganges and Brahmaputra, the tenants and cultivators are also Muhammadan, while the landlords are Hindu, with the exception of some prominent and meritorious gentlemen of the Muhammadan faith."

"Elsewhere in India the Muhammadans, being scattered, do much to leaven the mass of native opinion. Besides the discontent engendered among them by historic memories, there is one special circumstance affecting their contentment. Under native rule they enjoyed a large portion, perhaps the lion's share, of the State patronage, and at the outset of British rule were found in the front everywhere. But nowadays they are beaten by Hindus in the open competition of mind with mind. It is to this that the Muhammadans themselves attribute the fact that they are falling in wealth and status while the Hindus are rising."

"The temper and disposition, politically, of the Muhammadans form one of the many sources of anxiety in India. Some years ago the religious revival commenced by the Wahhabis in Arabia, the breeze of fanaticism which ruffled the surface of the Muhammadan world, and other causes difficult to define, excited the Indian Muhammadans considerably. Plots were discovered and State trials instituted; some grave and melancholy events occurred which need not here be recounted. Within the most recent years, however, the Indian Muhammadans have become comparatively well affected. Be the reasons what they may, the symptoms of disaffection among them have of late abated" (*India in* 1880, by Sir Richard Temple).

An offshoot of the Muhammadan community are the Khojahs, the real descendants of the famous assassins of the Middle Ages.

The English and Eurasians.

The dominant English race are still almost aliens in the land. They have nowhere formed any agricultural settlements or permanent trading communities, nor is it likely that any serious attempt will ever be made by them to colonise even the more healthy and temperate upland districts in the Himalayas or highlands of the Deccan. Numerous sanitaria have almost everywhere been established in the more favourable sites in these districts. But such places are merely visited periodically by the officials and military, who escape during the summer season from the almost intolerable heats of the plains. It may be questioned whether three generations of Englishmen are anywhere to be found in Simla, Darjiling, Mussurie, Utacamand, or any of the many other health-resorts dotted over the uplands of the peninsula.

Nor has any advance been made towards a fusion of the ruling and subject races. The Anglo-Saxon holds his head even higher than the haughtiest Rajput chiefs claiming descent from the gods and demigods of Hindu mythology. In former times alliances and other connections used to be formed between Europeans and Native females, but the result has not been such as to encourage a general spread of the practice. The offspring of European fathers and native mothers, called East Indians or Eurasians,[1] hold much the same position in relation to English society that the quadroons or octoroons do to the white classes in the United States. They do not exhibit any marked idiosyncrasy of race. Although both parents may belong to the Aryan stock, and although the English fathers are often distinguished by their physical qualities, and their Indian mothers by personal attractions, the Eurasians themselves do not generally display a striking appearance. They possess many intellectual endowments; but though quick of apprehension, they seldom acquire solid knowledge so well as Europeans, nor have they equal perseverance. From their mothers they seem to inherit gentleness and amiability. Among them individuals are found eminent in character and ability.

It fares still worse with the pure-blood European children, who are constitutionally unable to struggle against the enervating effects of the climate, especially in the Ganges valley. Till their sixth year they retain the high complexion of the race and seem healthy enough,

[1] That is, *European-Asians.*

but on entering their teens they begin to lose their fresh colour, their features grow pale and wan, the weariness of premature decay, or some unaccountable secret blight, steals over them. Without any decided outward symptoms of disease, they droop and pine away like hothouse plants deprived of light and air. This light and air must be sought in the home of their forefathers before they attain their sixth year, for nothing but a speedy removal to the fickle but invigorating climate of England will now save them from an early grave or from physical deterioration. Of the Anglo-Indian children thus brought up in Europe, many of the young men return to India before their twentieth year in order to make a career for themselves in the civil and military services, or else to fill positions secured for them in commercial houses or other employments.

8. *Topography.*

Although consisting mostly of agricultural and rural elements, the population of India is so enormous that enough still remains to overflow into many cities of the first magnitude. The number of large towns is also increased by the many administrative divisions and native feudatory States, each with its special capital, the centre of the government of the civil and criminal courts, and of other independent local interests. Thus it happens that besides nearly half a million rural villages there are no less than forty cities with populations of 50,000 and upwards. Many of these are mere aggregates of houses built of dried earth, with roofs of tile or thatch. But others not a few are of vast antiquity, the changeless capitals of shifting empires, reflecting in their monuments the varied tastes of many successive cultures, abounding in antiquarian and art treasures of every sort.

Srinagar.

In the upland regions of the Himalayas one of the most interesting places is Srinagar, the capital of Kashmir, which stands on the banks of the Jhelum in the midst of some of the loveliest scenery in the world. Like Venice, Srinagar is a city wherein the streets consist of numerous canals, or rather branches of the river, which traverse the place and connect it with the neighbouring Lake Dal. The canals are flooded by means of sluices from the lake when the Jhelum is low. But when the river rises above the level of the lake, the sluices are closed by the pressure of the back flow. There are shady poplar avenues in the neighbourhood. The lake is enlivened by the presence of water-fowl with brilliant plumage, while above its lotus-fringed banks majestic trees stand out against the azure sky. On the Dal itself, which is 5 miles long by 2 broad, the fantastic floating gardens recall the chinapas or swimming islands of Lake Tezcuco, near Mexico. Amongst the varied vegetable growths that here delight the eye, conspicuous is the thorny water-nut (*Traba bispinosa*), yielding a delicious flour and bread. About 60,000 tons of this substance are yearly produced at the larger Lake Wular in the same district. The Jhelum is spanned by several picturesque wooden bridges. Near the capital is the Takht-i-Suliman hill (Solomon's throne), from the top of which is seen the panorama of Kashmir, the finest landscape in the Indian Empire.

The only other large town in the Himalayan States is Katmandu, the present capital of Nepal, including the old capital of Patan close by, lying in a productive and well-watered valley in the heart of the country. The two capitals together make a very interesting locality, with good streets, pleasant houses, many temples of unique style and beauty, in its appearance betraying a certain mixture of Indian and Chinese elements.

TOMB OF RUNJIT SING, LAHORE.

Lahore—Delhi—Karachi.

In the Panjab, the most considerable places are Peshawar on the Afghan frontier, the present terminus of

CITY OF SRINAGAR.

the Indian railway system towards northern Afghanistan;

STREET IN PESHAWAR.

Lahore, capital of the province, on the Ravi, nearly due south of Srinagar; Amritsar, the old religious capital of the Sikhs, a few miles farther east; Multan, near the united Chenab and Jhelum, a few miles above the junction of the Satlaj; Delhi, on the border of the North-West Provinces in the Jamna valley. Lahore, which is a great railway centre, has in its neighbourhood many ruins of former brilliant epochs, and is still adorned with some fine palaces, mosques, mausoleums, and bazaars.

But none of these places can compare in interest with Delhi, which was for centuries the proud capital of the Mogul Empire, and the centre of the Moslem world in India. The present city occupies a circuit of little over 7 miles, in the midst of vast ruins, covering an area of 20 square miles. Yet its former greatness is still attested by several magnificent buildings, conspicuous amongst which are the Jama-Masjid, the largest and finest mosque in India, and the palace of the emperor Shah Jehan. The canal, 120 miles long, conveying water from the Jamna where it enters the plains, has been restored by the English. But Delhi never really recovered from the blow inflicted on it in 1739 by Nadir Shah, who carried off vast treasures in gold and precious stones, estimated at from eighty to over a hundred millions sterling. Amongst the prizes of conquest was the famous Koh-i-nûr diamond, the most highly esteemed heirloom in the family of the Mogul dynasty. After a series of almost fabulous accidents, this gem ultimately became an appanage of the Queen of England, who, as Empress of India, inherits all the possessions of that dynasty.

Haidarabad, which stands on the left bank of the Lower Indus, and the flourishing seaport of Karachi at the western extremity of the delta close to the Baluchistan frontier, are the chief places in the province of Sind. Karachi, which lies close to the Baluch

(From Photo. by Frith & Co.)

THE MAUSOLEUM OF AKBAR AT SEKANDRA, A SUBURB OF AGRA.

frontier, is the terminus on the Arabian Sea of the Indus Valley State Railway. Defensive works have here been undertaken, and much has lately been done to improve the harbour, which is somewhat obstructed by a bar, and affords room only for a limited number of large vessels.

Agra—Cawnpore—Lucknow—Allahabad—Benares.

Few regions in the world present such an array of splendid cities as those which line the banks of the main streams along the Ganges-Jamna valley for a distance of considerably over 800 miles. Between Delhi, capital of the old empire, now arbitrarily included in the Panjab province, and Calcutta, capital of the new Imperial India at the opposite extremity of this vast river basin, there follow in majestic procession such memorable places as Agra, Cawnpore, Lucknow, Allahabad, Benares, Mirzapur, Patna, Murshedabad, and Dacca.

At Agra, which, like Delhi, stands not on the Ganges but on its great tributary the Jamna, artistic interest must ever be centred in the Moti-Masjid and Taj-Mahal, two buildings of surpassing loveliness, in which Muhammadan architecture reached its acme under the Mogul emperor Shah Jehan. The Moti-Masjid, or "Pearl Mosque," stands within the enclosure of the old imperial palace, and though inferior in size, is perhaps superior in design and harmony of proportions to its rival, the Jama-Masjid of Delhi. It is built entirely of white marble, and, with its glorious cupolas, arcades, and lovely surroundings, presents a picture of enchanting beauty, surpassed only, if surpassed, by the peerless Taj-Mahal. This mausoleum, raised by Shah Jehan at a cost of three millions sterling over the grave of his beloved empress Mumtaz-i-Mahal, combines within itself more varied elements of beauty than almost any other building in the world. Site, size, general design, symmetry of parts,

(*From Photo. by Frith & Co.*)

THE TAJ-MAHAL AT AGRA.

exquisite finish of details, choice materials, play of colour, and all the delightful surroundings, afford a vision of supreme loveliness, which, seen especially when bathed in the liquid atmosphere of a clear moonlight night, leaves an undying impression on the memory of the spectator.

The Jamna is here spanned by a railway viaduct right opposite the Mogul fortress and city.

Cawnpore and Lucknow are names inseparably associated with the most thrilling events of the Indian

RUINS OF THE RESIDENCY, LUCKNOW.

Mutiny. The most sacred sight in India for men of English blood still must be the monument raised over the well at Cawnpore to the memory of the slaughtered innocents, whose piteous fate inspired their avenging fellow-countrymen with the heroism displayed in the defence and relief of Lucknow. Both places present in other respects many points of interest, although the ambitious palatial structures of Lucknow plainly mark a period of decadence in the Muhammadan architecture in India.

THE FORT, ALLAHABAD.

Standing at the confluence of the Ganges and Jamna, and nearly midway between Bombay and Calcutta on the Great Indian Peninsular Railway, Allahabad occupies perhaps the most central point of the empire. The Jamna is here spanned by a very fine railway viaduct.

commanding a view of the tongue of land at the confluence of the two rivers, which is held by a strong fortress containing an arsenal. The native city of Allahabad is not handsome, and has no buildings of note except the Muhammadan tombs in the Khusru gardens. But the European quarter has of late years become very fine, with its railway station, its military barracks, and its civil structures.

The first great city on the united Ganges and Jamna below Allahabad is Benares, which holds the same position in the Brahmanical that Delhi does in the Moslem world. It is crowded with palaces and Hindu temples, and, although none of these are of great size, the numerous towers, cones, spires, minarets and porticoes, and flights of steps, present an almost unrivalled river-frontage, nearly three miles in extent. The river view of Benares is one of the most characteristic in the empire.

But the interior of this city is far from inviting, with its close, dirty, and irregular streets, rickety houses, nauseous smells, repulsive mendicants, and stifling atmosphere. The great number of palaces is due to the fact that the Hindu chiefs and princes in every part of the empire endeavour to secure a residence in this sacred city, which during the festivals is crowded by pilgrims from all quarters. The innumerable little temples are compared by Bishop Heber to so many shrines "stuck in the angles of the streets and under the shadow of the lofty houses. Their forms, however, are not ungraceful, and many of them are covered over with beautiful and elaborate carvings of flowers, animals, and palm-branches, equalling in minuteness and richness the best specimens of Gothic or Grecian architecture." This description, though written many years ago, is applicable to this day.

Benares has always been a chief centre of Hindu learning, and the Sanskrit College founded here in 1792 is still the principal seat of native instruction in India.

Muhammadanism is also largely represented in Benares, where there are said to be as many as 300

BENARES.

mosques, including a structure with lofty minarets, erected by Aurengzeb on the site of a demolished Hindu temple.

Patna—Murshedabad—Dacca.

Lower down the river is the great trading city of Patna, capital of Berar, where the produce of the poppy is collected in order to be prepared as opium, to be sent to Calcutta for exportation to China. The city proper, within the crumbling old fortifications, occupies a comparatively small space; but the handsome suburbs, with their numerous mosques, temples, streets, and gardens, stretch nearly 10 miles along the river-bank. Patna is much more a Moslem than a Hindu town, and its Mussulman inhabitants have the reputation of being amongst the most fanatical in India. It is also a large industrial centre, and many of its linens, lacquered and other wares, find a ready sale at the great annual fair held at Hajipur on the opposite side of the river.

Near Patna is the railway viaduct over the Sone River, an affluent of the Ganges, one of the longest to be found in any country.

Between Patna and Calcutta the most important town is Murshedabad, on the Bhagirati near the head of the delta. It is a very large place, extending some 8 or 9 miles along both sides of the river. But though it has a large trade, and is in some respects flourishing, those parts which depended on the former court and camp of the Muhammadans present an appearance of decay. The Nawab's palace, however, is a fine structure, built in the European style.

Dacca has always been the centre of the Muhammadan world of Eastern Bengal. It has a flourishing trade, though some of its fine and delicate manufactures have decayed. Its climate is unfavourable. It presents a handsome frontage towards the river.

Calcutta—Jaganath.

On the Hugli, westernmost and largest branch of the Ganges delta, and about 100 miles from its mouth, stands

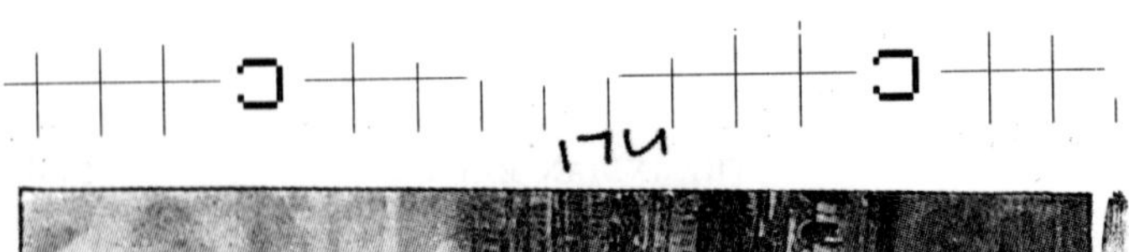

Calcutta, the modern capital of the Indian Empire. It is divided into a European and native city, jointly covering an area of some 15 square miles. The European quarter, which is inhabited chiefly by the English, has a Western aspect, being laid out with fine spacious thoroughfares, which in the Chowringhee or aristocratic quarter are lined with many fine public buildings and large private residences. It is the frontage of Chowringhee which has caused Calcutta to be called "The City of Palaces." The houses are built in an architectural style peculiar to Bengal and suitable to the climate. This style is handsome as well as commodious, and may be regarded as an instance of originality on the part of the English. The native city also, in which the native population is collected, has broad straight streets, well laid out, and in that respect differs from the aspect of an ordinary Eastern town. It is interspersed with fine public buildings and some native houses built in the English style; otherwise the native houses are poorly built, quite inferior to those of the other capitals in India, the climate of Bengal being unfavourable to native architecture.

But all alike have easy access to the pleasant Eden gardens, which with their tropical vegetation and refreshing ornamental waters form a charming foreground to the surrounding government buildings. Here the winding waters, the varied foliage, the amphitheatre of handsome edifices, the forest of masts from the shipping in the near distance, the guns of Fort William overlooking the animated scene, produce a very pleasing impression.

Calcutta is fairly well supplied with water pumped from the Hugli into filtering beds, whence it is conveyed through pipes for a distance of 14 miles to the city. Above the harbour the river is crossed by a pontoon bridge, which is one of the best works of its kind existing in any country, and gives easy access to the large and rapidly-increasing suburb of Howra on the opposite bank. Below this bridge the ships are moored together with strong chain cables along the quays and jetties, an arrangement adopted as a precaution against the tremendous cyclones to which the delta is exposed. The intricate navigation of the Hugli, with its treacherous sands and constantly-shifting shoals, is conducted by a pilot service admirably organised by Government, and composed exclusively of Europeans with their headquarters in Calcutta. Hence this great capital may be regarded as tolerably safe from the attacks of hostile fleets, which would be wrecked were they to venture into the river without competent pilotage.

Jaganath (Juggernauth), the most celebrated shrine in India, lies on the Orissa coast not far from the Madras frontier, and about 50 miles south of Kattak. Twelve great festivals, attended by over a million pilgrims, are here annually held in honour of Vishnu. Here is to be seen the huge car which, according to tradition, was supposed to be dragged over the bodies of devotee-victims. The great temple, which was finished at enormous cost in the twelfth century, stands at the head of the main thoroughfare.

Near here, on the sea-shore, stands the grand Hindu ruin known as "The Black Pagoda."

Nagpur—Jabalpur—Bhopal—Indore—Gwalior—Jaipur—Udaipur.

In the Central Provinces the only places with populations exceeding 50,000 are Nagpur and Jabalpur. Both are connected with Bombay by north-eastern extensions

of the Great Indian Peninsular railway system, one branch of which has a central station at Nagpur. Not far from the northern frontier of these Provinces are the important towns of Indore and Bhopal, in the political system which is termed the Central India Agency. Near Bhopal are the Buddhist remains known as the

JAIPUR.

Bhilsa Topes. Indore is the capital of Holkar's possessions, with one of the finest British Residencies in the peninsula.

Due north of Bhopal and about 70 miles south of Agra stands the famous fortress of Gwalior, one of the largest and strongest in the empire. It occupies the level summit of a steep rocky hill 350 feet high, rising abruptly from the surrounding plain, and completely commanding the city of Gwalior, capital of Sindhia's dominions, which lies at its base. Perennial springs, reservoirs, and cultivated grounds are enclosed within the walls of the stronghold, which is accessible only by steps hewn in the perpendicular side of the rock on which it stands. Yet this apparently impregnable fastness was twice stormed by the British—in 1780 by Major Bruce with a handful of native troops, and again in 1858 by Sir H. Rose, when held by a strong body of mutinous sepoys.

Jaipur is the principal town in Rajputana. It is quite modern and well laid out. In respect to arrangement of streets, it is superior to any native city in the empire. It is the seat of much wealth and commerce.

Udaipur is the very focus of heroic and chivalric traditions. Its palace-crowned hills, its tombs, its lakes and islets, make it the most picturesque city in the empire.

Madras—Bellary.

In the Madras Presidency there are few cities of large size, but many of great historic and antiquarian interest. None have a population of more than 60,000 except the capital, Trichinopoly and Calicut. There are many disadvantages in the site occupied by the city of Madras on the open, surf-beaten shores of the Coromandel coast, exposed for months together to the full fury of the north-eastern monsoons. Nor is the climate much more favourable, being intensely hot in summer and not entirely free from the malaria so prevalent along the eastern seaboard. Yet in spite of these adverse outward conditions, Madras has under British rule expanded into a flourishing city of nearly half a million inhabitants, with many stately public buildings, literary and scientific institutions, educational and charitable foundations. Something has even been done to improve the harbour, or rather to create one, by the construction of a large pier of great strength and size, which is capable of further extension. But with nothing but an open and shelving roadstead Madras can never become a great seaport, and must depend for its future expansion almost entirely on the system of railways by which it is already connected across the peninsula with Bombay, northwards with the Nizam's Dominions and Central Provinces, southwards with Pondicherry, Mysore, Trichinopoly, Madura, and Tuticorin. Madras is protected by Fort St. George, one of the earliest strongholds of the East India Company, and at present one of the arsenals of the empire. In its vicinity, on the sea-shore, are the rock-cut temples of Mahabalihuran, celebrated by Southey's poetry.

Near the main line from Madras to Bombay are the native rock-fortress and the European cantonment of Bellary. Near Bellary, again, are the wonderful extensive ruins of the Hindu city of Bijayanagar.

Trichinopoly—Madura—Tanjore.

Trichinopoly, the next largest place in the Presidency, lies in the fertile Kavari valley, a few miles west of Tanjore, and close to the famous temple of Srirangam. These Hindu buildings, which are amongst the most remarkable of their kind in India, occupy the western extremity of a large island in the Kavari, where the chief pagoda stands in the centre of seven separate square enclosures, with a total circuit of nearly four miles. It is a vast structure, surmounted by a gilded dome, beneath which is the statue of the presiding deity, one of whose glittering eyes, abstracted in the last century by a

TRICHINOPOLY.

French deserter, proved to be a diamond of almost matchless purity. This gem, known as the Orloff diamond, now figures as the chief ornament in the imperial sceptre of Russia. Trichinopoly is commanded by a strong fort, perched on a steep granite peak, 500 feet high. It is noted for its peculiar style of gold work. In the Protestant Church of St. John repose the remains of Bishop Heber, interred here in the year 1826

South of Trichinopoly, and connected with it by rail, lies the ancient city of Madura, with its truly magnificent temples, and other monuments of Hindu art. The palace, built by Tirumal Naik, a former ruler, the finest structure of its kind in India, never fails to excite the astonishment of visitors, who stand amidst its vast arcades, courtyards, vestibules, reception chambers, and halls, with their vaulted roofs and arches. Tanjore, at the head of the Kavari delta, also possesses some famous Hindu monuments, including a sacred bull 20 feet high, hewn out of a single granite block. It stands in one of the palace courtyards, but even modern engineers still marvel how it was carved and transported to its present site.

Calicut—Mysore—Seringapatam—Bangalore.

Calicut, on the Malabar coast, lies in one of the most fertile districts of the peninsula, yielding pepper, ginger, cotton, cardamoms, and other tropical products in vast profusion. This was the first Indian seaport visited by Vasco da Gama in 1498, and from the peculiar cotton fabric here formerly manufactured the *calicoes* of the modern European looms take their name.

A line drawn from Calicut north-eastwards will very

nearly intersect Mysore, Seringapatam, and Bangalore, in every respect the three most interesting places in the State of Mysore, which has again been placed under the administration of the native Raja. The city which gives its name to the State forms a pleasant aggregate of regular streets, avenues, gardens, and temples, the whole commanded by a strong fort, constructed from European designs. This stronghold, which is separated by an esplanade from the city, encloses within its precincts the Raja's palace, besides the dwellings of many wealthy citizens and other private buildings. But the British Residency lies some 5 miles farther south, on the summit of Mysore Hill, 1000 feet above sea-level.

Seringapatam, on the main head-stream of the Kavari, is chiefly noted for its fortress, which figured so prominently in Indian history during the closing decade of the last century. This formidable stronghold of Tippu Sultan occupies the west side of a large island in the river, and although considered quite impregnable, was finally stormed by the British in 1799. Its streets, houses, and fortifications remain, but it is now a city of the dead.

Bangalore, which lies almost exactly midway between Madras and Mangalore on the opposite coast, and nearly 200 miles from both points, is by far the largest city in the interior, south of the Kistna valley. Yet it is quite a modern place, having been founded by Hyder Ali about 1780 as a bulwark against the English. The fort has long been disused; but, thanks to its central position in the midst of an extremely fertile district, the town soon acquired a rapid expansion. Lying at an elevation of 3000 feet above the sea, on the Mysore plateau, it enjoys a delightful climate, and is consequently a favourite resort of Europeans. Here is a large British cantonment, with extensive barracks, library, public gardens, racecourse, and other attractions. From a combination of happy circumstances, Bangalore has thus become, in a few decades, the chief centre for the diffusion of Western ideas amongst the Dravidian inhabitants of the interior of Southern India.

Haidarabad—Secanderabad.

Haidarabad (Hyderabad), capital of the Nizam's Dominions, occupies a somewhat central position in a fine climate, the eastern terminus of the native state railway running thence to join the Madras main line. It is the largest city in the whole of the Deccan, with a present population of over 415,000, and with a picturesque situation. There is a handsome British Residency, one remarkable mosque, and one fine gateway. Haidarabad is much more a Moslem than a Hindu city, Pathans, Arabs, and Rohillas being here numerous.

The neighbouring town of Secanderabad may be regarded as a European quarter, this being the headquarters of the British subsidiary force in the Nizam's territory. Here are some of the largest and best-constructed cantonments in India, with extensive barracks, hospital, Protestant and Roman Catholic churches, Masonic Lodge, promenades, public libraries, racket courts, lawn-tennis grounds, and racecourse.

Near here are the old citadel and the mausolea of Golconda.

In the north-west corner of the Nizam's Dominions are the rock-cut temples and caves of Ellora (Hindu) and Ajanta (Buddhist), also the hill-fortress of Daolatabad.

Bombay.

Bombay, capital of the Presidency, is not only the most flourishing city in the Indian Empire, but possesses probably more elements of future greatness than any other city in Asia. It occupies the south-east end of the island of like name, which is 8 miles by 3, and which is connected by a mound with the larger island of Salsette. These, with Elephanta and two or three others, form a

TOWN HALL, BOMBAY.

little group close to the Konkan coast, in 18° 53′ N., 72° 48′ E., jointly enclosing with the mainland one of the most commodious and expansive harbours in the world.[1] The space available for shipping is nearly 14 miles long and about 5 broad, with an average depth of 10 to 12 fathoms. This splendid natural position has been greatly improved by artificial works, including extensive quays, wharves, and several docks, the finest of which is the

[1] The word Bombay—*i.e.* Bom Bahia—means in Portuguese "Good Harbour," although some take it to be a corruption of Mumbai, a small island named after the goddess Mumba.

Prince's Dock, with an area of 30 acres, completed at a cost altogether of over a million sterling.

The city consists properly of two parts, a native and European quarter, the latter stretching along the shore of the bay, where a line of magnificent buildings presents an imposing view when seen from Malabar Hill, at the south-west point of the island. The native city has several long streets, which are the finest in the Indian Empire. It is well supplied with good water, brought through pipes from two large artificial lakes embosomed in the picturesque wooded hills forming the advanced spurs of the Western Ghats, which here approach to within 20 miles of the coast. Although the scheme of defences is still incomplete, Bombay is already defended by several formidable batteries, as well as by two ironclad turret-ships permanently stationed at this port. In case of danger the whole of the shipping might also find absolute security in the inner waters behind the island of Elephanta. When we add that Bombay is the first important place reached by vessels from Europe and the Suez Canal, and that it is directly connected by several railway systems with every part of the peninsula, it will be seen that this great seaport lacks none of the elements calculated to secure it a foremost position amongst the cities of Southern Asia. Between it and Calcutta there is an honourable rivalry for the first position. Each city has advantages peculiar to itself, and it is hard to say which of the two will ultimately prevail.

Ahmadabad—Baroda—Surat.

The Presidency also contains several other large cities, the most important of which are Ahmadabad, Baroda, and Surat in the Northern, Puna and Sholapur in the Central, Dharwar and Belgaum in the Southern Division.

Ahmadabad, which is a very large place at the neck of the Gujarat peninsula, equidistant from the Rann of Katch and the Gulf of Cambay, contains many beautiful specimens of Muhammadan architecture. Unfortunately some of these monuments, including the great mosque of Sultan Ahmad, were shattered or destroyed by the terrible earthquake which seriously injured the place in 1819. Still many fine structures remain to delight the student of architecture. The city is now a great centre of Oriental art, producing exquisite specimens of damascened metal-work, gold and silver plate, mother-of-pearl objects, rich trappings and caparisons for the native princes.

The late Gaikwar of Baroda was deposed by the paramount power for maladministration. He was held by the Government of India to have been guilty of an attempt to poison the British Resident by a dose of diamond-dust. Baroda itself, which lies nearly midway between Ahmadabad and Surat, has prospered in the sunshine of Maratha royalty, and is a fine city, though not remarkable for architecture.

Surat occupies a convenient position near the mouth of the River Tapti, about 160 miles by rail due north of Bombay. It is the natural emporium of the rich Kandeish valley, and covers a large space some 8 miles in circumference on the left bank of the river, 20 miles from the Gulf of Cambay. In the early days of the East India Company it was the principal trading-place on the west coast, but during the last century it has become quite secondary to Bombay.

Puna—Sholapur—Bijapur.

No place in the Central Division of the Bombay Presidency can compare in importance with Puna, which is delightfully situated about 80 miles south-east of Bombay on the Deccan plateau, some 2000 feet above the sea. With its large British cantonments, hospitals, libraries, churches, colleges, and missionary schools, this famous capital of the Peishwas, or heads of the great Maratha confederacy, has in our days become the chief centre for the spread of European culture among the brave but somewhat turbulent Maratha races of Western India. The palace of the Peishwas, built of teakwood, a noble specimen of Maratha architecture, was burnt in 1879.

South-east of Puna, and close to the Nizam's frontier, lies the town of Sholapur, a former stronghold of the Marathas, with two distinct lines of fortifications.

A far more interesting place is Bijapur, which lies some 60 miles farther south on a small tributary of the Kistna; it was the capital of the Muhammadan kingdom, which comprised the Western Deccan, before the establishment of the Mogul Empire. The extent and splendour of the ruins attest the former greatness of this "Palmyra of the Deccan," as it has been called. These ruins,—which are remarkable especially for their great solidity and simple grandeur, and yet a suitable degree of ornamentation,—consist of Muhammadan palaces, mosques, and other structures, many of the domes, spires, and minarets of which are still standing. Among these is a mausoleum, with a cupola, the admiration of architects and the largest yet constructed in the world.

Satara—Ahmadnagar.

Satara and Ahmadnagar are the only other places in the Presidency which call for special mention. They both lie on the Deccan tableland and on small head-streams of the Kistna, the former 70 miles south, the latter 80 miles north-east of Puna. Ahmadnagar has a few good streets and substantial buildings enclosed by a wall, beyond which are a strong stone fort of historic celebrity, a finely-built palace, and on the crest of a neighbouring hill the tomb of Salabat Jung. Satara—much associated with stirring passages of Maratha history—is clustered round the base of a rocky eminence rising 800 feet above the surrounding plain, and crowned by the ruins of an ancient citadel. In the neighbourhood are European cantonments, which enjoy a favourable and healthy climate.

9. *Highways of Communication: Canals—Roads—Railways.*

Under the British administration a system of internal communication has been rapidly developed, which in this respect places India nearly on a level with the most civilised regions of the globe. Apart from the natural channels of the great rivers and their affluents, affording over 10,000 miles of navigable water highways, the irrigation canals, which are constantly increasing, are often navigable by small craft for hundreds of miles. Many of the larger ones have been specially adapted to this purpose, and by a wise provision have thus been made to serve a twofold object. The canals near Calcutta and in Orissa, and those of the Madras Presidency, are largely utilised in this way. The irrigation system has already assumed magnificent proportions. The chief scenes of these operations are the country between the Jamna and the Ganges, several parts of the Panjab, and the deltas of the Mahanadi, the Godavari, the Kistna, and the Kavari on the east coast, and the delta of the Indus on the north-west coast. Of main channels, great and little, there are no less than 13,000 miles completed, besides countless distributing rills with a total length of over 10,000 miles in the north alone. Thus have been brought under irrigation about 8,000,000 acres, mostly of extremely fertile land, at a total expenditure by the State of £22,000,000 sterling, yielding an average interest of 6 per cent.

Although occupied for ages by settled communities, which had attained a high degree of culture long before Britain had emerged from barbarism, India seems to have possessed scarcely any roads before the advent of the English. Neither the ancient Hindu dynasties nor their Moslem conquerors paid any attention to this primary condition of true civilisation. Many of the petty rulers were even directly opposed to the development of easy lines of communication, which would have the immediate effect of opening up the country to the attacks of hostile neighbours.

Now all this is changed, and although the system is still far from complete, over 33,000 miles of metalled or macadamised highways have been constructed, mostly within the last sixty years. Thus all the great cities have been brought into direct communication with each other, and the uttermost limits of the land have been made accessible to trade and to defensive or offensive warfare.

The great trunk lines are those running from Calcutta for 1000 miles to Delhi, and thence through

Lahore to the frontier at Peshawar; from Bombay for 900 miles to the last-mentioned through Agra; from Bombay for 800 miles over the Western Ghats and across the Deccan to Madras; from Bombay through Gujarat; from Madras northwards to Bengal, southwards to Trichinopoly and Madura, westwards through Bangalore to the Malabar coast. Important sections of the system are also the routes running in the Himalayas from Amballa to Simla and beyond it towards Chini, and from the Bengal plains to Darjiling and thence to the Chola range on the Tibetan frontier; in the Deccan the roads connecting Mirzapur on the Ganges through Jabbalpur and over the Satpura range with Nagpur in the Central Provinces; the line running from Poona southwards to Mysore, and that ascending from Coimbatore to Utacamand in the Nilgiris.

Most of these highways are solidly constructed, and often present splendid specimens of engineering skill in their gradients, cuttings, causeways, and bridges. Like the old Roman roads, they are in many places carried right over the Ghats, Vindhyas, and other ranges, and through such difficult passes as the Thal and Bhore in the Western Ghats. The section between the Jhelum and Indus, in the extreme north-west, consists of an almost continuous series of cuttings and embankments for a distance of over 150 miles.

The Indian railway system, carried out mainly on the wise plans laid down by Lord Dalhousie some thirty years ago, has already assumed considerable proportions, and in 1895 a total mileage of nearly 19,000 miles had been completed. The base of the system is the great trunk line running from Calcutta for 1500 miles up the Ganges valley through Allahabad and Lahore, and across the Indus at Attock to its present terminus at Peshawar on the Afghan frontier. From Allahabad, on this base, the

Great Indian Peninsula runs first over the Bandelkand hills, down the Narbada valley and through the Satpura range and Western Ghats for 700 miles to Bombay, and thence again over the Western Ghats, through Puna, and across the Deccan for 800 miles to Madras. Thus the three capitals are brought into direct communication with each other and with all the more central and populous parts of the empire.

Some of the other lines are of great length and of much commercial and strategical importance. Of these perhaps the most vital is the Indus valley line connecting Lahore with the sea at Karachi, the nearest port to England, and with a projected branch of 400 miles from Sakkar to Kandahar, already completed far beyond Quetta in the Pishin valley. A second line to this place by the Bolan Pass was opened in 1895. These lines have much political importance in reference to the completion of the Russian Trans-Caspian line to the Oxus and Samarkand.

Another great section runs from Bombay along the west coast, across the Lower Tapti and Narbada valleys, across Gujarat and Rajputana to the northern trunk line at Agra with a junction to Delhi. Several minor branches ramify from these main lines northwards to the Himalayas at Kurseong for Darjiling, southwards to Gwalior, eastwards to Nagpur and Secanderabad close to Haidarabad, capital of the Nizam's Dominions. From Madras two independent lines radiate, one right across the Deccan through Vellore and Coimbatore to Baipur near Calicut on the Malabar coast, the other southwards through Tanjore, Trichinopoly, and Madura, to Tuticorin and Tinnevelli near the apex of the peninsula. On these lines there are branches to Bangalore, Pondicherry, and Negapatam for Karikal.

Most of these lines, the materials for which had to be

brought mostly from England, were built at an average cost of about £14,000 per mile, the total capital already expended amounting to over £220,000,000. They are constructed partly on a narrow, but chiefly on a broad gauge, the former mostly by the State, the latter by private companies, to whom a rate of 5 per cent interest is guaranteed by the Government of India. Railway travelling is growing in popularity, and the various lines already convey (1893) nearly 130 million passengers yearly.

The telegraphic system, originally planned by Sir William O'Shaughnessy, may be regarded as complete, comprising a total length (1892) of nearly 39,000 miles. This system is connected by various submarine cables with the whole world.

10. *Administration: The Native States—Social Progress—Education.*

After the mutiny of the Native army of Bengal in 1857 the administration of the country passed from the old East India Company to the Crown, and on 1st January 1877 India was constituted an empire, the Queen of England assuming the title of Kaisar-i-Hind, or Empress of India. The sovereign is represented on the spot by the Viceroy and Governor-General, whose headquarters are at Calcutta, but who ordinarily resides during the summer months at Simla in the Himalayas. The Governor-General and the Governors of Madras and Bombay are each aided by Executive Councils, which are like Cabinets on a small scale. There is one Legislative Council of the Governor-General for legislation regarding imperial matters. There are also three local Legislative Councils, sitting at Calcutta, Madras, and Bombay respectively. The members of all these Legislative Councils are appointed by Government and not elected. The Government of India—that is, the Governor-General in Council—is subordinate to Her Majesty's Government in England, represented by the Secretary of State for India in London, who is assisted by a Council. Subordinate to the Governor-General are the Governors of the two Presidencies of Madras and Bombay, the Lieutenant-Governors of Bengal, the North-West Provinces, and the Panjab, the Chief Commissioners of the Central Provinces, Assam, and Lower Burma, the Resident at Haidarabad, the Agents to the Governor-General in Rajputana, Central India, and other Residents and Political Agents of the first rank. There are four High Courts of Judicature at Calcutta, Madras, Bombay, and Allahabad respectively, and one Chief Court for the Panjab. There are three armies, belonging to the Presidencies of Bengal, Madras, and Bombay respectively, each army having a Commander-in-chief; but the Commander-in-chief of the Bengal army has a supreme command over all the Royal troops in the empire, as contradistinguished from the Native. There are five Bishoprics, exclusive of the Missionary Bishoprics. The Bishop of Calcutta is also Metropolitan in India. At each of the three Presidency towns there is a bank connected with the Government.

This is an outline of the machinery by which England from a distance of 8000 miles administers the affairs of 287,000,000 people, including a large number of Native States, which recognise the supremacy of the paramount power. As, on the other hand, there are not more than a few thousand Europeans, exclusive of the military, and only a few hundred European civil officers in the whole empire, it will readily be imagined how arduous must be the task imposed on the Government of keeping order amongst such a mass of human beings, consisting of heterogeneous elements. Not perhaps unnaturally, the imperial race, to which such an inheritance has fallen, feels at times more oppressed by a deep sense of its overwhelming responsibilities than elevated by the commanding position it thus takes amongst the nations of the world.

All the Native States (some three hundred in number, great and small) may be regarded as placed under the protection of the suzerain power, the only really independent elements being some of the wild and often troublesome hill tribes on the frontiers. Of these States there are three categories—the allied, the tributary, and the protected. The allied are provided by the British Government with a regular contingency of subsidiary troops, for which a fixed charge is made. These represent a total population of over 25,000,000. In the tributary States the Government maintains no regular troops, but undertakes to defend them from any possible attacks from without, receiving in return a regular tribute. Of such States there are about fifty, with some 15,000,000 inhabitants. The protected States, exempt from tribute, stand in the same relation to the supreme authority, and number upwards of ninety, with a joint population of perhaps 26,000,000.

All three have renounced the right of self-defence and of independent diplomatic representation abroad, England guaranteeing them from attack, and acting as mediator in all the differences arising among them. They also maintain troops numerous enough to preserve peace within their borders. The English Government, moreover, reserves to itself the right of interfering in the internal administration whenever the native rulers become the oppressors instead of the protectors of their subjects. In fact, however, the Native States are becoming well governed.

The chiefs, princes, and other representatives of these

various Native States appear from time to time at the "durbar" or public audience of the Viceroy, for the purpose of paying homage to the Empress through her representative.

Under this administration, ensuring the blessings of peace at home and presenting a firm front to any possible assaults from without, the country has made astonishing progress both materially and morally in recent times. A far more radical transformation has taken place than might be suspected at a cursory glance. The removal of the centre of authority from the old inland capitals to the seaboard,—the general disarmament of the people, and the establishment of lasting peace and security in the remotest corners of the empire,—the suppression of savage rites, such as human sacrifices, amongst some wild hill tribes, and of Suttee[1] amongst the Hindus,—the surveys, trigonometrical, topographical, and geological,—the enlightened legislation, and the establishment of a system of civil and criminal justice,—the releasing of trade from transit duties and other fetters,—the assessment of the land-tax for long terms of years, and the recognition of proprietary right in the land,—the construction of highways, railroads, and telegraphic lines, and the extension of artificial irrigation,—the introduction of education on English principles—are all unmistakable evidences of social progress.

Some of the old native manufactures are dying out in many places, partly through the competition of the English looms, and partly through the introduction of modern machinery, while many of these manufactures continue to flourish. On the other hand, nevertheless, thousands are employed in the jute, cotton, and sugar factories, in

[1] *Suttee*, or rather *Ṣati* (that is, "the pure one"), properly means the widow who immolates herself on the death of her husband, but is commonly applied in English to the act itself.

the coal-mines, and in the plantations of tea and coffee. Nor has the traditional skill of native craftsmen and the hereditary genius of native artists succumbed to Western influences. Thus in Orissa and Southern India the hand-loom still maintains its place, and the most delicate muslins in the world may still be procured from the Dacca weavers, although at very high prices.

Another result of the English rule is the increased sense of unity that has been developed amongst the various nationalities. The same tendency is shown in the cultivation of the native languages (both classical and living), which formerly received little encouragement from the various Persian, Hindi, or Marathi speaking conquerors, but which are now fostered in the national education. Five Universities have been established (much upon the model of the London University) at Calcutta, Madras, Bombay, Panjab, and Allahabad respectively, to each of which several colleges, belonging to Government and to private bodies, are affiliated. Three Medical Colleges of the best possible kind, with several Medical Schools and two Colleges of civil engineering, also several technical and industrial schools, have been established. Much satisfactory progress has been made in the popular instruction, although much remains still to be done in this direction, and especially as regards female education. In the Panjab the schools and attendance have greatly increased; yet 50 per cent of the children are said to be still unprovided with instruction of any sort. In Bengal and Madras also the elementary schools have been greatly multiplied, and here the wish to learn English is increasing among the middle classes. In Madras the greatest development is in the primary schools and amongst the native Christians. A really sound beginning has also been made with female education. As regards the highest education, the number of students who matriculated at the five Universities rose from 4905 in 1889 to 6358 in 1892.

For Lower Burma an Educational Syndicate was established in 1881 for the purpose of controlling the public examinations, which under new regulations will be especially designed to encourage the study of law, medicine, engineering, and the technical arts.

State expenditure is largely incurred in the shape of grants-in-aid. About half of the educational expenses in the interior of the country are defrayed by the State, and the other half by the people.

The progress of Christian missions of all Protestant denominations is considerable, and the native Christians of all denominations exceeded 2,284,000 in 1891. The Protestant communities have increased at the rate of about 50 per cent in each decade during the last half century. There are about 450 mission stations and 500 European missionaries, 3 missionary bishops, and 300 native ordained clergymen. Several colleges and training institutions belong to the missionary bodies. The Vernacular Education Society conducts extensive operations in the publication and colportage of books for Christian instruction. The total income of the Protestant missions has been computed at something between £300,000 and £400,000 annually.

The Roman Catholic Church has real vitality, and includes Europeans, East Indians, and natives. It has archbishops, bishops, vicars apostolic, and lady superiors. It has many missionary stations, besides colleges, schools, convents, and other religious establishments, and the members of the Roman Catholic Church exceeded 1,315,000 in 1891.

Further proofs of material and social progress will be revealed in the subjoined tabulated statements of population, trade, education, etc.

11. *Statistics of British India.*

General Results of the 1891 *Census.*

The second general census of India, taken on 26th February 1891, showed a population of 287,223,431 for the British possessions, and including the French and Portuguese enclaves, 289,187,316, or about one-fifth of the population of the whole world, being an increase of nearly 28,000,000 since the previous census of 1881. Although this increase is nearly equal to the present population of England and Wales, Mr. J. A. Bains points out in his "General Report on the Census of India, 1891," that the percentage of increase is only 10·96, a comparatively low rate, for in a list of twenty-eight of the chief countries of the world, India ranks only twentieth in this respect, New South Wales being the highest and France the lowest. He considers that the present rate of increase is well within the means of subsistence of the people of India, showing a density of population of only 184 per square mile, though very unequally distributed, and ranging for instance from 522 in Oudh to 31 in Kashmir. In fact there appears to be no actual overcrowding, except in a few special localities, such as parts of Behar, of the Deccan and Gangetic valley, and a small tract on the Bombay coast.

The vast predominance of agricultural over all other interests is shown by the fact that the landowners, tenants, and general labourers, graziers, shepherds, and wool and cotton workers comprise over 70 per cent of the entire population. In the same report the ethnological and religious relations are thus summed up. In the extreme north-west the dominant element comes from West Central Asia, arriving at different times and probably from different sources. Adjoining it on both sides of the frontier is a group of still more western origin, but much mingled with the former. The Himalayan districts, skirting the plains of the Panjab and North-West Provinces, also preserve a considerable element of northern origin; but here the people farther east and north come from the eastern side of Central Asia, which seems to have peopled the whole range along the British frontier, and in the eastern sections of Upper India the greater part of the Gangetic valley also. Across the Ganges basin the dominant element numerically is a lower race of darker colour and different features, stretching with few interruptions to the extremity of the Peninsula and over the north part of Ceylon.

In general the element from West Central Asia is found in a comparatively pure state in the Indus valley. It rapidly deteriorates through fusion with dark blood in the direction of the east, until it meets the north-eastern strain in the Ganges delta. The strain of northern blood south of the Central Belt of hills is of the thinnest, and hardly extends into the plains at all. But on the west coast some pure specimens may probably be found, and the general average of the dark type throughout the Peninsula has apparently been raised by prosperity and a long period of peace considerably above the level of those tribes that have remained in a somewhat wild state in the hills and forests.

In respect of religion the Hindu sects of all kinds number 207,000,000 and the Muhammadans 57,000,000, or 72 and 19 per cent of the whole population respectively.

Areas and Populations.

BRITISH TERRITORIES.

Provinces and Tracts.	Area in sq. miles.	Pop. 1881.	Pop. 1891.
Ajmir	2,711	460,722	542,358
Assam	49,004	4,881,426	5,476,833
Bengal	151,543	66,750,520	71,346,987
Berar	17,718	2,672,673	2,897,491
Bombay Presidency — Bombay	77,275	14,057,284	15,985,270
Bombay Presidency — Sind	47,789	2,413,823	2,871,774
Bombay Presidency — Aden	80	34,860	44,079
Upper Burma	83,473	...	2,946,933
Lower Burma	87,957	3,736,771	4,658,627
Central Provinces	86,501	9,838,791	10,784,294
Madras	141,189	30,827,113	35,630,440
Kurg	1,583	178,302	173,055
North-West Provinces	83,286	32,762,766	34,254,254
Oudh	24,217	11,387,741	12,650,831
Panjab	110,667	18,843,186	20,866,847
Quetta (British Baluchistan)	170,000	...	27,270 145,418
Burma Frontiers	20,000 (?)	...	120,000(?)
Andamans	2,700	14,628	15,609
Total British Provinces and Tracts	1,157,693	198,860,606	221,438,370

NATIVE STATES AND AGENCIES.

	Area in sq. miles.	Pop. 1881.	Pop. 1891.
Haidarabad	82,698	9,845,594	11,537,040
Baroda	8,226	2,185,005	2,415,396
Mysore	27,936	4,186,188	4,943,604
Kashmir	80,900	...	2,543,952
Rajputana	130,268	9,959,012	12,116,102
Central India	77,808	9,387,119	10,318,812
Bombay States	69,045	6,926,464	8,059,298
Madras States	9,609	3,344,849	3,700,622
Central Provinces States	29,435	1,709,720	2,160,511
Bengal States	35,834	2,786,446	3,296,379
North-West Provinces States	5,109	741,750	792,491
Panjab States	38,299	3,860,761	4,263,280
Sikkim	3,000	...	30,458
Shan States	30,000 (?)	...	375,992
Total Native Territory	628,167	54,932,908	66,453,906
Total British India	1,785,860	253,793,514	287,892,276
Portuguese Possessions in India	1,605	444,617	572,290
French Possessions in India	200	285,022	280,303
Total India	1,788,665	254,523,153	288,744,869

Ceylon and other Islands .	30,000	2,780,000	3,040,000
Straits Settlements and Protected States . .	37,000	600,000	1,100,000
Hong-Kong . . .	30	160,400	221,440
Total, British Asiatic Possessions[1] . .	1,853,870	257,333,914	293,106,309

TOWNS WITH UPWARDS OF 50,000 INHABITANTS (1891).

	Population.		Population.
Calcutta	862,000	Faizabad	79,000
Bombay	822,000	Shahjahanpur . .	78,000
Madras	452,000	Farukhabad . . .	78,000
Haidarabad . .	415,000	Rampur	77,000
Lucknow	273,000	Multan	75,000
Benares	220,000	Mysore	74,000
Delhi	192,000	Rawal Pindi . . .	74,000
Mandalay	189,000	Darbhangah . . .	73,000
Cawnpore	188,000	Moradabad . . .	73,000
Bangalore	180,000	Bhopal	70,000
Rangun	180,000	Bhagalpur . . .	69,000
Lahore	177,000	Ajmir	69,000
Allahabad	175,000	Bhartpur	68,000
Agra	169,000	Salem	68,000
Patna	165,000	Jalandhar . . .	66,000
Puna	161,000	Calicut	66,000
Jaipur	159,000	Gorakhpur . . .	64,000
Ahmadabad . .	148,000	Saharanpur . . .	63,000
Amritsar	137,000	Sholapur . . .	62,000
Colombo	127,000	Jodhpur	62,000
Bareilly	121,000	Aligarh	61,000
Mirut	119,000	Muttre	61,000
Srinagar	119,000	Bellary	60,000
Nagpur	117,000	Nagapatam . . .	59,000
Howra	117,000	Haidarabad (Sind) .	58,000
Baroda	116,000	Bhaunagar . . .	58,000
Surat	109,000	Chupra	57,000
Karachi	105,000	Monghyr	57,000
Gwalior	104,000	Bikaner	56,000
Indore	92,000	Patiala	56,000
Trichinopoly . .	91,000	Maulmain . . .	56,000
Madura	87,000	Sialkot	55,000
Jabalpur	84,000	Tanjore	54,000
Peshawar	84,000	Combaconum . .	54,000
Mirzapur	84,000	Jhansi	54,000
Dacca	82,000	Hubli	53,000
Gaya	80,000	Alwar	52,000
Ambala	79,000	Firozpur	51,000

[1] Exclusive of Afghanistan and other spheres of influence.

POPULATION ACCORDING TO RACES AND LANGUAGES.[1]

ARYANS.	Hindi and Urdu	110,000,000
	Bengali	46,340,000
	Mahrathi	20,000,000
	Panjabi	17,700,000
	Gujarati	10,600,000
	Kanarese	9,750,000
	Uriya	9,000,000
	Sindhi	2,590,000
	Kashmiri	2,500,000
	Assamese	2,430,000
	Marwari	1,420,000
	Pushtu	1,080,000
	Gypsy	400,000
	English	238,000
DRAVIDIANS.	Telugu	20,000,000
	Tamil	15,500,000
	Malayalim	5,430,000
	Gondi	2,000,000
	Tulu	500,000
	Oraon	400,000
KOLARIANS.	Santali	1,700,000
	Kol	650,000
	Kond	320,000
	Bhil	200,000
TIBETO-BURMESE.	Burmese	6,000,000
	Bod-pa	2,000,000
	Karen	670,000
	Kacchi	450,000
	Naga	300,000
	Chins, Lushai, etc.	250,000
	Shans, Mons, etc.	1,000,000

POPULATION ACCORDING TO RELIGION (1891).

Religion	Number	Religion	Number
Hindus	207,731,000	Sikhs	1,908,000
Muhammadans	57,321,000	Jains	1,417,000
Pagans	9,280,000	Parsis	90,000
Buddhists	7,131,000	Jews	17,000
Christians	2,284,000	Sundries	43,000

CHIEF RIVERS OF INDIA.

River	Length. Miles.	River	Length. Miles.
Brahmaputra and San-po	1800	Tapti	440
Indus	1800	Kavari	470
Ganges	1500	Penner	355
Irawadi	1000	Luni	320
Godavari	900	Sitang	230
Kistna	800	Brahmani	410
Salwin	750	Mahi	350
Narbada	800	Baitarani	345
Mahanaddi	520		

CANALISATION (1891).

Capital expended	£22,000,000
	Miles.
Main Canals and branches in the three Presidencies	5,000
Panjab and Sind	2,700
Tanjore or Kavari system	900
Distributing Canals in North India	9,500
	18,100

[1] That is, so far as race can be determined by language; but it is obvious that this grouping practically resolves itself into a classification according to linguistic families.

		Acres.
Area of Irrigation	Madras and Bombay	2,000,000
	Panjab and Sind	2,800,000
	N.W. Provinces	1,000,000
	Behar and Orissa	400,000
		6,800,000

FINANCE.

Year.	Revenue.	Expenditure.
1882	75,685,000 Rx.[1]	72,090,000 Rx.
1888	78,760,000 ,,	80,788,000 ,,
1890	85,000,000 ,,	82,473,000 ,,
1892	89,143,000 ,,	88,676,000 ,,

Chief Heads of Income (1893).		Chief Heads of Outlay (1893).	
Land Revenue	25,157,000 Rx.	Army	23,012,000 Rx.
Opium	7,316,000 ,,	Railway Revenue Account	21,546,000 ,,
Salt	8,588,000 ,,	Civil Salaries	14,472,000 ,,
Stamps	4,434,000 ,,	Charges of Collection	8,685,000 ,,
Excise	5,146,000 ,,	Buildings and Roads	6,091,000 ,,
Provincial Rates	3,707,000 ,,	Miscellaneous Civil Charges	5,538,000 ,,
Customs	1,617,000 ,,	Interest	4,066,000 ,,
Assessed Taxes	1,687,000 ,,	Irrigation	2,860,000 ,,
Forests	1,589,000 ,,	Post-Office, Telegraph, and Mint	2,610,000 ,,
Civil Departments	1,636,000 ,,	Refunds, Compensations	1,703,000 ,,
Irrigation	2,338,000 ,,	Famine Relief, etc.	1,160,000 ,,
Railways	19,552,000 ,,		
Post-Office, Telegraph, and Mint	2,721,000 ,,		

ARMY (1893-94).[2]

	Officers.	Rank and File.	Total.
British	3,470	70,511	73,981
Native	1,578 (British) 2,757 (Native)	141,301	145,636
Total	7,805	211,812	219,617

[1] Rx. = 10 rupees, approximately equivalent to one pound sterling before the year 1873; but owing to the depreciation of silver the rupee has since then gradually fallen to less than 1s. 4d. in 1894. The loss to India in the exchanges is correspondingly great, and a chief cause of the financial difficulties from which the country has suffered during the last two decades.

[2] Since 1893, when the former Presidency commands were abolished by Act of Parliament, the forces of India have been constituted in four local armies, each under a Lieutenant-General, subordinate to the Commander-in-chief.

DISTRIBUTION OF THE BRITISH FORCES (1893-94).

	Bengal.	Bombay.	Madras.	Total.
Artillery	7,716	3,315	2,287	13,318
Cavalry	3,786	631	1,262	5,679
Engineers	187	79	73	339
Infantry	33,453	10,130	10,130	53,713
Miscellaneous Officers	1,311	140	261	1,712
Total	46,453	14,295	14,013	74,761

EDUCATION (1892).

Colleges	139	Attendance	16,460
Secondary Schools	4,907	,,	477,500
Primary Schools	97,180	,,	2,842,000
Technical Schools	560	,,	22,000
Universities	5[1]	Matriculated	6,358

State Expenditure on Public Instruction . . . . . Rx. 3,073,000
Proportion of Population receiving Instruction . . 12 per 1000

LITERATURE.

Average yearly publications—	English	960
	Vernacular languages	5000
	Classical languages of India	900
	In more than one language	800
		7660

RAILWAYS.

Lines open (1895), 18,855 miles.
Lines in construction or sanctioned, nearly 3000 miles.
Total capital expenditure, Rx. 234,464,000.
Passengers carried (1892), 127,457,000.
Goods, material, and live stock carried, 26,334,000 tons.
Total working expenses, Rx. 10,900,000.
Total net earnings, Rx. 12,400,000.

POSTAL AND TELEGRAPH SERVICES.

Year.	Post Offices and Letter Boxes.	Letters, etc., Carried.	Revenue.	Expenditure.
1888	16,967	274,399,000	1,214,000 Rx.	1,375,000 Rx
1890	19,196	311,988,000	1,301,000 ,,	1,377,000 ,,
1892	21,465	347,133,000	1,446,000 ,,	1,496,000 ,,

Year.	Telegraph Lines.	Miles of Wire.	Despatches.	Revenue.	Expenditure.
1888	31,894 miles	93,500	2,808,000	764,000 Rx.	787,000 Rx.
1890	35,280 miles	106,140	1,133,000	767,000 ,,	731,000 ,,
1892	38,625 miles	120,000	3,309,000	919,000 ,,	839,000 ,,

Telegraph Offices open (1892), 1001.

[1] Calcutta, Madras, Bombay, Panjab, and Allahabad. Not more than one-fourth of the matriculated students become graduates.

TRADE.

Year.	Imports.	Exports.
1889	86,657,000 Rx.	105,367,000 Rx.
1891	84,155,000 ,,	111,460,000 ,,
1893	83,275,000 ,,	113,554,000 ,,

	Imports (1893).	Exports (1893).
Bengal	25,486,000 Rx.	42,200,000 Rx.
Burma	5,466,000 ,,	9,236,000 ,,
Madras	5,365,000 ,,	11,263,000 ,,
Bombay	39,743,000 ,,	46,580,000 ,,
Sind	3,555,000 ,,	4,186,000 ,,

	1888.	1890.	1892.
Exports to Great Britain	£30,764,000	£32,669,000	£30,513,000
Imports from Great Britain	32,539,000	33,641,000	27,902,000

CHIEF CUSTOMERS OF INDIA.

	Indian Exports to (1893).	Imports from (1893).
Great Britain	32,267,000 Rx.	44,005,000 Rx
China	14,402,000 ,,	2,843,000 ,,
France	9,083,000 ,,	1,040,000 ,,
Italy	3,641,000 ,,	356,000 ,,
Straits Settlements	4,441,000 ,,	2,372,000 ,,
United States	4,513,000 ,,	1,135,000 ,,
Egypt	4,832,000 ,,	159,000 ,,
Belgium	4,414,000 ,,	1,645,000 ,,
Austria	2,607,000 ,,	1,032,000 ,,
Australia	1,107,000 ,,	240,000 ,
Japan	1,611,000 ,,	91,000 ,,
Germany	6,517,000 ,,	1,451,000 ,,
Holland	670,000 ,,	207,000 ,,
Persia	509,000 ,,	687,000 ,,
Spain	626,000 ,,	17,000 ,,

CHIEF IMPORTS AND EXPORTS (1893).

Imports.	Value.	Exports.	Value.
Cotton goods	25,626,000 Rx.	Rice	12,407,000 Rx.
Hardware, cutlery	6,600,000 ,,	Wheat	7,440,000 ,,
Silks	2,818,000 ,,	Cottons	20,800,000 ,,
Sugars	2,626,000 ,,	Opium	9,255,000 ,,
Woollen goods	1,523,000 ,,	Oil seeds, etc.	11,631,000 ,,
Liquors	1,447,000 ,,	Hides and skins	5,592,000 ,,
Oils	2,919,000 ,,	Jute	11,150,000 ,,
Machinery	2,359,000 ,,	Tea	6,292,000 ,,
Coal	1,142,000 ,,	Indigo	4,141,000 ,,
Provisions	1,862,000 ,,	Coffee	2,067,000 ,,
Clothing	1,384,000 ,,	Wool	1,117,000 ,,

Foreign Trade of the Chief Seaports (1893).

Bombay . .	65,675,000 Rx.	Madras . .	9,067,000 Rx.
Calcutta . .	63,640,000 ,,	Karachi . .	7,546,000 ,,
Rangun . .	12,525,000 ,,	Tuticorin . .	1,803,000 ,,

Shipping (1893).

Vessels Entered.	No.	Tons.	Vessels Cleared.	No.	Tons.
British . .	2047	3,158,000	British . .	2010	3,073,000
British Indian	1035	150,000	British Indian	1063	156,000
Foreign .	734	508,000	Foreign .	681	481,000
Native . .	1568	86,000	Native . .	1585	80,000
Total .	5384	3,902,000	Total .	5339	3,790,000

Steamers Entered and Cleared *via* Suez Canal.

	No. Entered.	Tons.	No. Cleared.	Tons.
1883	711	1,153,000	1645	2,586,000
1890	677	1,332,000	1608	3,055,000
1891	752	1,487,000	1717	3,309,000
1893	782	1,638,000	1711	3,525,000

Agricultural Returns (1892).[1]

Land under cultivation and fallow	166,176,000 acres.
Land untilled but cultivable	99,306,000 ,,
Land not available for cultivation	110,000,000 ,,
Land under forests	48,124,000 ,,
Land for which no returns are yet available . .	603,518,000 ,,

Chief Crops (1892).

	Acres.		Acres.
Rice	27,225,000	Cotton	8,860,000
Wheat . . .	18,574,000	Oil seeds	8,498,000
Other grains . .	76,452,000	Indigo . . .	541,000
Sugar-cane . . .	1,940,000	Tobacco	327,000
Tea	266,000	Coffee	128,000
Other food crops .	3,884,000	Jute and other fibres .	301,000

Vital Statistics (1892).

	Births per 1000.	Deaths per 1000.
North-West Provinces and Oudh . .	33·26	31·14
Panjab	34·02	29·13
Central Provinces	43·09	35·54
Lower Burma	20·74	15·93
Assam	28·59	29·91
Madras	34·4	26·2
Bombay	36·27	27·26
Bengal	21·46 (?)	26·94 (?)

[1] Exclusive of Bengal, for which the returns are not available.

Occupations of the Population (1891).

Agriculture	171,735,000	Wood, cane-making	4,293,000
General labour	25,468,000	Transport, storage	3,953,000
Food, drink, etc.	14,576,000	Metals and jewellery	3,821,000
Weaving	12,611,000	Stock-breeding, provisions	3,646,000
Domestic service	11,220,000	Light, firing, forage	3,522,000
Professions	5,672,000	Leather, horns, etc.	3,285,000
Administration	5,600,000	Glass, pottery, stoneware	2,361,000
Independent	4,774,000	Buildings	1,438,000
Trade	4,686,000	Defence	664,000

Emigration.

Total emigrants (1870-80), 180,000.

Coolie emigrants (1886), 7980; (1888), 6450; (1890), 16,874; (1892), 16,567; chiefly to Demarara, Trinidad, Mauritius, and other British Colonies.

Crime (1891).

Tried, 1,525,000; convicted, 749,000; fined, 572,000.

Sentenced to death, 427; to transportation, 1873; to imprisonment, 174,922.

Convictions for murder, 940; for cattle-lifting, 8015; for theft, 55,443; for housebreaking, 17,682.

Police, 144,420; central gaols, 36; district and other gaols, 710.

Ceylon.

Provinces.	Area in sq. miles.	Population (1891).
Central	2,304	474,487
Uva	3,725	159,155
North Central	4,047	75,319
Western	1,371	763,187
Sabaragamuwa	2,085	258,605
North-Western	3,024	320,032
Southern	1,980	489,761
Eastern	3,657	148,727
Northern	3,171	319,193
Total	25,364	3,008,466

Population according to Race and Religion.

Singhalese	2,041,000	Buddhists	877,000
Tamils	724,000	Hindus	616,000
Moormen (Arabs) and others	214,000	Muhammadans	212,000
Descendants of Europeans	21,000	Christians of all denominations	302,000
English	6,000		

Chief Towns: Colombo, pop. (1801), 127,000; Kandy, 20,000; Galle, 34,000; Trincomali, 11,000; Jaffna, 43,000.
Revenue (1892), 18,509,000 Rupees; Expenditure, 17,762,000 Rupees.
Imports (1892), 70,688,000 Rupees: Exports, 62,272,000 Rupees.
Chief Exports: Tea, 32,527,000 Rs.: Coconut products, 9,567,000; Plumbago, 4,307,000; Coffee, 3,294,000; Cinchona, 822,000; Areca Nuts, 887,000.
Imports from Great Britain (1892), £945,000: Exports to Great Britain, £3,945,000.
Schools (1892), 3872; Attendance, 170,000.
Railways (1893), 230 miles; Telegraphs, 1550 miles.

CHAPTER III

INDO-CHINA AND MALACCA

1. *Boundaries—Extent—Area.*

THE south-eastern section of Asia, commonly spoken of collectively as Indo-China, Further India, Trans-Gangetic India, or the "Golden Peninsula," really consists of two distinct peninsular regions — Indo - China proper and Malacca — which differ profoundly from each other in their physical conditions no less than in the ethnical affinities, culture, and religion of their inhabitants. In all these respects the northern peninsular mass of Indo-China still belongs to the Asiatic mainland, whereas Malacca, projecting southwards parallel with the neighbouring island of Sumatra, forms, strictly speaking, an integral part of the great Oceanic world. Amongst the great schemes of canalisation projected since the piercing of the Isthmus of Suez, not the least ambitious is the recently-proposed connection of the Bay of Bengal with the China Sea by a canal across the narrow Isthmus of Krah, by which Malacca is at present connected with Indo-China.

Meantime, Further India, taken in its wider sense, lies almost exactly within the northern torrid zone, stretching from about 3 degrees beyond the tropic of Cancer southwards to Cape Romania, which is the southernmost point of the continent, and which approaches almost to within 1 degree of the equator. This gives an extreme length of about 1800 miles, with a breadth varying from 700 miles in the north to a little over 60 miles in the Isthmus of Krah, a total area of over 870,000 square miles, and a population vaguely estimated at from 35 to 40 millions.

The coast-line of Indo-China is far more diversified by bays, bights, gulfs, islands, and headlands than is the somewhat monotonous seaboard of British India. The west coast is watered by the Bay of Bengal, which here forms the Gulf of Martaban, while contracting southwards to the Malacca Strait, between Sumatra and the mainland. East of Malacca the coast is washed by the storm-swept China Sea, which here develops the great Gulfs of Siam and Tongkin, between which the continent is rounded off by the graceful curve of the Cochin-Chinese seaboard.

2. *Relief of the Land: Mountain Systems—Cochin-Chinese Coast Range.*

The interior of Indo-China is one of the least-known regions in Asia, and here are concentrated some of the most interesting orographic and hydrographic problems that still await solution from modern research. The surface is covered with a number of parallel mountain ranges running mainly north and south, with intervening longitudinal river valleys broadening southwards to extensive alluvial plains, where are developed some of the largest deltas on the globe. We are not only still ignorant of the real character of these mountain ranges, but we do not even know exactly where the large intervening rivers take their rise. Nor does the solution of these difficulties depend on a wider knowledge of the interior of the peninsula itself so much as on the further exploration of the Tibeto-Chinese frontier lands, whence the mountains radiate and whence the rivers flow southwards. Do these mountains form an independent highland system? or are they, as many suspect, simply the south-eastern continuation of the Tibetan plateau, cut up into so many separate ridges by fluvial action? In any case, it is clear that the sources of the great rivers must be sought for on the Tibetan plateau itself, and probably in the great lacustrine region where rise the Hoang-ho and Yang-tse, and where Nain Sing and Bower have discovered a vast system of lakes, some of which appear to drain eastwards. From the explorations of A. K. (Pundit Krishna), it seems evident that here also the Salwin and the Mekhong have their rise.

In the east of the peninsula the Cochin-Chinese coast range separates the Mekhong basin from the numerous short streams flowing to the China Sea. But the term "range" is somewhat inaptly applied to an intricate system of moderately elevated ridges crossing each other at all angles, and giving rise to a number of rivers, flowing some to the Mekhong, others to the coast. This upland system runs north and south, throwing off numerous spurs and offshots which project seawards, breaking the Cochin-Chinese seaboard into a number of bays, bights, and inlets. It terminates in the extreme south with the headland of Cape St. James, at the entrance of the Saigon River.

This little-known water-parting between the coast streams and the Mekhong basin was partly surveyed in 1892 by Dr. A. Yersin, who crossed it by a hitherto unvisited route from east to west, his main object being to discover the sources of the Sebong (Se-bang), a large affluent of the Mekhong near the Khong rapids. The region traversed is occupied by the Benongs and numerous other wild tribes, who hold no intercourse with the

Annamese, and who had never before seen or heard of Europeans. They have no kind of political unity, each village forming an independent petty republic, at constant war with its neighbours. The country forms rather a plateau than a coast range, rising to a mean altitude of about 1500 feet above sea-level, and scored by numerous river gorges. It is covered by almost continuous forest stretching from the narrow strip of low-lying Annamese coastlands all the way to the Mekhong. This forest contains many gigantic trees festooned with a great variety of parasitic orchids, and inhabited by elephants, buffaloes, tigers, bears, the wild boar, rhinoceros, many species of monkeys and other animals. Yet some districts are densely peopled, villages of from 100 to 400 inhabitants occurring at intervals of six or eight miles, while other parts are absolutely uninhabited; sometimes the explorer journeyed for a whole week without meeting a single human habitation. He traced the course of the Sebong from its source to the confluence, and found that its basin is separated from that of the Don-nai (river of Saigon) by a high mountain range, which can be crossed only during the dry season.

Coal is widely diffused throughout the uplands of Tongkin and Annam, and British capital has already been invested in the mines of Nongson (Nong-sin), 40 miles south-west of Turan, and of Hatu, near Hongai, on Along Bay. These have hitherto yielded the best results, and at Hatu Mr. Curzon was shown the unique spectacle of a solid seam of black coal exposed to a thickness of 180 feet down the entire front of a hill, below which it still continued to an unknown depth. Extensive beds have also been discovered at Yenbai, Laokai, Kwangyen, and other parts of the interior. But opinions differ greatly as to the quality, which evidently varies considerably. The best kinds are bituminous, and burn well, while others require a great draught and readily crumble to dust, to utilise which a briquette factory has been established at Hongai. The Nongson Concession, visited in 1892 by Consul Parker, covers a space of about four square miles, and contains an abundance of coal, chiefly anthracite.

The chain running parallel with the coast range between the Mekhong and Meinam valleys merges southwards in low plateaux west of the Tonle-sap (properly Tale-sap) or Great Cambodian Lake. But it again acquires a considerable elevation at Chantabun, near the coast, whence it runs south-east along the Gulf of Siam, and then trends round to the north-east, here culminating with the Pursat or Krevanh (" Cardamom ") hills south of the Great Lake.

Between Arakan and Burma runs the Arakan Yoma range, which, north of Sandoway, has a mean altitude of from 4000 to 6000 feet, but farther south falls rapidly in the direction of Cape Negrais. Beyond this point the system is continued seawards by the Preparis and Coco Islands, the Andaman and Nicobar groups. The prevailing formations are lime and sandstones of the cretaceous and tertiary epochs, with some igneous rocks, but no erupted lavas or any volcanoes, although Ramri and Cheduba Islands, as well as the neighbouring coast districts, are dotted over with numerous mud volcanoes. These are still very active, periodically ejecting vapours, mud, and even stones; petroleum springs also occur in the district, which is occasionally subject to earthquakes.

The Yoma range is crossed by a few passes, of which the easiest and most frequented appears to be that of Aeng or An (4700 feet). This pass was followed in 1891 by Lieutenant Walker, when he crossed from Burma to Arakan, for the purpose of reporting on the possibility of establishing railway communication between Arakan and Burma. He returned by the Sawbwa's route, which traverses the Yoma range some 80 miles farther north, and which is much more difficult and higher (about 6000 feet, if not more, at the highest point), and altogether unsuitable for a railway, for which the An Pass offers by far the greatest facilities.

3. *Hydrography: The Irawadi, Salwin, Meinam, Mekhong, and Song-ka Rivers—Lake Tonle-sap.*

North of the Bramaputra valley the eastern section of the Tibetan plateau seems certainly to be geologically continued eastwards far into Yun-nan and Se-chuen. But in this little-known region, inhabited by the Mosso, Lolo, Si-fan, and other semi-independent aboriginal tribes, there are pressed together an extraordinary number of separate ridges, possibly produced by the action of running waters. Parallel with the Tant-la run several chains, mainly north and south, nearly at right angles with the Kuen-lun system, and these "Cross Ridges," as Blakiston calls them, penetrate far into Burma and Siam, where they form the Indo-Chinese mountain system. Their general direction is indicated by the course of the great rivers, some of which at all events take their rise on the Tibetan plateau, and which flow first north-east, parallel with the Tant-la and the other cross ridges on the Tibeto-Chinese frontier. All these rivers, amongst which must be included the Irawadi, Salwin, and Mekhong, as well as the Yang-tse and Min, then trend gradually round to the south, flowing in this direction for hundreds of miles in the closest proximity. Nowhere else is there any instance of so many large streams flowing in independent parallel valleys, separated only by single ridges, without uniting into one general water system. All the large rivers which reach the coast between the Irawadi and

Yang-tse deltas, a distance of at least 5000 miles, are confined in their upper courses within the comparatively narrow tract which lies between the eastern tributaries of the Brahmaputra and the head-waters of the Hoang-ho.

Margary, Gill, M'Carthy, Szechenyi, Walker, Errol Gray, Prince Henry of Orleans, and others who have traversed the ground between Yun-nan and Upper Burma, all speak of the numerous river valleys running north and south which they had to cross between Lake Ta-li-fu and Bhamo. Justus Perthes's map (1881) of Szechenyi's route from Sayang to Bhamo, a distance of about 160 miles as the bird flies, lays down, besides numerous tributaries, no less than four main streams identified by that explorer as the Mekhong, Salwin, and the two great forks of the Irawadi. The same phenomenon is described by Desgodins, whose route lay far to the north between Se-chuen and Tibet, and by A. K., who in 1882 also traversed the remarkable region of contiguous river valleys between the Yang-tse and the Brahmaputra.

Going eastwards, and keeping within the limits of Indo-China, first comes the Irawadi, formed by the junction of the Nam-kiu in the west, and the Phung-mai in the east, neither of which head-streams has yet (1895) been traced to its source. Hence the problem of the Irawadi still remains unsolved, although Mr. Needham, the Abbé Desgodins, and A. K. (Pundit Krishna) have finally disposed of its claim to be regarded as the continuation of the Tibetan San-po. During a three years' residence in the valley of the Lu-kiang (Upper Salwin ?) at a convenient place for studying the question, Desgodins ascertained that the two rivers could not possibly be connected, and that consequently the San-po must flow to the Brahmaputra, as has since been placed beyond doubt by Needham and others.

Much of the confusion regarding the hydrography of

the Upper Irawadi and Salwin basins is due to the perplexing nomenclature of a region where rivers and mountains often bear as many as half a dozen different names, which it is not always easy to discriminate. Thus the western and eastern branches of the Upper Irawadi are respectively the Myit Gyi ("Great River") and Myit Gney ("Little River") of the Burmese, the Nam-kiu (Nam-cheo) and Phung-mai of the Shans, and the Mali-kha and Nmai-kha of the Kachin aborigines, while the eastern and really larger branch appears to be identical with the Kuts or Chitom of the Tibetans and the Kinsha-ho of the Chinese. Although the sources of these two branches have not yet been reached, the recent explorations of Captain Borwick by steamer from Bhamo (1890), of Eliott, Hobday, and Blewitt by land, also from Bhamo (1890-91), and of Errol Gray by land from Assam (1892-93), have made it evident that both rise within the limits of the Indo-Chinese peninsula, the eastern and longest scarcely farther north than 28° 30′ or 29° N. lat., the western certainly on the southern slopes of the Khamti hills, about 28° N., 97° E. Borwick's steamer was arrested on both forks by rapids 4 or 5 miles above the confluence, where Lieutenant Blewitt roughly estimated the discharge of the Phung-mai and Nam-kiu at 33,500 and 23,000 cubic feet per second respectively. Lieutenant Eliott also infers that the Lu-kiang (Nam-kong) must be the upper course, not of the Irawadi, as some even still hold, but of the Salwin, as Desgodins always held. In fact this point was regarded as settled till a doubt was raised by M. Loczy, of the Szechenyi expedition, who argued that a stream of such feeble volume as the Lu-kiang could not have its source hundreds of miles away in the heart of the Tibetan plateau. But its feeble volume may be due to a very narrow catchment basin hemmed in between those of the Mekhong and Upper

DEVA-FACED CLIFF ON THE IRAWADI.

Brahmaputra, while the more copious Irawadi, although so much shorter, draws its supplies from a broader expanse between Yun-nan and the eastern bend of the Brahmaputra. Here also the rainfall from the moisture-bearing clouds rolling up from the Indian Ocean is far more copious than in the region farther inland through which the Salwin flows. It may be added that during his second journey through Tibet (1892) Mr. Rockhill crossed a stream, the Chang-tang-chu, about lat. 33°, long. 91° E., which he supposed might be the westernmost feeder of the Upper Jyama-nu-chu. Lower down (lat. 32°, long. 94°) he reached the Su-chu, held by the natives to be the same Jyama-nu-chu, which is supposed to be the Upper Salwin. At this point the Su-chu, flowing at an altitude of 13,700 feet on the Tibetan plateau, was already 75 yards wide and from 8 to 10 feet deep.

At the confluence of the two head-streams the Irawadi is already a majestic stream 500 yards broad, and navigable by steamers of considerable size throughout the whole of its course from this point to the delta, a distance of nearly 900 miles. There are no rapids or obstructions of any sort, except about 20 miles above Bhamo, where the stream suddenly narrows from 1000 to 150 yards, rushing with great velocity through a dangerous rocky gorge, in which the navigation is much impeded and at times arrested by swift eddies and backwaters. Below the junction of the two forks, the main-stream is joined on its left bank just above Bhamo by the Ta-ping (Ta-ho) descending from the Yun-nan uplands, and on its right bank below Ava by the Khyendwen (Chindwin), which drains the greater part of the Chin-Lushai Hill Tracts. The Irawadi has a total length of probably over 1000 miles, with a mean discharge in the delta of nearly 500,000 cubic feet per second, and of about 2,000,000 cubic feet during the floods.

To the Irawadi system belongs also the Sittang, which, after a course of over 360 miles, unites with the main stream in a common delta. The Sittang drains a region about 22,000 square miles in extent, between the Pegu-Yoma range, separating it from the Lower Irawadi and the Panglung hills forming the divide towards the Salwin. In its broad estuary it collects the running waters of Lower Pegu from the west, and is joined on the east by the Bilin, which also communicates with the Salwin estuary through several creeks and channels.

Since the completion of the new Myit-kyo canal the Irawadi delta, which begins below Mianan and terminates south-westwards at Cape Negrais, presents a continuous tidal waterway, extending from the Bassein branch for about 370 miles round the Gulf of Martaban to the Salwin estuary at Maulmain. During the floods a great part of this low-lying alluvial region is transformed to an inland sea, and at other times it is intersected in all directions by an intricate system of creeks and channels, most of which are continuously shifting their beds, and are consequently useless for navigation. But the eastern branch of the Irawadi delta proper is permanently accessible for 25 miles to large vessels as far as the port of Rangun, from which it takes the name of the Rangun river. This branch, however, although now communicating by several channels with the main stream. is properly speaking the estuary of the Hlaing, which winds along the west foot of the Pegu-Yoma hills, and which before the creation of the alluvial delta must have reached the sea in a channel altogether independent of the Irawadi system. At that time the Gulf of Martaban penetrated inland nearly to the parallel of Prome, and the heights now rising in the midst of the plains above the present head of the delta, some 200 miles from the coast, were formerly so many rocky islets washed by the marine waters. As the alluvial deposits gradually encroached on the sea, the Irawadi proper continued to follow its normal southerly trend; hence it is that the main deltaic branch, flowing about midway between the eastern (Rangun) and the western (Nawun or Bassein) mouth, still properly bears the name of Irawadi, that is, *Airavati*, "Elephant River," as it was named by the Hindu missionaries who introduced Indian culture into this region over 2000 years ago. The delta proper, which has altogether nine main branches, and an area of nearly 20,000 square miles, is still advancing seawards, and a submarine bank with a depth of 240 feet already extends beyond the present coast-line over 60 miles into the Gulf of Martaban. Since the British occupation costly engineering works have been carried out at various points to control the discharge of the flood-waters, which nevertheless at times break through the dykes, laying extensive fertile tracts under water.

Assuming that the Lu (Nu, Lutze, Nam-long) is the true upper course of the Salwin, this river must have a length of probably not less than 2000 miles. Rising on the Tibetan plateau, it enters Indo-China through a deep, narrow rocky bed, which may be described as almost a continuous mountain gorge descending from the Langtan (Gulong-Sigong) hills on the Yun-nan frontier. Lower down the Salwin flows first along the eastern frontier of Burma, and then between Pegu and Siam, to its mouth in the Gulf of Martaban. Near the confluence of the Thung-yang, one of the few affluents that join its lower course, the stream is contracted to a width of little over 100 feet, and farther down the navigation is almost entirely arrested by numerous reefs and rapids. In this region it rises from 30 to 35 feet during the floods, when the discharge is as much as 600,000 or 700,000 cubic feet per second. But even at its mouth the approaches to Amherst and Maulmain are obstructed by shifting sandbanks, so that despite its great length the Salwin is practically useless as a highway to the interior.

Still farther east comes the Meinam, the "Mother of Waters," the only large Indo-Chinese river whose course lies entirely within the geographical limits of the peninsula. It flows through the Laos States and Siam proper, mainly southwards, to the head of the Gulf of Siam, which it enters through three channels. Of these the easternmost is the most navigable, but even this is obstructed by a bar with scarcely 4 feet of water at ebb and 12 at flow. Hence large vessels proceeding to Bangkok, 38 miles from its mouth, discharge most of their cargoes in the roadstead. Throughout most of its course the Meinam is fringed by forest trees, behind which the low-lying rice and sugar plains are regularly flooded during the inundations. A few miles above its mouth the Meinam communicates with the Mekhong, which is really an independent river, although from this circumstance often represented as a branch of the Meinam.

The Mekhong Basin—Lake Tonle-sap—The Song-ka Basin.

Beyond the Meinam follows the Lantsang or Kin-lung-kiang, better known as the Mekhong, or great river of Cambodia.[1] Thanks to the famous French expedition of 1864 and several subsequent explorations, there no longer remains much doubt as to the true course of this, the longest of all the Indo-Chinese rivers, which rises in East Tibet and flows through Yun-nan and between Siam and Cochin-China to its delta in Cambodia, at the south-eastern extremity of the continent. Its upper course is

[1] Mekhong is the Lao name of this river, which the Cambodians call the "Tonle Thom"—*i.e.* the "Great River"—whence the European expression, the Great River of Cambodia, or simply the Cambodia River.

separated by a single narrow ridge from that of the Kinsha-kiang or Yangtse-kiang, the two streams here flowing for a long distance in parallel meridional valleys along the eastern scarp of the Tibetan plateau.

In its lower course the Mekhong is connected with the Tonle-sap or Great Lake of Cambodia, which drains to the river at low water, but which during the inundations receives a back current from the river. The Tonle-sap, which is almost the only lake in Indo-China, has a mean area of about 1000 square miles. But during the summer floods its level is raised nearly 40 feet, by which its length is increased from 70 to 120 miles, and its area is tripled. It abounds to such an extent in fish of every sort, that their capture and cure for exportation forms one of the chief industries of the country.

"The entry of the great Cambodian lake is at once grand and beautiful. It presents the aspect of a vast inland strait, with its low banks covered with dense and half-submerged forest growths, but encircled in the distance by a vast mountain range, whose farthest crests merge in the azure sky or disappear in the hazy atmosphere" (Mouhot).

"The Tonle-sap may be regarded as a remnant of the marine inlet which formerly penetrated up the present Mekhong delta as far inland as the 13° 30′ parallel. This inlet has been filled in partly by slow upheaval of the land, partly by the alluvia of the great river, by which the delta is still steadily advancing seawards" (Aymonnier).

A similar process is going on along the east coast, where extensive shallow lagoons have been enclosed by long narrow strips of sand from the sea. Such are the western and eastern lagoons about the Hue River estuary, with their southern continuation the Kohai lagoon, offering an uninterrupted waterway of difficult navigation, which extends for over 30 miles from Hue to about halfway to the Bay of Turan. One of the two seaward outlets of these inland waters is already choked by the sands, and during the periodical floods they overflow their low banks far and wide, so that the lagoons and surrounding rice-fields become merged in a single sheet of water. It is evident that in course of time another "Great Lake" will be developed about the Hue River estuary, similar in all respects to the Cambodian Tonle-sap.

The last important river reaching the coast in an independent channel is the Song-ka (Sang-koi, Nhi-ha, Hong-kiang), or Red River of Tongkin, which flows from the south Chinese highlands south-eastwards to the head of the Gulf of Tongkin. This river, the navigation of which is open to the Chinese frontier town of Lao-kai in Yun-nan, possesses considerable commercial importance, and the French have already made several attempts to open up a trade with the southern provinces of China through this channel. But a portion of its course is occupied by independent wild tribes, while it is obstructed at several points by difficult rapids. Hence the Song-ka does not offer the great advantages which the French at first expected to derive from it. Nevertheless, two steamers built specially for this traffic have succeeded in reaching Lao-kai (1891).

Both the Mekhong and the Song-ka are joined by several important tributaries, many of which have been carefully explored by Aymonnier, Neis, Harmand, and other French naturalists. Dr. Neis completed in 1881 the survey of the Don-nai, fixing the position of its source in the rugged highland region, whence it flows to the Mekhong delta. Here he determined two distinct lofty ranges, with an extensive intervening plateau, which it took seven days to cross.

In 1877 Dr. Harmand explored a considerable portion

PLAIN OF BAC-NINH IN THE SONG-KA DELTA.

of the Si-bang-hieng, an important affluent of the Mekhong, famous for its magnificent tropical scenery.

The whole of the Mekhong and Song-ka deltas, as well as all the intervening low-lying tracts extending between the Annamese coast range and the sea, ranging from about 5 to 15 or 20 miles in width, consist exclusively of alluvial matter washed down by the running waters from the surrounding uplands. The process is still going on, and owing to a combination of favourable conditions, winds and marine currents co-operating with the action of the coast streams, the land is here encroaching on the sea at a very rapid rate. The phenomenon was carefully studied and well described by the Hon. G. N. Curzon during his visit to the French Indo-Chinese possessions in 1892. "On the maritime fringe the brick or gruel-coloured streams, surcharged with alluvium, leave their detritus, which the tide is not sufficiently powerful to remove, and which gradually solidifies, and gives birth to a rank vegetation of mangroves and other aquatic plants. Sometimes for a while these form floating islets, which eventually coalesce and find a common anchorage. Sometimes they are covered with saline swamps, in which case they are utilised by the natives as salt-pans. A little later, as the sea recedes, they can be drained and planted, and in a few years what began as a muddy lagoon is transformed into a rice-field of cloth of gold" (*Geograph. Jour.*, August 1893).

Hanoi, capital of Tongkin, now 60 miles inland, stood on the seashore in the seventh century A.D., and in the seventeenth century the Dutch traded with the seaport of Hongyen, which is now 35 miles from the coast. Owing to these conditions the river mouths, almost without exception, are obstructed by bars, where the waves break with great fury, and prevent large vessels from gaining access to the estuaries for months together.

Another phenomenon of great geological and geographical interest is the peculiar rock-formation, composed of a limestone or marble overlying the Devonian schists, and presenting the most picturesque and fantastic con-

STREET IN HANOI.

tours both in the sea at no great distance from the mainland, and also at several points along the river valleys, which have in earlier days been similarly situated. It consists of detached blocks 50 to 500 feet high, with scarped sides, but with summits and ledges overgrown with a superb vegetation. The action of the sea has hollowed the interior of these rocks into vast caverns, and carved their surface into strange and fantastic shapes. These formations have received their greatest development in Along Bay and its prolongation Fai-tsi-long Bay, where thousands of such rocky islets fringe the coast for over 100 miles from the east side of the Red River delta nearly to the Chinese frontier at Cape Pak-lung. "One may spend days sailing in and out of the islets of this

THE BAY OF ALONG.

astonishing inland sea, which I do not hesitate to characterise as one of the wonders of the world, and which far excels the better-known beauties of the Inland Sea of Japan. Low-tunnelled passages, accessible only at low water, conduct to hidden basins or remote caverns in the heart of the rocks, and till recent times afforded an impenetrable retreat to the corsairs who devastated these waters. Near Turan, maritime port of Hue, a cluster of similar rocks, called by the French the Marble Mountains, rises abruptly from the sand-dunes on the seashore, and is perforated with grottoes, which have been utilised by

Buddhist monks for the establishment of one of those retreats wherein they appear uniformly able to combine æsthetic attractions with devotional needs. Their altars are enshrined in the bowels of the earth, and the ecstatic face of the gilded god shines faintly from the cavernous gloom" (Curzon, *ib.*).

4. *Natural and Political Divisions: Manipur—Lower Burma—Upper Burma—Siam—Annam—Cambodia—French Cochin-China—Malacca and Straits Settlements.*

The political condition of Indo-China has been largely determined by its prominent natural features. To the great river valleys of the Irawadi, Meinam, and Mekhong correspond the ancient historical kingdoms of Burma, Siam, and Annam (Cochin-China), while the still more ancient empire of Cambodia, founded by the primitive Caucasic race of the peninsula, has been gradually restricted to the broad alluvial plains and delta of the Lower Mekhong by the later Mongoloid intruders from the north. Malacca, also, almost physically detached from the mainland, has from prehistoric times been occupied by petty States, founded by peoples of Malay stock, either here indigenous or more probably intruders from the neighbouring archipelago of Malaysia.

Manipur.

East of Cachar lies the vassal State of Manipur, which acquired some notoriety owing to the rebellion and massacre of 1891. It comprises a valley about 8000 square miles in extent, at a mean altitude of 2500 feet above the sea, draining to the Chindwin affluent of the Irawadi, and encircled and partly traversed by several mountain ranges between Assam, Eastern Bengal, and Upper Burma. These ranges are generally disposed in the direction from north to south, decreasing in height towards Chittagong and the Lushai territory. They are crossed by three main routes, one from Cachar, one from Kohima in Assam, and one from Tammu on the Upper Burma frontier, all converging in the Manipur valley, which appears to be the bed of an old lacustrine basin, now represented by the small Logtak Lake.

Owing to the general altitude the climate is relatively temperate, with cool nights and mornings even in the hottest season. The mean annual rainfall also scarcely exceeds 39 or 40 inches, the moisture-laden clouds being intercepted by the encircling hills, which on the north and north-east frontier have a yearly discharge of 120 inches and upwards. Hence the slopes are clothed with magnificent forests of teak, fir, bamboo, and many other species, including the tea shrub, indigenous here as it is generally in the uplands between India and China. These forests afford cover to the elephant, tiger, leopard, wild cat, bear, rhinoceros, buffalo, and several kinds of deer. Snakes, though represented by many species, appear to be for the most part harmless. There is a strong, hardy breed of ponies, similar to that of Burma, used as pack animals and mounts, as in the game of polo, which, before its recent spread in India, was known only in Manipur and Ladakh and neighbouring districts at the opposite extremity of the Himalayas.

The inhabitants form two socially distinct groups—the hillmen, scarely removed from the savage state, and the Manipuri proper, that is, the settled and civilised population of the valley—collectively numbering about a quarter of a million. In the north the hillmen are not to be distinguished from the neighbouring Nagas, while in the south they form a branch of the widespread

Kuki family. The Manipuri themselves are probably members of the same family, though claiming Aryan descent, and long Hinduised in religion. The language is not Sanskritic, but of Tibeto-Burman stock, cultivated

MANIPURI HUT.

and written with a peculiar alphabet, derived, like the Burmese itself, from the Devanagari.

British supremacy may be said to date from the year 1823, when the Indian Government restored Gumbhir Sing, a member of the deposed Manipur family.

Lower Burma.

From the Chittagong district forming the south-eastern limit of Bengal proper there stretches a territory for about 1000 miles between the Bay of Bengal on the west, and on the east Upper Burma and Siam southwards to about the 10th parallel. This region, which before the annexation of Upper Burma formed the province of British Burma, consists entirely of Burmese territory at various dates during this century ceded to the English. It comprises three divisions, Arakan, Pegu, and Tenasserim, which completely shut off Upper Burma from the sea, and about the 12th parallel nearly reach across to the Gulf of Siam, between Siam proper and its lower province at the neck of the peninsula of Malacca. In Malacca itself the Straits Settlements, terminating at Singapore at its southern extremity, continue the British domain almost uninterruptedly round the Bay of Bengal to the Eastern Archipelago.

Arakan, the northernmost division, presents from the coast a fine appearance. The mountains, forming a southern continuation of the Lushai hills, and clothed to their summits with a rich forest vegetation, rise in a succession of parallel ridges from the plains to a height of from 5000 to 6000 feet. The plains themselves are of small extent, being mostly either limited by offshoots of the lower coast ranges, or else hemmed in by wooded tracts, which on the coast consist exclusively of mangrove trees. The lowlands are intersected by countless streams from the hills, while the spring tides flood extensive low-lying districts, forming a labyrinth of channels and backwaters. These watercourses take the place of highways, serving as a means of rapid intercourse between the towns and villages. Mud volcanoes occur both along the coast and on the neighbouring islands, and coal, iron, and petroleum are found in many places, while salt of a fine quality is obtained by evaporation in the numerous tidal estuaries.

Pegu comprises the region of the Lower Irawadi and Sittang rivers, which here form a common wide-branching delta, in its main features resembling that of the Brahmaputra-Ganges. The land is mostly low, sandy, or muddy, and in the wet season exposed to destructive floods. But it is well suited for the cultivation of rice, which is here produced in superabundance. Its trade and industries are also furthered by the railway running from Rangun through Prome, and the line along the Sittang to the Upper Irawadi valley near Mandalay.

The mountain system throughout the whole of this coast region is of a very simple character, consisting of regular and parallel ridges running uniformly north and south, and forming water-partings between all the large rivers, which thus find their way independently to the Bay of Bengal.

For a portion of its lower course the Salwin forms the border line between Siam and Tenasserim, the southernmost of the three great divisions of Lower Burma. This division, whose southern extremity approaches the insular region of Malaysia, is itself fringed along its entire length by a vast number of islands forming in the north the Moscos, in the south the much larger Mergui Archipelago. A few only of these little-known islands are inhabited, chiefly by Burmese and Karens from the opposite mainland. They are all hilly, with peaks 2000 to 3000 feet high, and are often densely wooded with the caoutchouc and other valuable trees. They are said to abound in minerals, and are tenanted by the tiger, rhinoceros, deer, and a great variety of reptiles.

South-east of Dumel Island, one of the largest members of the Mergui Archipelago, is situated the remarkable Bird's Nest or Elephant group, the Ye-ei-gnet-thaik of the Burmese, which consists of six marble rocks, the largest and highest rising to an altitude of

1000 feet. The precipitous sides of these limestone islets are partly clothed with vegetation, including a species of tree-fern projecting at all angles to a considerable distance high above the surrounding waters. But the chief feature are the birds' nest caves, generally opening into the sea below high-water mark. At the head of a deep cave in the large island, a tunnel, fretted with large stalactite knobs, leads into another circular basin of crater-like aspect. Similar basins occur elsewhere, and Commander Alfred Carpenter of the Geological Survey of India thinks that these basins were the floors of vast caverns at a time when the islands were far higher than at present. The work of disintegration by moisture is still in progress, pulling down these marble monuments of a giant age (*Geo. Proc.*, 1888, p. 303).

The eastern frontier of Tenasserim is formed by a mountain range 5000 feet high, which again acts as a water-parting between the Tenasserim and the Siamese river systems. On the British side the chief river is the Tenasserim, named from the capital, and flowing between the hills and the coast, mainly south, for over 230 miles, of which about 100 are navigable.

Upper Burma.

At the beginning of the present century the Burmese empire was by far the largest and most powerful in Farther India. It occupied nearly the whole of the Irawadi, Sittang, and Salwin basins, with a coast-line stretching for about 900 miles from the head of the Bay of Bengal to the Isthmus of Krah. Since then a series of disastrous wars with the English has caused the gradual loss of all the coast regions—Arakan, Pegu, and Tenasserim—which now constitute the flourishing province of Lower Burma.

Thus entirely cut off from the sea, the country remained in a state of chronic trouble until the close of the year 1885, when King Thebaw, last of the native rulers, was dethroned, and the administration of Independent Burma taken over by the British authorities.

Within its limits at the time of the annexation the late kingdom of Ava, as it was often called from one of its ephemeral capitals, was hemmed in on the west and south-west by British Burma, on the north-east by the Chinese province of Yun-nan, on the south-east by the kingdom of Siam. With an extreme length north and south of about 500 miles, and a mean breadth of 300 miles, it had a total area of about 190,000 square miles and a population of less than 4,000,000. It is divided into three distinct sections—Burma proper, between 24° 30′ and 18° 50′ N. latitude, inhabited by the pure Burmese people; North Burma, occupied by the Sing-fu and other semi-independent hill tribes; and the tributary Shan States to the east. All the Shan or Laos States stretch eastwards to the Mekhong valley; but those subject to Burma lie mainly between the 24th and 20th parallels and between 97° to 101° E. long.

The Burmese rule, which was severely felt by the districts in the proximity of Mandalay, the present seat of government, became continually less oppressive as we proceeded eastwards. In the north-east it was, so to say, overlapped by the Chinese authority, so that it was here often difficult to say where the one ceased and the other began. In some districts the triennial tribute due to the Burmese Court consisted of such trifles as gilded wax tapers, a little salt and tea, or perhaps a pair of embroidered shoes, a gold drinking-cup, a silver plume, or suchlike tinsel, and these presents were sent by several of the Shan districts both to China and Burma. The dignity of "tsauwab" or "thabwa"—that is, feudal lord

—is hereditary in all the ruling families, but the Burmese Court conferred the investiture on each successive lord, and designated the next heir. In the principalities, ruled jointly by the Chinese and Burmese, both suzerains generally came to an understanding in the choice of the next heir; but, in case of disagreement, two chiefs were appointed, and fought it out. But since the annexation of Burma several of these Shan States have been transferred to Siam, and those lying east of the Mekhong bend, which were to constitute the "Buffer State" between the British and French possessions, have been ceded to France by the Anglo-French treaty of January 1896. The claims of China have also been settled by the Anglo-Chinese Convention of 1894, in virtue of which she abandons her pretensions to the region north of Bhamo, receiving in return the two important Shan States of Monglem and Kianghung, between the Salwin and Mekhong rivers, which were formerly subject to Burma. In this convention the so-called tribute mission is not referred to, the subject being allowed to drop by mutual consent.

As in other Indo-Chinese States, the white elephant ranked in Burma next to royalty itself. This elephant had a palace to himself, with a personal chamberlain and estates in the most fertile cotton districts, besides four gold umbrellas and a suite of thirty courtiers. At the same time the expression "white elephant" is extremely elastic, the colour being often of a dirty yellow, or even brown, if only a few light specks can be shown behind the ears, on the forehead or trunk.

The famous ruby mines of Burma, surveyed by Mr. Robert Gordon in 1887, lie in the Mogok district, about 40 miles from the left bank of the Irawadi, between Mandalay and Bhamo, the nearest station by the water route being Thabyetkin, about 50 miles above Mandalay. Mogok stands at an altitude of 4100 feet, and the surrounding ruby hills rise to heights of from 4800 to over 6000 feet. The mines are of three kinds: workings in fissure veins of soft material embedded in the crevices of the harder rock caused by shrinkage at a remote geological epoch; the *Myaw*, or washing, which corresponds on a small scale with the hydraulic mining in California and elsewhere; lastly, the deposits on the flat-bottom lands at depths of from 10 to 30 feet. It is difficult to account for these deposits of nearly pure corundum, from a few inches to a few feet in thickness, lying on a bed of earth in which no stones occur, and covered by a similar layer of porous earth. When brought to the surface the corundum layer sparkles with myriads of rubies mostly too small to be of any value, while the larger stones are rarely free from flaws. Good, flawless stones, from three to five carats in weight, are much more valuable than the best diamonds of like weight; a five-carat ruby may be worth £3000, while a similar diamond will scarcely fetch more than £300. The present value of the Burmese ruby-fields has not yet been accurately determined. From this source the late King Thebaw is believed to have derived a revenue of from £12,500 to £15,000.

Siam.

From Burma and the region of the Irawadi we pass eastwards to the basin of the Meinam, politically comprising the kingdom of Siam or Tai—that is, of the "Free." Siam occupies the heart of the Indo-Chinese peninsula between Burma and British Burma on the west, Yun-nan on the north, Annam and Cambodia on the east. Southwards it includes the strip of territory between Tenasserim and the Gulf of Siam, as far as the Isthmus of Krah, in 10° N. latitude; and beyond this point all the northern section of Malacca nearly to Perak.

Till recently the eastern frontier coincided, or was supposed to coincide, with the ill-defined crest of the Annamese coast range, giving the kingdom a breadth of about 400 miles between the British and French possessions, but narrowing to 60 in the Malay Peninsula,

CHULALONGKORN, KING OF SIAM, AND SECOND QUEEN.

with an extreme length of at least 1000 miles north and south, an area of 360,000 square miles, and a population variously estimated at from 8,000,000 to 10,000,000. But in the year 1893 "a budget of fictitious grievances," raised by the French Chauvinist party, and supported by a blockade of the Siamese coast, the appearance of gunboats at Bangkok and the seizure of Chantabun, compelled the king to accept a treaty which involved a partial dismemberment of the kingdom in favour of his powerful and aggressive neighbour. The eastern frontier was shifted westward to the right bank of the river Mekhong, between Cambodia and Luang-Prabang, Siam thus ceding the whole of the territory between the river and the coast range, a region about 80,000 square miles in extent, with a population vaguely estimated at some two millions. From Luang-Prabang to the British frontier the Mekhong, here running west and east, becomes the northern boundary of Siam. But the little-known region stretching northwards to China, and extending from Tongkin westwards to British Burma, is not appropriated by France, but through the intervention of England was reserved to form a future neutral zone or " Buffer State " between Great Britain, China, France, and Siam, the last-named thus sacrificing another slice of territory, estimated at 20,000 square miles, with a population of perhaps 1,000,000. Siam also withdraws all armed vessels and military posts from the provinces of Battambang and Siemreap, that is, from the Siamese section of the lake district, Cambodia, and also from within 15 miles of the right bank of the Mekhong below Luang-Prabang, France to hold Chantabun until all the terms of the treaty are complied with. The last-mentioned clauses are a practical surrender of the Siamese Cambodian provinces and of the 15-mile zone along the right bank of the Mekhong from Luang-Prabang southwards. A borderland that cannot be held by military tenure is obviously either given over to anarchy or to the control of the more powerful conterminous State. Such an arrangement could only be, as it was meant to be, temporary, and the continued advances of the French in the Upper Mekhong basin resulted in the Anglo-French treaty of January 1896, by which the proposed buffer state disappeared, and the Mekhong became the boundary between the British and French possessions from the Nam-Huok confluence right up to the Chinese frontier. At the same time both contracting powers agreed to abstain from further encroachments on the kingdom of Siam, which, however, is thus practically reduced to the Meinam basin and the northern section of the Malay Peninsula, with a superficial area of about 200,000 square miles and a population of at most 5,000,000.

Siam proper consists mainly of the low-lying alluvial basin of the Meinam and its numerous tributaries, branches, and backwaters, which form an extensive and intricate delta, like that of Cambodia, continually advancing seawards. The northern background occupied by the Lao or Shan States is a more or less hilly country, which by the 1896 treaty has ceased to form an integral part of the kingdom. Both sides of the Meinam basin are skirted by densely-wooded terrace-like ranges, forming the water-partings towards the Salwin and Mekhong, but whose structure, form, and general elevation were little known before the recent official surveys of Mr. M'Carthy, and the expeditions of Carl Bock, Lord Lamington, Prince Henry of Orleans, and a few other travellers. Even still much of the interior remains an almost unexplored wilderness, mostly covered with dense tropical forests, and thinly inhabited partly by semi-civilised Lao peoples, partly by semi-independent wild tribes.

It results from Mr. M'Carthy's Government surveys that Siam, within its rectified frontiers, is a far less mountainous region than had hitherto been supposed. The greater part, not only of the Meinam, but also of the Mekhong basin within the Siamese borders, is described as "mostly flat, diversified by isolated hills and broken jagged ridges of limestone mountains." The former is an extensive alluvial plain representing the waste of the northern and western mountains, washed down by the

Meinam and its numerous head-waters, and gradually rising from the Gulf of Siam northwards to a height of less than 300 feet. Thus Saraburi, on the Nam-Sak, 70 miles north-east of Bangkok, is still only 46 feet above sea-level, while Pechai, on the Nam-Pat, about 280 miles north of the Gulf, has an altitude of not more than 260 feet. The rise to the foot of the northern hills would therefore appear to be at the rate of rather less than one foot per mile. Hence to a traveller ascending the Meinam towards Rahang the country presents the aspect of an uninterrupted level plain all the way to Chainat, 100 miles above Bangkok, where the monotony of the scene is first broken by a few isolated hills.

The still more extensive eastern region draining to the Mekhong is a moderately elevated plateau, at a mean altitude of less than 600 feet, and nowhere rising above 700 feet except near the western hills, which form the divide towards the Meinam basin, and at some points on that section of the Mekhong, which flows west and east below Chieng-Kan. Thus Korat, towards the southern margin, and Nongkhai below Wieng-Chan on the Mekhong, stand at the respective heights of 765 and 727 feet, while the intervening space of 283 miles falls towards the centre, as at Pathai-Soang, as low as 450 feet. Hence this Siamese section of the Mekhong basin is somewhat in the nature of a depression, where the sluggish streams wander about with uncertain flow, developing in many places extensive saline wastes and swamps, or flooded lagoons, such as those of Nong-Han, about 17° 10′ N. lat. During the rainy season the whole region "must be a perfect sea, judging from the water-marks on the trees 4 feet up the trunk. There are numerous swamps all over the country, and salt is procured in large quantities. It appears in the form of an efflorescence on the surface of the ground, and on a cool morning has all the appearance of a hoar frost" (J. M'Carthy, *Geo. Proc.*, 1888, p. 124).

Thus, in its present reduced limits, Siam consists mainly of two vast level tracts, the alluvial Meinam basin in the west and the slightly elevated Mekhong plateau in the east. Westwards the Meinam plains are limited by the long chain of mountains which forms the frontier towards Burma, and which is continued in an unbroken range to the southern extremity of the Malay Peninsula. Some of the peaks between Siam and Burma have an altitude of 7000 feet, and the system culminates in the Siamese section of the Malay Peninsula in a summit 8000 feet high.

Northwards the Meinam is separated from the Mekhong basin by an irregular orographic system, which breaks away from the western range at a point north-east of Chengmai. Here the divide is so ill-defined that before Mr. M'Carthy's surveys the head-waters of the Meinam were commonly supposed to intermingle with the western affluents of the Mekhong. In many places the sources overlap, or approach so near as to appear continuous even on a large scale map. The hills themselves are of low elevation, scarcely anywhere rising to a height of 3000 feet. Nor do they form a clearly marked range running west and east, but rather a series of short parallel ridges disposed north and south, some of the intervening valleys opening north in the direction of the Mekhong, others draining south to the Meinam.

At present the chief wealth of Siam is derived from the rich and well-watered alluvial plains of the Lower Meinam, which yields magnificent crops of tobacco, sugar, cotton, maize, indigo, and especially rice, the staple product and chief export of the country. Pepper is largely grown in the Shantabun district, and the forests of the western and northern uplands contain many valuable species, such as teak, sapan-wood, ebony, and rose-wood.

In Malacca the dependent States are Quedah, Kalantan, Patani, Ligor, Talung, Tringanu, whose Malay sultans pay merely a nominal tribute of a gold or silver tree or flower, sent every third year to Bangkok.

Annam and Cambodia.

East of Siam the remainder of the Indo-Chinese peninsula is occupied by the kingdom of Annam in the east, and the French possession of Cochin-China with the neighbouring vassal kingdom of Cambodia in the south.

The French, who have taken direct possession of the Mekhong delta, have during the past twenty years gradually extended their influence throughout the whole of this region. For Annam itself has by the recent treaties of 1885 become a French vassal State. It consists of a comparatively narrow strip of coast-lands stretching nearly due north and south between the China Sea and the coast range skirting the left bank of the Mekhong. It consists of three distinct sections—Tongkin, or Dang-ngoi—that is, the "Eastern Land," watered by the Song-ka, or Hong-kiang; Cochin-China, or Dang-kong—that is, the "Interior Land"; and Chiampa, or Tsiampa, in the extreme south-east corner of the peninsula. To these must also be added the domain of the semi-independent wild tribes (Moi), and the section of the Lao nation settled on the left bank of the Mekhong, who are subject to Annam. Formerly Annam claimed jurisdiction over Cambodia and the Mekhong delta, but even within its present restricted limits it stretches north and south across thirteen degrees, between 23° and 10° N., with a total area of some 200,000 square miles, and a population vaguely estimated at from ten to twenty-one millions

Annam proper, or Dang-kong, a narrow strip from 10 to 20 miles wide, extends from about 12° N. on the Chiampa frontier northwards to Tongkin. It is enclosed on the west by bare hills covered with a very sparse vegetation. The domain of the Moi stretches west of this province from about 10° to 16° N. Under the general designation of Moi, the Annamese comprise all the numerous hill tribes known to the Siamese as Kha, and differing widely from each other in speech, type, and usages. The land of the Laos subject to Annam lies north of Cambodia and the Mekhong, and varies in breadth from 20 to 24 miles. In the south and west are several settled districts, but the east is an arid waste. The plains are enclosed on the north by two ranges, and the rivers are mere mountain torrents. Yet all the accounts of recent travellers represent the country as in a prosperous state, inhabited by a peaceful and industrious people living under the authority of patriarchal chiefs. They cultivate the land, and have some silk and earthenware manufactures.

Tongkin is very mountainous in the north, where it presents the same general features as the neighbouring Chinese provinces of Yun-nan and Kwang-si. The eastern districts are almost flat, merging seawards in an extensive alluvial plain. Most of the streams, which flow mainly in a south-easterly direction, contain large quantities of auriferous sands, the washing of which employs thousands of hands. An old wall running along the southern frontier towards Annam from the hills to the coast has been rendered useless since the union of the two States under one sovereign. Next to Korea, Tongkin had persisted most obstinately in the exclusive system as regards foreigners, and the policy even of the central government had been to keep Tongkin in complete seclusion from outward influences. But it was at last thrown open to the trade of the world by the commercial treaty concluded between France and Annam in 1874. In 1885 the French protectorate was formally recognised by China.

Tongkin is extremely fertile, and abounds in mineral wealth of all kinds. As many as 14 gold, 7 silver, 3 copper, 1 tin, 17 iron, and 3 salt mines have already been opened, while the extensive coal-fields, some reaching down to the coast, still remain almost untouched. But the vast resources of the country must remain undeveloped, while it continues a prey to anarchy and to the attacks of regularly-organised bands of marauders. A writer in a recent number of *Globus* tersely remarks that "before an active foreign trade can be developed the banditti must be exterminated."

The ancient kingdom of Cambodia, which formerly comprised a large portion of Indo-China, has long been restricted to the lower course of the Mekhong between Lake Tonle-sap and the delta. Till recently it even stood for some time in the position of a vassal State to Siam. But since the French occupation of Lower Cochin-China the king of Cambodia has transferred his allegiance to France. His territory forms an extensive and exceedingly fertile alluvial plain watered by the Mekhong, the great Lake Tonle-sap, and their numerous affluents, branches, and connecting channels. The plain is diversified in the west by isolated hills and short ridges, and is confined in the north by the Phnom Dongrek range. The space between this range and the northern shores of the lake is strewn with the stupendous ruins of Angkor Wat (properly Nakhon Wat) and many other remains, which still attest the former greatness of the Cambodian empire, when it formed one of the chief centres of Hindu culture in the East. These monuments, the finest of which are now included within the limits of Siam,

ANGKOR WAT.

contain vast archæological and architectural treasures, which have in recent years been carefully studied by Aymonnier, Delaponte, and other French antiquarians. From their researches, as well as from the inscriptions, some of which have already been deciphered, it appears that these monuments, covering a space of no less than 20 square miles, date from different periods between 700 and 1100 A.D. They were originally dedicated, not to snake worship nor to Buddhism, as had long been supposed, but to pure Brahmanism, although after the introduction of Buddhist teachings many images of Hindu deities were replaced by statues of Buddha. The materials employed are a coarse porous red stone for the substructures, a hard fine-grained sandstone for the sculptures, and good durable bricks for the later buildings. Collectively they form "the most remarkable collection of ruins in the world, whether we regard the prodigious magnitude of the ground-plan, the grandiose dimensions of the principal palaces and temples, or the artistic beauty and delicacy of the bas-reliefs and sculptures" (Curzon).

The erection of these monuments, of which there are seven distinct groups, is attributed by some French authorities to the Khmers, a people of Aryan speech, who arrived overland from India, conquered the country, and founded the Cambodian empire, afterwards disappearing, or becoming absorbed in the present Cambodian population. But there is no record of any such Aryan invasion of this region, and in all probability the invaders were simply Hindu missionaries, who settled in the Mekhong basin, and introduced Aryan culture and orderly government amongst the rude inhabitants of Cambodia, as they did in other parts of Indo-China, as well as of Malaysia, as still attested by the marvellous remains of Ayuthia in Siam, of Pagan in Burma, and of

Borobodo in Java. It is to be noted that *Khmer* is and always has been the national name of the Cambodians themselves, and that they were almost certainly the builders of these monuments under the guidance of their Hindu teachers. They belong to the same indigenous stock as the neighbouring Kuy, to whom they give the name of *Khmer-dom*, that is, "original Khmers," and who dwell chiefly to the north of the Great Lake about the frontiers of Cambodia and Siam. These Kuy, though regarded as wild tribes, possess a considerable degree of culture, and are specially noted for their skill as workers in iron. They speak various dialects of a primitive language, which is also common to the Cambodians, but which is neither "Aryan" nor yet in any way related to the Siamese, Annamese, or to any other member of the Indo-Chinese linguistic family. These are all toned languages, and mostly monosyllabic, with no grammatical inflections, whereas Khmer is polysyllabic and spoken without tones. No "Khmers of Aryan speech" are known to history or tradition, and none are needed to account for the monumental remains of Cambodia, any more than for those, for instance, of Java.

Cambodia has an extreme length of 240 miles north and south, with a breadth of 180 miles, a total area of 40,000 square miles, and a population of under 1,000,000.

French Cochin-China.

The French colony of Lower Cochin-China, which has become the centre of political power east of Siam, comprises the whole of the hot and marshy Mekhong delta. It consists of six provinces detached in 1863 from Annam, and now administered by the French marine department. It has an area of some 22,000 square miles of extremely fertile land, with a population of 1,600,000.

The Malay Peninsula—The Straits Settlements.

The Malay Peninsula stretches from 13° 45′ N. first southwards to 8° 50′ N., and thence south-eastwards to Cape Romania in 1° 35′ N. The angle thus formed is marked by the Isthmus of Krah, which is about 3000 feet above sea-level. South of this point rise the Malacca mountains, whose irregular masses fill the whole of the peninsula to its southern extremity. It has an extreme length of about 900 miles, with an area of 83,000 and a population of about 1,000,000. The northern section, with an area of 14,000 square miles, and a population of 400,000, is held by petty Malay States tributary to Siam. All the rest is occupied partly by the British possessions, collectively known as the Straits Settlements, partly by a number of Malay States, which have accepted the British protectorate.

The Straits Settlements, which lie scattered along the west coast, include the province of Wellesley, Malacca, and the two important islands of Singapore and Pulo-Penang. Of the protected States the most important are Perak (Perah), Selangor, Pahang, Negri Sembilan, and Johor.

The interior of the peninsula, which consists of magnificent wooded ranges, intersected by numerous fertile river valleys, is still imperfectly known. But a systematic survey of Perak and the neighbouring districts was begun in 1879 by H. S. Deane, who determined the height of the Shin (6000 to 7000 feet) and Titi Wangsu (7000) ranges, and ascended the navigable River Plus, a tributary of the Perak, for 50 miles to the Jeram Dina rapids.

In Perak the culminating point appears to be Gunong Kerban (7127 feet), which was ascended in 1892 by Mr. G. A. Lefroy, chief surveyor of the State. Following the course of the Kinta River, here flowing at a level of about 1000 feet above the sea, the explorer made the ascent from Batu Shallek, which stands about 3400 feet above the sea. Up to a height of about 5000 feet Kerban consists of grey granite rocks and boulders, with a shallow soil. Above 5000 feet the granite is overlaid by a schistose formation, which is being rapidly removed by denudation. The evidence of numerous

PEKAN IN PAHANG.

landslips in the neighbourhood would point to the fact that the summit was much higher in past ages than it is at present. The crest, in fact, is the remains of a secondary formation, traversed by several quartz veins, which has now disappeared, except on the tops of the higher mountains, and on some isolated hills in the low-lying Larut district. The change in the geological formation at 5000 feet is plainly marked by the characteristics of the flora, which at this altitude becomes very stunted.

There are altogether over 1200 miles of water transport, and this region excels in the extent of its forest lands available for plantations. In a total of 5,000,000 acres, fully 200,000 are well suited for the cultivation of tea, coffee, cinchona, and indigo. The chief mineral is tin, of which there seem to be vast deposits. Tin also abounds in other parts of the peninsula, and the rich mines of the island of Thalang (Junkseylon) on the west coast, about 8° N., employed over 30,000 Chinese hands till 1872, when the works were partly abandoned.

But a much more important island is Pulo-Penàng ("Prince of Wales"), which lies off the coast of the British province of Wellesley at the northern entrance of Malacca Strait. Although scarcely 15 miles long by 7 broad, such is the fertility of its soil that it produces large quantities of rice, pepper, cloves, nutmegs, betel, cotton, tobacco, tea, coffee, sugar, and cocoa-nuts.

At the southern entrance of Malacca Strait, and close to the mainland, lies the little island of Singapore, which, since its occupation by the English in 1819, has become one of the great centres of trade in the East. Formerly almost entirely uncultivated, it is now covered with pepper, sugar, rice, sago, and gambier plantations, on which a large number of Chinese coolies are employed. But the rapid progress of Singapore is due not so much to its agricultural produce as to its geographical position on the great trade route between India and China, combined with the enlightened policy and liberal institutions of its present rulers.

The Straits Settlements have a total area of about 1500 square miles, with a mixed Malay and Chinese population (1891) of 513,000.

Johor.

For a very full account of the Sultanate of Johor with the dependent territory of Moar (Muar), we are indebted to Mr. Harry Lake, engineer in the service of the Sultan, by whom the whole region was surveyed and mapped in the years 1890-92. Johor, which occupies the extremity of the peninsula as far north as about 3° 1′ N. latitude, has an area of over 9000 square miles, but being still mostly under primeval forest, the population scarcely exceeds 300,000. As in the rest of the peninsula, the coast-lands are bold and rocky on the east side facing the China Sea, low and swampy on the west along the Strait of Malacca. The interior forms in many places a continuous mass of dense jungle, the low-lying marshy tracts, rising grounds and uplands, being all alike clothed with a rank growth of tropical vegetation, "so thick and closely interlaced with thorny creepers and rattans, that it becomes almost impossible to move a yard in any direction without previously cutting a path. Even on the smaller rivers the foliage and creepers will stretch entirely across from bank to bank, and a way must be cut for canoes to pass under, whilst the river-bed is full of snags and fallen timber, which have to be hacked through or moved before a passage can be effected" (Lake, *Geographical Journal*, April 1894, p. 283).

The prevailing geological formation is granite, traversed here and there by quartz veins and dykes of intrusive diorites, quartz felsites, and trachytes, and overlaid by clays, clay shales, and beds of laterite, the clays being non-fossiliferous, and probably of palæozoic origin. Gold occurs in Moar, and iron ores are widely diffused, while the tin deposits, less abundant than farther north, are entirely alluvial. It is noteworthy that, as elsewhere in the Malay Peninsula, the hills of Johor give no indication of the presence of tin oxide *in situ*, that is, in distinct veins traversing the native rock. The ores yield from a few ounces to as much as 50 or 60 lbs. per cubic yard; but mining operations are carried on in a somewhat primitive manner almost exclusively by Chinese.

Unlike other parts of the peninsula, Johor proper is not a mountainous region, the main central ranges here breaking into small isolated groups of hills, and culminating in the Blumut group (3100 feet) in the centre of the territory. But the neighbouring district of Moar, politically attached to Johor in 1877, is much more hilly and elevated. Here "Mount Ophir," properly Gunung Ledang, rises to a height of 4150 feet in the midst of a small group of hills, at one time supposed to be the highest land in the peninsula. But Gunung Tahan, and several other peaks farther north, are now known to tower some thousand feet above Ophir, which was so named by the early European adventurers everywhere in search of the Biblical Mount Ophir. The district is certainly auriferous, and has yielded considerable quantities of gold; but the alluvial deposits are now all but exhausted, while reef-mining, lately introduced, has hitherto yielded but poor returns. It has been proposed to make Ophir more accessible by a bridle-path, and establish a health resort for the residents of Singapore, Johor, and Malacca, on the summit, which enjoys a good climate, abundant water, and a splendid view of the surrounding region.

Of the inhabitants of Johor over 200,000, or fully two-thirds, are Chinese, engaged mainly in mining and agricultural pursuits. They chiefly cultivate pepper and gambier, which, with rubber and gutta, form the staple products of the country. Gambier, which is used both for tanning purposes and as a brown dye, is prepared from the foliage of *Uncaria gambir*, a climbing plant widely diffused throughout Malaysia. After boiling for some hours the liquor is strained and hardened in cooling to a bright yellow clay-like mass, which is cut into cubes, and in this state exported. The trade of Johor, which has steadily increased in recent years, is carried on almost exclusively with Singapore, the great distributor of the produce of the Malay Peninsula. In 1890 over 21,500 tons of gambier and 9230 of pepper were exported, besides copra, coffee, tea, areca nuts, rattans, resins, timber, gutta, rubber, and tin, the chief imports being rice (about 50,000 tons annually), salt, fish, sugar, tobacco, hardware, and Manchester goods.

Although theoretically autocratic, the government of Johor has assumed under the present enlightened ruler almost a constitutional character. Abubakar, hereditary Temenggong (chief) of Johor, was recognised by the British Government in 1855 as the *de facto* sovereign of the State; in 1868 he took the title of Maharaja, which by the treaty of 1885 was changed to that of Sultan. He thus represents the rulers of the former powerful State of Malacca, whose records date from the beginning of the thirteenth century, when the reigning prince, Mahmud Shah, adopted the faith of Islam, and took the title of Sultan. Abubakar, who visited England in 1866, 1878, and 1886, may, in fact, on historic grounds, claim to be the heir of those princes, for Sultan Ahmed, when expelled by the Portuguese from Malacca in 1511, withdrew first to Moar and then to Johor. During the long struggle of 140 years between the Portuguese and the Dutch for the possession of Malacca, Johor took a prominent part, the seat of government being at that time established at Johor Lama on the Johor River. But in 1855 the capital was removed to Johor Bahru, "New Johor," a town of about 15,000 inhabitants, pleasantly situated on the Strait of Tebrau, 15 miles north-west of Singapore. "Facing the sea is the Istana Laut, the principal residence of the Sultan, a long two-storied building, fitted up with every European comfort and luxury, and looking deliciously fresh and cool in the glaring sunlight. Well-laid-out roads, an esplánade over a mile long, large airy hospitals, water-works and wharves, all testify to the enlightened and energetic administration of the present ruler" (Lake, *loc. cit.*).

The Sultan is assisted by a Council of State, composed of chiefs and nobles, and the various departments are organised on the model of a British Crown Colony. Justice is meted out by the Sultan with an impartial hand, and above the various minor tribunals the Council of State itself acts as a supreme court of appeal, to which all subjects of the State have access.

5. *Climate.*

Owing to its position between two oceans, and almost entirely within the tropics, the climate of Further India may be described as normally hot, moist, and relaxing, and about the large deltas distinctly enervating and malarious. There is scarcely any cold season except on the northern uplands towards the Tibeto-Chinese frontier, and the rainfall, due mainly to the south-west monsoons, from April to October, ranges from about 90 inches in Singapore to over 200 in parts of Burma. Owing to this abundant rainfall the Irawadi is a copious river even before emerging from the last Tibetan highlands, and this, more than any other circumstance, has lent a colour to the theory of its connection with the San-po.

the Lower Meinam and Mekhong valleys, notwithstanding the excessive heat, the atmosphere remains charged with an unusual quantity of moisture both day and night throughout the greater part of the year. To this cause are due the many diseases, such as dysentery and typhoid fevers, which are here endemic. Exposure to the solar rays is also frequently attended by fatal sunstrokes, although the cholera, which is also endemic, commits less ravages than might be expected. The climate of French Cochin-China and Cambodia seems even to have become less fatal to Europeans of late years. Here the rainfall is scarcely more than 54 inches, while the temperature at Saigon averages about 80° F., the extremes being 75° and 95°, as results from the observations recorded between the years 1874 and 1881.

Notwithstanding its proximity to the equator, Malacca would seem to be, on the whole, rather cooler and drier than Indo-China proper. Thus the temperature of the Malacca district at the southern extremity of the peninsula is described as salubrious and equable, the glass ranging between 72° and 85° F. Even in Singapore and the province of Wellesley it is said seldom to rise above 87° or 88° F., and although very damp, the climate of Singapore agrees well with Europeans. On the other hand, that of Pulo-Penang is very oppressive and enervating, with a rainfall varying from 60 to 90 inches. The winter months are here the driest, and the northern winds the coolest and most invigorating.

In the Malay Peninsula the climatic relations are largely determined by its marine position, exposing it to the influence both of the north-east monsoon, which blows steadily from November to April, and of the south-west monsoon prevailing from May to October. The latter is less regular, veering round to west and even to south-east, and at times shifting from day to day. It is also much warmer and somewhat moister than the north-eastern monsoon, which is relatively cold and healthy. In the extreme south (Johor and Singapore) the mean annual temperature is stated to be 80° F. by the Dato Abdul Rahman, who adds that although India is many degrees higher in latitude, Johor is a much cooler region. "Johor and Singapore have been called the Paradise of India" (*Geographical Journal*, April 1894, p. 298).

6. *Flora and Fauna.*

Owing to its hot and moist climate and naturally fertile soil, the vegetation of the Indo-Chinese peninsula scarcely yields in exuberance and variety to that of the neighbouring archipelagoes. A great part of the surface is still everywhere covered with dense primeval forests, in which teak, eaglewood, gum-trees, the gutta-percha plant, bamboo, dye-woods, cardamum, vanilla, and many other useful tropical plants, are found in great abundance. The staple of agriculture is rice, of which vast quantities are produced, especially in the Irawadi, Meinam, and Mekhong valleys. Other cultivated plants are cotton, tobacco, indigo, the areca palm, the sugar-cane, cloves, cinnamon, coffee, tea, sago, pepper, ginger, besides maize, wheat, and tropical fruits in endless variety and abundance. Such are the jack-fruit, mango, mulberry, tamarind, letchi, bread-fruit, orange, lemon, and pine-apple. Vegetables are less varied, the most generally cultivated being sweet potatoes, beans, radishes, and onions. In Cambodia and Cochin-China the hills are mostly overgrown with wild vanilla, various species of caoutchouc, many oil, resin, gum, and lacquer yielding plants, while the shores of the southern islands and many parts of Malacca are fringed with the cocoa-nut palm.

Of larger wild animals the most common are the

elephant, tiger, leopard, wild boar, rhinoceros, and crocodile. The gibbon and other large species of apes, snakes, and birds abound in all the wooded districts, while the rivers and especially Lake Tonle-sap teem with every variety of fish. The chief domestic animals are the buffalo, ox, and horse, besides the tame elephant, which, contrary to the generally received opinion, breeds in confinement (Dr. Harmand).

Of bird and insect life in Johor (Malay Peninsula) Mr. Harry Lake gives a vivid description. "Birds of every size and colour, from the tiny, bronze-green sun-bird, and the blue and orange kingfisher, to the big crimson-beaked black hornbill, rise from the trees. Insect life swarms in myriads; dragon-flies of bronze, blue, purple, and vermillion, and butterflies in every shade of yellow, from pale primrose to orange, delicate rose-pink and bright crimson, are in endless variety. Now and then a lizard or snake will glide away in the dense undergrowth, while troops of monkeys chatter and skurry off, crashing through the trees, and taking break-neck leaps from branch to branch in their haste to escape the intruders. On each side is primeval forest, huge trees loaded with creepers drooping in a thousand fantastic shapes, dark green foliage, yellow sand and clear water, overhead a blue sky and blazing sunshine. But let a cloud obscure the sun, and the whole aspect changes; the trees and water look sombre, the birds and butterflies vanish" (*Geographical Journal*, April 1894, p. 288).

Amongst the larger fauna of the Malay Peninsula are the tiger, of smaller size, but as powerfully built as his Indian congener; the leopard, or "starry tiger," as the natives call him, now exceedingly rare; the wau-wau, or white-handed gibbon, keeping to the tops of the highest trees; the *broh*, or pig-tailed monkey; the black lotang species (*Semnopithecus obscurus*); the elephant, rhinoceros, bison, tapir, sambur, and other deer,—some, such as the kijang, of diminutive size. On the other hand, very large monitor lizards are met, as well as crocodiles from 18 to 20 feet long.

7. *Inhabitants: The Shans, Laos, and Siamese—The Annamese and Cambodians.*

Till recently Indo-China proper was supposed to be exclusively occupied by peoples of Mongoloid stock, allied in speech to the Tibetan and Chinese branches of that family. But the systematic researches of the French naturalists carried on in the Mekhong valley during the last twenty years have determined the presence in that region of a second ethnical element apparently of Caucasic type, and speaking languages akin to those of the Malayo-Polynesian linguistic family. The Malay race itself has been settled probably from prehistoric times in Malacca, while the true aborigines of this peninsula seem to be a dark race akin to the Negritoes of the Eastern Archipelago, and of which a few surviving tribes still linger in the interior of the country. All the inhabitants of Indo-China, taken in its widest sense, are thus reducible to four distinct groups, as under :—

(*a*) Mongoloid Stock.

Kakhyen (Singfo)	Khakhun ; Maran Mariss ; Nkum Lataung ; Lepei	Upper Irawadi basin above Bhamo ; east to Momein.	
Kuki	Lushai (Dzo) Chin (Shendu)	Hill Tracts, North Burma.	
Rakhaingtha ("Mugs")		Arakan Plains	500,000
Karen	Sgau ; Pwo Bghai ; Tari Mopgha Tungthus	S. Arakan Hills, Tenasserim	? 250,000
Kumi	Awa Kumi Aphya Kumi	Koladyne River, N. Arakan	? 20,000

Talaing (Mon)		Pegu	500,000
Burmese		Tenasserim Pegu, Arakan	250,000
Tai Family. Shan	Tai-shan Tai-neua	Shan States, North Burma, North Siam, and Yun-nan	
Tai Family. Siamese		Meinam basin.	
Tai Family. Lao	Lau-pang-kah	Middle course Mekhong River.	
	Lau-pang-dun	North and East Siam.	
Tongkinese		Tongkin.	
Annamese		Cochin-China.	
Unclassified wild tribes	Moi[1] Kha[1] Phnom[1]	Siam, Cambodia, and Annam frontiers.	

(*b*) Caucasic Stock.

Khmer (Cambodians)	Cambodia.
Khmer-dom (Kuy)	Cambodia and Siamese frontier.
Samre	South Lao, near Cambodian frontier.
Charay Stieng	Annamese and Cambodian frontier.
Cham	Southernmost districts of Annam.
Chong	South-east Siam, Gulf of Siam.

(*c*) Malay Stock.

Orang Malayu ("Malay Men")	The civilised Malays, dominant throughout the Malay Peninsula.
Orang Benua ("Men of the Soil")	The Malay wild tribes, constituting an aboriginal element intermingled with the Negritoes in the more inaccessible parts of the peninsula.
Orang Laut ("Sea Men")	The semi-civilised floating populations of the straits and inland waters, formerly corsairs, now fishers and sailors; are the "Sea Gipsies" of early English writers.
Sam-Sams	Malayo-Siamese half-breeds about the borders of Siam proper and Malayland.

(*d*) Negrito Stock.

Samangs Sakeis Jakuns Besisi Mentras	A dwarfish black race, with woolly hair and extreme prognathism, scattered in small groups over the wooded uplands, mostly keeping aloof, but in places intermingled with the Orang Benua; are the true aborigines of the peninsula, and the only distinctly Negro people still surviving on the Asiatic mainland.

[1] *Moi* is the Annamese, *Kha* the Laotian and Siamese, *Phnom* or *Penom* the Cambodian collective name for all these aboriginal wild tribes, which seem to belong partly to the Mongol, partly to the Caucasic stock.

The Kakhyens, Chins, and Lushai.

From the recent researches of Jenkins, Sladen, Forbes, Errol Gray, Elliot, and others, in North Burma, it appears that the Singfo and Kakhyens, hitherto regarded as two distinct races, are really one and the same people. Although split up into a great number of small tribes, they everywhere call themselves Singfo, properly *Changpaw,* that is, "men," but are always spoken of by the Burmese as Kakhyen (Kachin). This is again the same word as Karen, another form of Rakhaing, whence the province of Arakan takes its name. The Singfo claim to be the elder branch of the Burmese family, and although before the annexation nominally subject to the "wun," or governor of Magong, they paid little heed to his mandates, and on all occasions showed contempt and aversion for their "younger brothers," the civilised Burmese. They reach eastwards as far as Momein, and are generally regarded as a savage, unruly, and treacherous race. Major Sladen, however, found them friendly and intelligent, although extremely suspicious of strangers. They are active traders, and would willingly abandon their lawless and predatory habits were regular commercial relations established across their country between Assam and China. Their religion consists mainly in the worship of good and evil spirits (*Nats*), to whom they offer sacrifices. Mountains, valleys, trees, rivers, the sun and moon themselves, are under the influence of these *Nats,* who seem to be sometimes confounded with the spirits of the departed.

Despite their claim to be regarded as Burmese, the Kachins are said to be a northern branch of the Karens, who originally dwelt near the Khamti country in a territory regarded as the cradle of the Karen race. The oldest descendants of this stock are stated to be the Mariss,

Lataung, Lepei, Nkum, and Maran tribes. But at present the most powerful are the Khakhus, that is, "River-head" people (Kha = river, Khu = head), so-called because their domain lies about the Nam-Tisang, Pungsan-Kha, and other head-waters of the western fork of the Irawadi. Formerly the Kachin villages were ruled by a hereditary *Sawbwa*, or chief, whose subjects were obliged to cultivate his lands without pay. They were also oppressed by so many other burdens that about 1870 a revolution broke out, resulting in the massacre or deposition of many Sawbwas, who were replaced by *Akyis* or *Salangs* (headmen). The villages without Sawbwas are now called *Kamlao* ("rebel"), in contradistinction to the *Kamsa* communities still under Sawbwas.

Much trouble has been caused by the change, the Kamlao villages being practically so many petty republics, in which the headmen exercise little or no power over the unruly classes. But the villages near the Chinese frontier are said to be already tired of the lawlessness prevailing under the new order, and are now treating for the return of their Sawbwas. On the other hand, the Chinese element is pressing hard on these aborigines, and immigrants from Yun-nan are settling in continually increasing numbers in the Kachin domain, both as traders and agriculturists. Some have even penetrated west of the Nam-Kiu branch of the Irawadi, and these are chiefly engaged in collecting rubber, making salt, and gold-washing. They are said to have arrived about 1890 from China at the invitation of a local Sawbwa (Lieutenant Eliott's *Report*).

To the same ethnical group as the Kakhyens undoubtedly belong the Chins and the Lushai, who give their names to the Hill Tracts. This region lies generally between latitudes 21° and 24° north, and longitudes 92°-94° east, the conterminous territories being Manipur and

Cachar in the north and north-west, Arakan and Chittagong in the west, and Burma proper in the south and east. A series of parallel mountain ridges extending about 400 miles north and south, with an extreme breadth of 260 miles, covers the whole surface, and in many places rises to an altitude of over 9000 feet. This excessively rugged territory, almost unknown before the annexation of Upper Burma, has since been traversed in every direction by the various military expeditions, which have been engaged in reducing the lawless hill tribes to order between the years 1886-94.

As might be expected from its general relief, the Chin-Lushai country "embraces every variety of physical features and climate, from the dense and deadly jungles below, through the tangled mazes of which the ponderous elephant and rhinoceros push their way, to the invigorating summits crowned with pines, where the sheen of the pheasant's wing catches the eye, as, with lightning speed, he skims down the mountain side. People this region with dusky tribes, almost as numerous in dialect and designation as the villages in which they live, owning no central authority, possessing no written language, obeying but the verbal mandates of their chiefs, hospitable and affectionate in their homes, unsparing of age or sex while on the war-path, untutored as the remotest races of Central Africa, and yet endowed with an intelligence which has enabled them to discover for themselves the manufacture of gunpowder."[1]

Despite the multiplicity of names by which these aborigines call themselves, or are called by their neighbours, they all appear to belong to the same ethnical family as the Nagas of South Assam, the Abors of North Assam, and the Kakhyens of the highlands on the Chinese frontier. The Kuki, that is, "Hill-men," as they are

[1] Surg.-Lieut.-Col. A. S. Reid. *Chin-Lushai Land*, Calcutta, 1893, p. 2.

collectively called by the Bengalese lowlanders, form two geographically distinct groups—(1) The *Dzo* (*Zao*), better known as *Lushai*, or head-cutters (*lu* = head, *sha* = to cut), west of the Koladyne; (2) The *Lai*, better known as *Chin* (*Khyen*), or *Shendu*, east of that river.

Before the annexation, the Lushai raided into British, the Chins into Burmese, territory, as far east as the Chin-dwin (Khyen-dwen) affluent of the Irawadi, and it would appear that this river was so called because it formed the limit of Chin incursions into the Irawadi valley. Hence it is that the Chins were unknown to the English till quite recently, whereas the Lushai had already been heard of in the days of Warren Hastings. "The first record of the raids of these savages dates from 1777, when the chief of Chittagong, a district which had been ceded to the British under Clive by Mir Kisim in 1760, applied for a detachment of sepoys to protect the inhabitants against the incursions of the Kukis, as they were then called" (A. S. Reid, p. 7). In later expeditions frequent mention occurs of the Baungshes, Soktes or Paites, Looes, Kabui, and many others, all of whom belong to the same Kuki group, although often differing in dress, customs, and even speech. As amongst the Nagas, a great variety of dialects prevails, and these, though probably derived from a common Tibeto-Burman stock, have diverged so widely that some of the Chins living only a few miles apart are mutually unintelligible, while few of the Lushai can make themselves understood by their eastern kinsfolk.

But the religious beliefs and practices present everywhere much the same characteristics, for all the Kuki peoples are still nature-worshippers, so far as they can be said to worship anything. "When they are prosperous they practically worship nothing at all. Sometimes, when wandering through the jungle on a hunting or trapping excursion, or in quest of roots and leaves to eke out the rice supply, they may here and there tie down the top of a bamboo shoot, and deposit a few humble offerings of flowers and fruit to the spirit guardian of the woods; or they may fix upon a particular tree, some great forest giant, or some rare species, as the abode of a *Nat*, and make propitiatory signs, or mutter a few prayers. But they have no regular religious services, though they are in constant fear of the ghostly inhabitants of the cliffs and trees and streams.

"It is only when they fall upon bad times that they indulge in any organised system of devotion. They ascribe their misfortunes to the anger of the demons, either on account of neglect, or of some particular offence against their dignity. The two spirits most generally recognised and appealed to on these occasions are the *Nat* of the forest and hills and the *Nat* of the village. To these pigs and fowls are sacrificed, libations of rice beer are poured out, flowers and fruit and cooked rice are offered, a tiny hut or two are built for them in the branches of a tree, and there are many prostrations and noisy incantations. Nevertheless, in spite of the intermittent character of their adorations, they are careful never to run the risk of offending any satyr or dryad. When they make a clearing, they always leave a tree or two standing, so that there may be a dwelling-place for the local *Nat*, who, if he were evicted, would inevitably revenge himself, not only on the reckless cultivator, but on the entire community. Not unseldom prayers and incantations are scratched on the tree-trunks to soothe the disturbed demon, and lull him into acquiescence in the new conditions of his haunts. But the spirits are not held in anything like real reverence. They are all considered to be evil-minded and malicious. The worship, what there is of it, is entirely deprecatory. No one wants the *Nats*, however easy-going they may be. If there were any method of getting rid of them it would forthwith be adopted, for the spirits never do any good except in the negative kind of way of preventing others of their kind, or strangers of earthly birth, from encroaching on their domains and possibly doing mischief" (J. G. Scott, *Burma*, p. 174).

The Burmese.

On the other hand, the civilised Burmese all profess Buddhism, which in Burma seems to have preserved

BURMAN.

itself freer than elsewhere from intruding divinities. Here also the monastic vows are more faithfully observed than in other Buddhist lands, and the bonzes have generally promoted the education of the people. A complete national system of public instruction has been developed, all youths being obliged by law to reside for three years in a "khyung" or religious house, where they minister to the "phungys" or priests, and are by them instructed in reading, writing, the elements of arithmetic and of religion. Hence a knowledge of letters is universal in Burma; and here also the women enjoy a remarkable degree of freedom.

In their character the Burmese have much in common with the Chinese. They possess a considerable degree of intelligence and independence, and are shrewd and enterprising, although somewhat indolent. Free from the spirit of caste and national prejudices, they readily acknowledge the superiority of the Europeans, and are eager to learn from them. While extremely tolerant, or rather indifferent, to other religious sects, they remain steadfastly attached to their own tenets. Owing to local maladministration there was a constant migration from Independent to British Burma, until the events of 1885.

The Talaings and Karens.

In Lower Burma the leading races are the Burmese, who are found everywhere in the open country, the Rakhaingtha, popularly known as "Mugs" in the Arakan plains, the Talaings or Mons of the Irawadi delta, and the Karens of the coast ranges in Pegu and Tenasserim. Bengali immigrants and Muhammadan Hindus are numerous in Arakan, where, however, the indigenous Mugs still constitute more than half of the population. They have a strong family likeness to the Burmese, but are of smaller stature and darker complexion. They speak a monosyllabic language accompanied with great emphasis and much gesticulation. Closely akin to them are the Kayans

(Khayengs), a rude but inoffensive hill tribe, who live mostly on game killed with poisoned arrows, and resemble the Chinese in their partiality for dog's flesh.

The Talaings or Mons, if not the aborigines, are at least the earliest known immigrants into Pegu, where they form an isolated linguistic group, now restricted to the east and south of the Irawadi delta in Martaban and North Tenasserim. Wearing the same dress, they differ little from the Burmese in appearance, but are generally of lighter complexion, with more delicate features and a slight growth of beard. But the two races live so intermingled, and alliances are becoming so frequent between them, that the time is perhaps not distant when the Talaings will have become absorbed in the dominant Burmese race. The two languages differ fundamentally, and affinities have been sought for the Talaing as far east as Cambodia and westwards amongst the Kolarians of Central India.[1]

The Tenasserim highlands are occupied exclusively by the aboriginal Karens, who still continue to live in the greatest seclusion. Having been formerly subjected to much hard treatment and oppression by the Burmese conquerors of the land, they now avoid, as far as possible, all intercourse with them. They, however, occasionally visit the towns in the lowlands for the purpose of procuring by barter the indispensable articles of domestic use. Settling in small communities of twelve or fourteen families near some stream in the higher woodlands, they clear the ground with fire, and cultivate rice, bananas, betel-nut, sweet potatoes, and other vegetables, on the reclaimed space. These products, with some poultry and game, suffice to supply all their daily wants. They are of a less robust build, with less prominent cheek-bones,

[1] Captain C. J. F. S. Forbes's *Comparative Grammar of the Languages of Further India.* London, 1881.

less oblique eyes, and a fairer complexion, than their neighbours, thus approaching in some respects to the South European type. The high colour of the cheeks, often suffused in the young women by a slight blush, is very striking in a region inhabited mostly by yellow or olive-brown races. Possessing no writing system for their rude uncultivated speech, and being destitute of all instruction, they lack all the higher religious conceptions. In the natural phenomena surrounding them, recognising agencies inexplicable to their untutored minds, they attribute them to the *Nats*, or good and evil spirits.

The Shans and Siamese.

The Shans and Laos, who are essentially one race under two names, stand in much the same relation to the Siamese proper that the Talaings and Kakhyens do to the Burmese. Their domain occupies the whole of North Siam and a portion of East Burma, whence it stretches along the Salwin valley far into Yun-nan, and down the Mekhong River to the frontier of Cambodia. Their allegiance is thus divided between Burma, China, and Siam. But their ethnical and linguistic affinities are entirely with the Siamese proper, all being so many closely related members of the Tai—that is, "Free" or "Noble" race—which seems to have occupied the Yang-tse basin before the arrival of the Chinese in that region. By the Chinese they were partly absorbed, partly driven southwards to their present homes in Yun-nan and Further India. Here many, especially of the Lao tribes, have become intermingled with, and often assimilated to, the Kha, as they collectively call the aboriginal wild tribes of the peninsula.

But the pure Tai stock, which has almost universally adopted Buddhism, is everywhere distinguished by its low stature, light yellow complexion, black hair and eyes, small nose, dilated nostrils, and somewhat dull, unintelligent expression in the eye. The teeth are often dyed an ebony black, while the thick lips acquire a deep-red colour from the universal custom of chewing betel. The Siamese shave the hair of the head, leaving nothing but a tuft on the crown, which is always carefully dressed, especially by the women. The type is on the whole decidedly ugly, although the children are often pretty, and the women retain a certain comeliness till their twentieth year. In the *Land of the White Elephant*, Vincent remarks that outwardly there is little to distinguish the two sexes, both wearing the "languti," or loin cloth of coloured silk or cotton, with an upper garment varying with the season. The women often add a vest or strip of cloth folded across the breast. But rings, charms, ear-rings, and other jewellery are reserved chiefly for the children of the upper classes, whose naked bodies are often profusely decked with gold and silver spangles and suchlike trinkets. Mouhot tells us that he saw in Bangkok a royal prince, some six or eight years old, so over-weighted with these objects that he was unable to stir.

Besides betel-chewing, tobacco-smoking is very general, and nearly every one has a cigarette stuck like a clerk's quill behind his ear. The staple food is rice and fish, varied with vegetables, fruits, and spiced soups.

The Lao domain, now subject to Siam, is divided into a great number of provinces, which are ruled partly by hereditary "Kiao," or princes, partly by governors appointed from Bangkok. The present Kiao of Bassac, on the Cambodian frontier, is the last survivor of the old Lao dynasty, which was deposed in 1828 by the Siamese. The national assemblies in the Lao States are usually conducted in the same way as in Siam and Cambodia. Under the Kiao are three dignitaries — the "Opalat," somewhat like the second king of Siam, the "Latsvong," and the "Latsbut." These positions are also held by members of princely families chosen by the Siamese Government, but all other officials are named by the ruling prince.

The development of the Lao States has been much retarded by the practice of slave-hunting, which till recently prevailed to a far greater extent than is generally supposed. Regular expeditions were organised by the Lao rulers themselves, or by their immediate subordinates. Constant forays were made amongst the wild tribes, and often even amongst the half-caste Lao communities, especially along the banks of the Mekhong, and the captured victims either distributed among the wealthy Lao families, or forwarded in gangs to Korat, to Cambodia, and even to Bangkok, where they were publicly sold as slaves. Dr. Harmand, an eye-witness, thus describes some of the scenes :—"The brother of the prince of Bassac told me without any reserve that he was about taking a trip to the left bank of the Mekhong in order to hunt down the Khas. It seems that when times are bad the Lao mandarins organise these expeditions against the savages. Under some slight pretext a favourable camping-ground is selected, whence forays are made against the surrounding villages.

"When a sufficient number of all ages and both sexes have been captured, they are bound together and led to Bassac, Stung-treng, and Attroppeu. Here they are purchased by native, Chinese, and especially Malay traders, who form them into gangs forwarded chiefly to Bangkok, Korat, and Phnom-penh, capital of Cambodia."[1]

[1] "Le Laos et les Populations Sauvages de l'Indo-Chine," in *Tour du Monde*, July 5, 1879.

The Annamese and Cambodians.

The inhabitants of Annam, while mainly of Mongoloid stock, present great differences both physically and mentally. The highlanders are, as a rule, of taller stature, lighter complexion, and ruder habits, than the lowlanders, and many of these tribes still lead a nomad life.

The settled and civilised inhabitants of Tongkin, Annam, and Cochin-China form a distinct branch of the Mongolic family, which is everywhere characterised by a remarkable uniformity of physical appearance, mental qualities, language, traditions, religion, and social usages. Living in the same geographical environment, and under the same political institutions for countless generations, the diverse elements originally occupying this region — Giao-shi, Muong, Cham, Shan, Khmer, perhaps Negrito and others — have been gradually moulded into a single ethnical group, somewhat more robust and vigorous in the extreme north (Red River delta), softer and more effeminate in the extreme south (Mekhong delta), coarser and more repulsive along the intermediate coast-lands, but everywhere the same in all essential particulars.

Although observers are far from unanimous in their appreciations of the physical and mental characteristics of the Giao-shi,[1] a name by which the race was known to the Chinese over 2000 years before the new era, it is generally allowed that they are outwardly the least prepossessing, or, in plain language, the ugliest, morally the most disagreeable, of all the Indo-Chinese peoples. Even

[1] *Giao-shi* or *Kiao-shi*, meaning "Bifurcated Toes," has reference to the abnormal space between the great toe and the others, which still distinguishes them, and which has been regarded as a racial characteristic. But it may be due to the universal habit of riding the hardy little Annamite pony with a narrow iron stirrup, which is gripped in the cleft between the big toe and the others.

Mr. Curzon, who is more favourably disposed towards them than perhaps any other writer, applies the epithet "ugliness" especially to the inhabitants of the coast districts between the Red River delta and Hue. He also admits that "the Annamites have the faults inseparable from an Oriental race that has never been divorced from its own surroundings. They are tricky and deceitful, disposed to thieve when they get the chance, mendacious and incurable gamblers, who never lose an opportunity of throwing a die or casting a lot." On the other hand, he declares that he has nowhere met "a more gentle and amicable race. They have the submissiveness without the nerveless apathy of the Hindu, while they possess industrial aptitudes, rendering them diligent workmen, and an artistic ingenuity which on the one hand makes them excellent cooks, and on the other inspires the various artistic productions, such as inlaid work in mother-of-pearl, embroideries, wood-carving, and jewellery. Though not a courageous people in the sense of inviting or voluntarily meeting danger, they are very tenacious in resistance, and make capital soldiers against an Asiatic enemy. They are, moreover, hospitable, polite, lively, sentimental, and of easy temper. The women present two types, the wife or concubine, who is merely the brainless instrument of her master's pleasure, and the active and business-like housewife, who toils hard either in the fields or at the oar, and who in the upper ranks of life frequently takes to business and manages all her husband's affairs."

Much of this is unquestionably true, and the artistic or at least mechanical faculty is certainly well developed, as shown in the surprising skill displayed in the construction of the embankments erected to protect the flat lands of the Red River delta against the periodical floods, works involving more patient labour and ingenuity than the Wall of China, or even the Grand Canal itself. But other observers are far less complimentary in respect of their moral character, and the Abbé Gagelin, who lived, not for a few months, but for many years in their midst, pronounces them at once insolent and dishonest, and dead to all the finer feelings of human nature. So M. Mouhot: "They are headstrong, revengeful, deceitful, thieves, and liars. Their dirty habits surpass anything I have ever seen, and their food is abominably nasty, rotten fish and dogs being their favourite diet." Hence Mr. J. G. Scott's famous exclamation: "The fewer Annamese there are the less taints there are on the human race."

The features are of a coarse Mongolic texture, characterised by a dirty yellow or sallow complexion, very prominent cheek-bones, square massive jaw, thick lips, oblique eyes, and long black hair shaven in childhood, but never cut afterwards, but worn chignon-fashion at the back of the head. These traits are even more pronounced in the women, most of whom are certainly ugly and even repulsive. Polygamy is universally practised, there being no limit except expense to the number of wives, and the Annamese are not a religious people, despite the innumerable shrines and pagodas crowning every wooded knoll and picturesque spot in the landscape. As in China, ancestry worship, which cannot be called a cult in the higher sense, is the dominant sentiment, pervading all classes and all orders, from the court through the officials down to the village commune and the humblest peasant family. In other respects the national faith "is a strange and incongruous amalgam of various superstitions, cults, and creeds. Buddhism is more or less widely diffused, but in a very attenuated form, barnacled with all manner of corruptions. Chinese ascendency has brought with it the ethics of Confucius and the worship of a host of demi-gods or glorified heroes. But for the most part spirit-worship or a crude demonology may be denominated the popular creed, the majority of the pagodas containing little beyond altars on which a censer smoulders to the *genius loci*, to the good spirit or evil spirit of the site, to the dreaded tiger, or to the *manes* of some celebrity of the past. The larger pagodas consist of two or three courts, in the hindermost of which stands the temple beneath a tiled roof, closed round with wooden doors. Fantastic animals, elephants with howdahs, kings and warriors on horseback, or tigers, are painted in fresco, or sometimes fashioned in high relief on the entrance gateway and around the courts; and a bizarre but brillant effect is produced by fragments of broken pottery stuck in plaster, and forming patterns of dragons on fabulous monsters on the walls and roofs" (Curzon, *ib.*).

The Cambodians, whose national name is Khmer, may be taken as the typical representatives of the Caucasic element in Further India, of which they are probably the aboriginal inhabitants. Although now restricted to the valley of the Lower Mekhong, their domain formerly occupied a large portion of the peninsula, where they are still represented in a fragmentary way by the Kuys, Charays, Chongs, Stiengs, and other tribes scattered over the hills and forests of East Siam and West Annam. But next to the Cambodians proper, by far the most important section of the race are the Kuys, or Khmer-dom, —that is, "Primitive Khmers,"—who are divided into several branches, occupying an extensive region round the northern and eastern shores of Lake Tonle-sap. Here was the original centre of Cambodian culture, the magnificent remains of which now lie mostly buried amidst a rank vegetation of a thousand years' growth. The human figures sculptured on these Hindu monuments, many of which must be some twenty centuries old, are not only of regular Caucasic type, but are often a faithful reflex of the features, dress, and ornaments of the present Khmer populations.

Some of these wild tribes are still distinguished by a gentle disposition, a certain innate politeness and courtesy, as well as a surprising artistic taste and skill lavished on their dress, ornaments, pipes, quivers, and other objects. These traits may well be the faint reflection of a now extinguished culture still cherished by these children of nature, lost for ages amid their dense woodlands, which they believe to be the centre of the universe, and which nothing can ever induce them to leave (Mouhot). But the Cambodians themselves seem to have retained little of their former greatness, except an overweening pride and arrogance. They are being gradually absorbed by the surrounding Annamese and Laos populations. A strange mystery hangs over this Cambodian race, which, fully 2000 years ago, built cities and raised monuments amid the swamps of Tonle-sap, vying in size and grandeur with those of the Mesopotamian and Nile valleys. Their culture is certainly of Hindu origin. But by what channel did Hindu influences penetrate to this remote corner of the continent? And whence came the race itself, with its European features and polysyllabic speech, totally distinct from that of the surrounding Mongoloid peoples, but showing marked affinities to the Oceanic linguistic groups? Lastly, by what barbaric hordes was their development arrested, their culture extinguished, their cities wasted, their stupendous shrines and monuments left to crumble in the midst of the exuberant tropical vegetation of Angkor and Battambang? These are questions which cannot yet be answered.

8. *Topography: Rangun—Moulmain—Prome—Mandalay—Bhamo.*

In Lower Burma the only large places are Rangun and Moulmain, the two great seaports of Pegu almost facing each other across the Gulf of Martaban. The Shway Dagohn pagoda at Rangun is one of the most remarkable structures of the kind in the Buddhist world. It stands on a wooded eminence, above which its gilded "htee" or umbrella shoots up to a height of 300 feet. From a distance it seems to flash in the sunlight above the dark foliage like a fiery meteor. The hills about Moulmain are also crowned with Buddhist pagodas, whence an extensive and varied prospect is commanded of the city, and the plains watered by three converging streams and enclosed eastwards by the distant Siamese frontier range. There is some literary activity in Rangun, which is gradually becoming the centre of intellectual life for the Buddhist world in Indo-China. Here are four vernacular presses, which have already issued a good many theological, literary, and scientific works, including dramas chiefly adapted from the Sanskrit, Buddhist tracts, often of a very polemical character between the rival Mahagandi and Sulagandi sects, an encyclopædia of Burmese knowledge (the Kawi Lekhana Dipani), many translations from English works, and some periodical literature.

In the middle valley of the Irawadi stands the old Burmese city of Prome, with gilded pagodas and wooden pinnacles on the summits of a cluster of hills near the river's bank.

During his expedition from Burma to Arakan in the year 1891 Lieutenant Walker visited Myohaung (Mrohaung), Ptolemy's Triglyphon, which was the ancient capital of Arakan. At present it has a population of scarcely more than 2500, but its former magnificence is still attested by the extensive remains of its fortified enclosures, and especially by two gigantic pagodas, containing 80,000 and 90,000 images respectively. Myohaung is reached by water from Minbya, which does a brisk trade with Akyab, and which stands on the Semra estuary, here nearly four miles wide.

As might be expected, the few large towns of Upper Burma all lie in the Irawadi valley, which is distinguished both for its picturesque scenery and great fertility. Here is Mandalay, one of the many places to which the seat of government had in recent times been shifted, partly in consequence of revolutions or changes of dynasty, partly through superstitious motives or royal whims. Ava, which had been the capital since 1364, gave place about 1740 to Mutshob, which yielded in 1782 to Amarapura. In 1819 the Court returned to Ava, whence it again passed in 1837 to Amarapura, and in 1857 to Mandalay, where it remained till the deposition of King Thebaw (1885). This place lies a few miles above Amarapura, on the same side of the Irawadi. The houses on the river and in the suburbs are of the usual Burmese type — frail structures of bamboo and matting erected on piles sunk in the mud. They are generally small and packed closely together, and their materials are inflammable. Consequently fires once breaking out become extremely disastrous. The main streets are lined mostly with brick houses, but even here the bricks are merely attached to the wooden framework. A pretty effect is presented by the Chinese shops, which are often two stories high. In every direction the eye lights on gilded or painted pagodas, temples, and "Khyungs," or cloisters with schools attached. The city is encircled by a stout brick wall, with a ditch crossed by one or two rude bridges. Here was the scene of the horrible

butcheries which accompanied the accession of the late King Thebaw in 1879. Mandalay was occupied by the British under General Prendergast on November 28. 1885.

AVA.

On the Upper Irawadi lies the important station of Bhamo, in 24° 15′ N., at the junction of the Tapeng. This is the starting-point of caravans proceeding eastwards to Yun-nan, and should a regular overland trade be established between British India and West China, Bhamo must from its position become one of the great emporiums of the East. At present it is a small stockaded town with a few hundred houses, occupied chiefly by Chinese and Shan traders.

Bangkok—Ayuthia—Chantabun.

Bangkok, the present capital of Siam, lies near the mouth of the Meinam a few miles below Ayuthia, the former seat of government. Bangkok presents a pleasant prospect from the water, whence a panoramic view is commanded of its glittering palaces and temples surrounded by dense masses of a gorgeous tropical vegetation. But the charm is soon dispelled by a closer inspection of the place itself, which consists of a confused aggregate of narrow, muddy lanes, stagnant canals, and wretched hovels, occupied by a mixed Siamese and Chinese population, estimated at from 400,000 to 500,000. As in so many other parts of Further India, the Chinese have here almost monopolised the local trade, while the foreign exchanges are mostly in the hands of English merchants.

Beyond Ayuthia, the site of which is marked by some magnificent Buddhist ruins and a large elephant park, the most noteworthy places in the interior are Raheng in the Meinam basin, north of Bangkok, much spoken of in connection with recent railway projects; Kheng-mai (Xieng-mai, Zimme), the great capital of the Siamese Shans, on the right bank of the Meping, comprising an inner and an outer town, both with separate fortifications, and a collective population of probably over 100,000, altogether the largest and most flourishing city in the Shan domain; Xieng-kong on the Mekhong, northernmost town in Siamese territory; Luang-Prabang, lower down on the left bank of the same river, important as an advanced station on the Franco-Siamese

BANGKOK : THE ROYAL PALACE AND QUAY.

frontier, but with a pestiferous climate, to which the naturalist Henri Mouhot fell a victim in 1861; Korat, also a future railway centre, on the plateau north-east of Bangkok, about the water-parting between the Meinam and Mekhong basins. On the south-east, not far from the Cambodian frontier, is the important seaport of Chantabun, which gives access to the rich mineral districts of the provinces of Chantabun and Battambang. Lately mining operations have been centred chiefly at Payrinh, midway between Chantabun and Battambang; but rubies and sapphires, as well as jade, topazes, and other costly stones, appear to abound in the whole region. In 1893 the French seized Chantabun, and still hold it (1896), as a pledge for the faithful execution of the treaty, although the Siamese Government has long since discharged all its obligations.

In Tongkin and Annam the towns everywhere present a somewhat uniform type, consisting for the most part of clusters of villages grouped together within an outer enclosure of walls and moats, and defended by a huge citadel, often spacious enough to contain the whole population. In the villages the houses are thatched with palm-leaves, and built with a watling of bamboos and mud, the interior being disposed in a number of low platforms used as tables by day and bedsteads by night. Every village has its communal hall set apart for municipal and general purposes, and a similar reception-room usually occupies the centre of the richer dwellings, being fitted with a central table, arm-chairs, a shrine at the back, and sleeping stands on either side. These houses are generally built of wood or bricks and roofed with tiles, and the streets even of the smallest hamlets are at least clean and orderly, presenting in this respect a favourable contrast to the filthy towns of China and Korea. The urban nomenclature, however, is exceedingly confused, many towns bearing as many as four distinct names, one for the particular hamlet or group of houses, one for the larger unit of which it forms part, and which

A NATIVE FAMILY AT HOME, TONGKIN.

may contain any number of similar groups, one for the *tram* or wayside station, and one for the market, which is the most characteristic feature of Annamite towns. Markets are held on stated days, either in an open space in the middle of the village, often tiled or thatched and let out in stalls and booths, or else in some central place between several villages. Marketing is entirely conducted by the women, who squat down amid their wares, and intersperse a ceaseless chatter with the chewing of the betel leaf. Here are exposed for sale "pigs, chickens, and ducks in hampers, fish, fresh, slimy, shell, and sun-dried, big prawns and tiny land-crabs, cabbages, radishes, the areca-nut, vermicelli, cakes, sweetmeats, and eggs. Elsewhere will be cheap articles of furniture or raiment, tin lamps for petroleum, pottery, brass ware, opium pipes, bracelets, necklets, amber buttons, palm-leaf hats, turbans, Bombay cottons, and scarves" (Curzon, *ib.*).

Hue, capital of Annam, occupies an important strategical position at the mouth of the River Hue on the Cochin-Chinese coast. It was strongly fortified with skilfully-planned outer and inner lines by French engineers at the beginning of the nineteenth century. With its extensive arsenal, magazines, walls 60 feet high, and moat 120 feet broad, Hue formerly ranked as one of the strongest military positions in Asia. The king resides in a spacious and fortified palace in an inner enclosure of the citadel 800 yards square; but all these defensive works have not prevented him from sinking to the position of a vassal of France, in whose favour he signed a treaty in 1874 practically accepting the protectorate of that country. The population of Hue, formerly estimated at 50,000, 100,000, or even 150,000, is reduced by Mr. Curzon to not more than 15,000 at the outside.

On the road from Turan to the Nongson coalfields lies the historical city of Faifo, which in the sixteenth century was a great trading centre. But when visited by Consul Parker in 1892 it was found to be a mere depôt for the transit of inland goods from the semi-civilised districts of the interior. At present its chief exports are sugar, silk, cinnamon, and edible birds' nests.

In Tongkin the chief Treaty port is Haiphong on the Lower Song-ka, which is navigable to this point by vessels drawing 8 to 10 feet. At Haiphong the French have a naval station for the ostensible purpose of suppressing piracy on the high seas, but which also serves to keep the country in subjection to the mistress of the Mekhong delta. Kesho, capital of Tongkin, which lies some miles above Haiphong, is accessible to large junks drawing from 4 to 6 feet.

Langson, scene of so many conflicts and even disasters during the early period of the occupation, has now become the chief bulwark of French power towards the Chinese frontier. Its position on a head-stream of the Si-kiang also promises to ensure its future prosperity as the most important centre of international trade in the interior. It is already connected by a good road with Phu-lang-thuong on a branch of the delta navigable by steamers, and from this road a narrow gauge railway now (1896) runs by way of Bac-ninh right through to Hanoi. At one point the line surmounts a pass 3000 feet high. Here also a junction has been effected since 1891 between the French telegraph system and a Chinese line running from Canton up the Si-kiang to the frontier. Meantime the trade of Langson is small, and the goods offered for sale are chiefly English and American cottons, soap, matches, umbrellas, besides mirrors, drugs, and crockery from Canton. Mr. Agassiz, who passed through in 1891, doubts whether "this line will be a means of opening up trade with western Kwang-si" (*loc. cit.*).

Phnom-penh—Saigon.

Cambodia some years ago shifted its capital from Udong to the neighbouring Phnom-penh, which is more

PHNOM-PENH.

conveniently situated at the junction of all the navigable waters in the kingdom.

The French have chosen for the capital of their Cochin - Chinese settlements the town of Gia - dinh, renamed Saigon, picturesquely and conveniently situated on the Dong-nai. Although lying at some distance from the sea, Saigon is accessible to large vessels, and is much frequented by English and Chinese traders engaged in the foreign and coast trade of Cambodia and Lower Cochin-China. The neighbouring island of Pulo Condor is well adapted to become a depôt for the produce of the surrounding districts. The staples of the export trade are cotton, pepper, raw silk, tobacco, betel leaves, sugar, gums, hides, horns, and fish.

Perak—Singapore.

In Siamese Malacca the chief place is Toneah on the island of Junk Seylon, which is said to have a population of 30,000. Of the towns in British Malacca the largest are—Perak, capital of the protected State of that name; Malacca, on the south-west coast of the peninsula to which it gives its name; Penang, on the same coast; and Singapore, unquestionably the most important place in the whole of Indo - China. Notwithstanding its proximity to the equator, the "Lion City," as its name means, enjoys a fairly healthy climate, and this flourishing free port has during the British occupation been transformed from an almost uninhabited rock into one of the great centres of Eastern trade. Very few places in the old or new world present such an example of rapid development as does the city of Singapore, which now counts its population by many tens of thousands and its exchanges by millions. It possesses a magnificent harbour, well sheltered and of easy access, lying at the converging point of the great trade routes between India, China, Japan, the Dutch East Indies, and Australia.

Singapore, which is well laid out, with a Chinese, Malay, and European quarter, is the capital of all the Straits Settlements, with a governor's residence, Protestant cathedral, hospital, schools, and several benevolent institutions.

9. *Highways of Communication.*

In Further India there are few regular roads, and most of the trade routes follow the course of the great rivers and their affluents. Thus travellers and explorers wishing to penetrate from the west into China, ascend the Irawadi to Bhamo, where they follow the caravan route up the Tapeng River valley through the Kakhyen highlands to Manwyne, and so on to Momien on the Yun-nan frontier. Since the reduction of the hill tribes, who formerly levied blackmail on passing caravans, this route is much frequented.

Before the annexation of Upper Burma, the railway system in the Irawadi basin was limited to the line running from Rangun to Prome, and to another nearly completed from Rangun up the Sittang valley to Tungu near the frontier of the Karen country. Since then the Sittang Valley line has been completed to Mandalay, and the section from Shwebo to Wuntho and Mogaung, with a branch from Meitkila Road to Meitkila (nearly 13 miles), has been opened. The surveys have been completed for farther extensions from Mogaung to Myitkina on the Irawadi (40 miles), from Mandalay to the Kunlong Ferry, and from Chittagong through the Aeng Pass (Arakan-Yoma Range) to Minhla on the Irawadi.

The chief highway to the interior of Siam follows the course of the Meinam from Bangkok to Ayuthia and Prabat. Here it trends east over the Dong Phya

Phai range to Korat, whence it runs due north through the Laos country to Luang-Prabang on the Mekhong. Bangkok is also connected with Cambodia and the Lower Mekhong by a track running eastwards through the Shong country, and over the coast range to Battambang and thence along the south side of the Great Lake to Udong and Phnom-penh. From Battambang a branch runs round the west end of the lake to Angkor, thus giving access to most of the magnificent ruins of the ancient Cambodian Empire.

In Annam all the chief towns are connected by a coast route running from Kesho southwards through Thanh-hoa-noi, Koang-tri, Hue, Kwang-nai, Quin-hon, and Binh Thuan, to Saigon. This is the so-called "Mandarins' Road," a famous historical highway, which was thoroughly restored and extended to Langson on the Chinese frontier by the Emperor Gia-Long at the beginning of the nineteenth century. In Annam it runs close to the sea-shore, and in some places surmounts the projecting spurs by means of steps cut in the face of the rock, and *trams* or wayside stations for the convenience of travellers and the public service occur at intervals of from 8 to 15 miles along the whole route. "In the parts where I travelled upon the Mandarins' Road, between Ninh Binh and Hue, it had been in places repaired during the early part of the year (1892) for the passage of the Governor-General, and was commonly a flat track about 20 feet wide, either running upon a bank between the rice plots, or often over hot sands, whilst elsewhere, after rain, it was converted into a horrible bog. The poles of the French telegraph are planted along its side" (Curzon, *ib.*):

From Kesho a track runs also west to the Mekhong at Kiang-kheng above Luang-Prabang, and thence across the Siamese Shan States, through Kiang-mai to the Lower Salwin valley and Martaban.

In French Indo-China a short line of railway between Mytho and Saigon was opened to the public in May 1894.

10. *Administration.*

The Burmese Government was a pure despotism, his majesty of "the Golden Feet" ruling as an absolute monarch. Even the British Envoys were required to appear barefooted [1] in the presence of these despots, whose chief characteristics have too often been cruelty, licentiousness, and arrogance. Hence while court etiquette was rigidly maintained, the progress of the country was retarded by the policy of its rulers. Anarchy and decay were the prevailing features of the regions subject to the influence of the central government, while the more remote Shan States enjoyed a comparative degree of peace and immunity. The resources of the land were also further drained by the steady migration of the more intelligent and wealthier classes, who were glad to find a refuge in the neighbouring province of British Burma.

All this is now changed, and since the unification of Burma under British rule the country has undergone a vast change for the better. The population is increasing twice as rapidly as that of India proper, and material progress is advancing even at a still more rapid rate. In Upper Burma the unruly element, represented by numerous bands of dacoits or brigands, has been almost everywhere reduced to order, and nearly all the frontier hill tribes have tendered their submission. According to the official reports crimes of violence are less frequent in the Upper Irawadi districts even than in the older province of Lower Burma. Since the annexation the internal

[1] The great "Shoe Question," which never was settled, is fully discussed in Col. Laurie's work on *Ashé Pyee; or, The Superior Country*. London, 1882.

trade has more than doubled, while the sea-borne commerce of Rangun, the chief outlet of the whole region, has advanced from 120 million rupees in 1886 to 180 millions in 1892, and the revenue from 25 to nearly 55 millions in the same period.

The Siamese and kindred peoples form the only nation that has two kings, although this old custom has at present lost all its significance. The "Wangna," or second king, usually chosen from amongst the nearest relatives of the first king, resides quietly in his palace without at all interfering in State affairs.

Siam itself is an electorate, the succession being determined by the nobles, who, however, are bound to elect a "Chao Fa"—that is, a prince of royal blood—on both sides. Hence to secure a posterity capable of succeeding to the throne, the first king is bound to select a consort from amongst the daughters of the second reigning king, or of some former first or second king, or at least a princess from the ancient royal families of the Lao States. But in the latter case some of the public jurists doubt whether the issue of such alliances are entitled to the succession.

In theory all the inhabitants of the land stand in the relation of serfs or slaves to their sovereign. At the same time, there exist certain substantial class distinctions denoted by special tattoo marks on the wrist, all being so branded except the nobles and officials, with their families. In the artisan classes the son is compelled to follow his father's trade. During war military service is of universal obligation, but in peace the army is recruited from the working corporations under the control of the war office. But notwithstanding the ancient traditions of oriental despotism, Siam has made considerable social progress under the enlightened administration of the reigning sovereign. Free scope has been given to the action of European influences, and while Buddhism remains the State religion, Christianity is allowed to be preached and practised without any restrictions. To Bishop Pallegoix, for many years head of the French missions in Siam, science is indebted for some of the most valuable contributions to the history and ethnology of that State.

The Government of Annam is so far absolute that the whole authority is centred in the king or emperor, and emanates from him. The sovereign has his privy council, besides a ministry for the administration of justice, a war office, a ministry of public worship, a board of works, and a home and foreign office. But all this machinery of government has lost its significance since Annam has, by the treaties of 1874 and 1885, become a dependent State of France.

The ancient kingdom of Cambodia in 1864 accepted the French protectorate. The two provinces of Battambang and Angkor, seized by the Siamese at the end of the last century, but the occupation of which had never been acknowledged by Cambodia, were finally ceded to Siam by the treaty of 1867, when the new boundary-line was fixed between the two States. Thus were separated from Cambodia the two historical provinces, in which are situated the ancient capital, Angkor, and most of the grand monuments of former Cambodian culture.

The king of Cambodia is in theory absolute master of the life and property of his subjects, as well as of all the land. The only check to his despotism are certain long-established usages, combined with the fear of causing troubles which might require the intervention of the French. The nobles and officials are divided into four classes, at the head of which are the prime minister (*Shanfea*), the ministers of justice, finance, and trade. For administrative purposes the kingdom is divided into five *dey* or departments, and fifty-six *khet* or districts, the dey being governed by the ministers of State, the khet by inferior mandarins named by the king. Under the governor are a *balat* or lieutenant, two *snâng* or prefects, besides other smaller functionaries, and the *me srok* or village magistrate.

Formerly attached to British India, the Straits Settlements now form a Crown Colony, with a separate administration vested in a Governor at Singapore, a Lieutenant-Governor at Penang, and a Resident Councillor at Malacca, appointed by the Queen. The Governor is assisted by Executive and Legislative Councils, of which the Lieutenant-Governor and the Resident Councillor, above mentioned, are members. The lieutenant-governors of the several settlements are also members of these bodies. Seat of government, Singapore.

11. *Statistics.*

AREAS AND POPULATIONS.

		Area in sq. miles.	Population (1891).
Upper and Lower Burma		172,000	7,606,000
Siam, less territory ceded to France, 1893		200,000	5,000,000 (?)
FRENCH INDO-CHINA	Tongkin	35,000	9,000,000
	Annam	46,000	5,000,000 (?)
	Cochin-China	23,000	2,000,000
	Cambodia	39,000	1,800,000
	Wrested from Siam, 1893	80,000	2,000,000 (?)
Zone between Burma and Tongkin		20,000	1,000,000 (?)
Straits Settlements and British Malaya		35,000	1,225,000
Total Indo-China and Malay Peninsula		650,000	34,631,000

SIAM.

		Area in sq. miles.	Population.
Siam proper and Lao States		160,000	4,000,000
SIAMESE MALAYA	Ligor and Senggora	17,000	250,000
	Kedah	4,000	250,000
	Patani States	6,000	100,000
	Kelantan	7,000	300,000
	Tringganu with Kemaman	4,000	100,000
Total		198,000	5,000,000

Chief Towns: Bangkok (capital), pop. 400,000 (?); Ayuthia, Korat, Chantabun, Luang-Prabang.
Revenue, £2,000,000. No Public Debt.
Army, 12,000; Navy, 1 ram-ship, 2 steam corvettes.
Imports (1892), £1,296,000; Exports, £1,387,000.
British Imports (1892), £110,000; Exports to Great Britain, £52,000.

Chief Imports (1892).		Chief Exports (1892).	
Cottons	£293,000	Rice	£956,000
Hardware	84,000	Teak	63,000
Kerosene	50,000	Pepper	54,000
Silks	49,000	Salt fish	38,000
Sugar	48,000	Woods	32,000
Jewellery	47,000	Cattle	27,000
Cotton yarn	42,000	Hides, dried fish	32,000

Bangkok Shipping (1892): 292 vessels of 209,745 tons entered. Of these 248 of 182,354 tons were British.
Railways: Bangkok to Paknam, 14 miles, opened 1893. Line from Bangkok to Korat (165 miles) in progress.
Telegraphs: 1780 miles in Siam; 710 miles in Malay Peninsula.
Post-Offices (1890), 98; letters forwarded, 410,000.
Since 1885 Siam has been a member of the International Postal Union.

French Indo-China.

Total Imports (1891), £2,700,000; Exports, £2,720,000.

Tongkin: Capital, Hanoi; pop. 150,000.
Revenue and Expenditure, about £700,000.
Army (1892), 18,550, including 6500 natives.

Annam: Capital, Hue; pop. 30,000.
Army (1892), 23,330, including 11,830 natives.

Cambodia: Capital, Pnom-Penh; pop. 30,000.
Revenue and Expenditure, about £120,000.
French troops, 300.

Cochin-China: Capital, Saigon; pop. 60,000.
Revenue and Expenditure about £1,200,000.
Army, 4600, of whom 2800 natives.
Railways (1892), 51 miles; Telegraphs, 1840 miles.

British Malaya.

	Area in sq. miles.	Population (1891).	
Straits Settlements—			
Singapore	1,472	184,554	512,342
Penang		235,618	
Malacca		92,170	
Protected States—			
Perak	10,000	214,254	718,527
Selangor	3,000	81,592	
Negri Sembilan	2,660	65,219	
Pahang	10,000	57,462	
Johor	9,000	300,000 (est.)	
Total	34,660		1,230,869

Straits Settlements (1893).

Revenue, £517,500 ; Expenditure, £604,500 ; Debt *nil.*
Total Imports, £20,036,000 ; Total Exports, £19,038,000.
Imports from Great Britain, £3,159,000 ; Exports to Great Britain, £3,554,000.
Registered Tonnage, 44,000 ; Tonnage entered and cleared, 10,200,000.
Vernacular and English Schools, 192 ; Attendance, 11,310.
Railways (in Protected States), 102 miles open ; 137 in progress.

Racial Elements.

	Malays.	Chinese.	Klings (Indians).
Singapore . . .	35,992	121,908	16,035
Penang	106,756	87,920	36,245
Malacca	70,325	18,161	1,647
Total . .	213,073	227,989	53,927

WESTERN ASIA: MUHAMMADAN STATES

(TURKEY IN ASIA, ARABIA, AND PERSIA)

CHAPTER IV

ASIA MINOR

1. *Boundaries—Extent—Area.*

THE Asiatic portion of the Turkish Empire comprises with Arabia the whole of the south-western section of the continent west of the Tigris. It is thus conterminous eastwards with Persia and Russia, the three empires converging about Mount Ararat, the culminating point of the Iranian plateau towards the north-west. Elsewhere Asiatic Turkey is, except towards Egypt, everywhere surrounded by water—Persian Gulf on the east, Arabian Sea on the south, Red Sea, Mediterranean, and Ægean on the west, Black Sea on the north. It comprises four well-marked natural and historical divisions—the two peninsulas of Arabia and Asia Minor, the basin of the Euphrates and Tigris, and the upland region of Syria and Palestine. These four main divisions will here be treated under four separate heads, a general survey of the empire being reserved for the end of the section.

Asia Minor, or the "Lesser Asia," is so named relatively to the Greater Asia of which it forms the westernmost projection. But relatively to Europe it is the

"Anatolia" of the Greeks, and the "Levant" of the Italians—that is, the "Orient," or "Land of the Rising Sun."[1] Projecting far into the Mediterranean, it is washed on three sides by inland seas—the Euxine on the north, Marmora and Ægean with their connecting straits on the west, the eastern section of the Mediterranean on the south. Eastwards it is limited by a somewhat arbitrary line running from Alexandretta Bay east to the great bend of the Euphrates, and thence follows the course of this river to its source, where it trends northwards to the Euxine, mainly along the valley of the Choruk-su. Anatolia thus lies between 36° and 42° N. latitude, consequently between the same parallels as the southern sections of the three European peninsulas and the northernmost portion of Barbary. Its greatest length from Cape Baba to the Euphrates, west and east, is about 700 miles, and its extreme breadth from Cape Anamur opposite Cyprus to Cape Injeh near Sinope on the Black Sea is rather over 400 miles. Within these limits its area is roughly estimated at about 220,000 square miles, or nearly 20,000 more than France, but with scarcely one-fourth of the population of that country.

2. *Relief of the Land: Taurus, Anti-Taurus, and Amanus—Passes—Plains—Volcanic Agencies—Geological Formation.*

Geographically Asia Minor must be regarded as a western extension of the Armenian and Kurdistan high-

[1] Anatolia, from ανατέλλω; Levant, from *levare*, both of which terms mean "to rise," hence are the exact equivalents of the "Orient," from *oriri*, the corresponding Latin word. But while *Anatolia* is by the Greeks strictly limited to Asia Minor, *Levante* is by the Italians extended to all the lands lying east of the Mediterranean, and *Orient* is applied to the East in general. *Anadoli*, the Turkish form of Anatolia, is more usually restricted to the western and northern provinces of Asia Minor, while the rest of the country is known as Karamania.

lands, from which it can nowhere be separated by any hard-and-fast line. The plateau formation prevails throughout, the interior of the peninsula forming an extensive tableland at a mean elevation of from 3500 to 4000 feet above sea-level, and stretching north-east and south-west for a distance of over 200 miles, with an average breadth of about 140 miles. Above this tableland rise several loosely-connected mountain ranges, while over its surface are scattered a number of salt-lakes, morasses, and watercourses, without any visible outflow seawards, besides several streams which find their way mainly northwards to the Euxine and westwards to the Ægean.

The plateau is skirted south and north by two broken mountain ranges, which radiate from the Armenian uplands, and to which the terms Taurus and Anti-Taurus were somewhat vaguely applied by the Ancients. The Taurus or southern branch, which forms a continuation of what Kiepert calls the "Armenian Taurus," rises close to the Euphrates, where one of its peaks attains an elevation of 10,000 feet. From this point it pursues a very irregular course, under the more specific name of the Amanus, down to Karamania, and thence along the Mediterranean coast to the Ægean, with ramifications projecting northwards and southwards at various points. These branches, like the several sections of the main range itself, bear special names, such as the Ala-dagh, the Karmez-dagh, the Bulgar-dagh, the Sultan-dagh, the Jebel-kum, and others, ranging from 7000 to 10,000 and even 13,000 feet high. The large island of Cyprus here lies, like a detached fragment of the mountain mass, opposite the angle formed by the Anatolian and Syrian coast-lines, while the south-western extremity of the peninsula is continued seawards by the lofty island of Rhodes, facing which the Massacitus spur terminates and culminates with Mount Takhtalu, 7820 feet high. But elsewhere the escarpments of the tableland fall westwards down to the Ægean, whose southern islands may be regarded as their advanced terminal peaks. Between the hills and the coast space is left in many places for lower valleys, and even for alluvial plains, varying in width,

TAURUS RANGE, NEAR TARSUS.

but mostly of great fertility, and sloping gently in all directions seawards.

The Anti-Taurus,[1] now perhaps better known as the Agha-dagh, forms a western extension of the Lazistan highlands, running in two and occasionally three nearly parallel chains from the neighbourhood of Batum along

[1] This term is by some geographers applied to the Amanus or north-eastern section of the Taurus between the Armenian highlands and Adana.

the coast of the Euxine, and at no great distance from the sea, as far as the Bosphorus. Here it throws off a southern branch to the great western network, culminating with the Keshish-dagh (Olympus), the Morad-dagh, and the Kas-dagh (Ida), which rises 5700 feet above the plains of Troy at the head of the Gulf of Edremid. The Anti-Taurus forms a water-parting for the streams rising on the southern slopes of the Armenian uplands, and flowing some westwards and others towards the Euphrates. Like the Taurus, it also throws off several side ridges seawards and to the interior. Here a number of smaller and more isolated chains run in various directions, and often attain considerable elevations, culminating with the volcanic Ergish-dagh (Argaios), which is 13,100 feet high, and apparently the culminating point of the peninsula. This cone, which is nearly isolated from the Taurus, forms a striking landmark on the plains of Kaisarieh (Cæsarea) which here attain an elevation of over 3000 feet. It consists altogether of igneous matter, and its summit terminates with two craters, through which in former times the underground forces found an outlet. In its central section the Taurus itself varies in height from 2700 to 5500 feet, while the Asi-Kur (Niphates), one of its loftiest summits, rises above the snow-line.

Both the Taurus and Anti-Taurus are crossed at various points by passes generally at low elevations and of moderately easy access. Of these the most important strategically and commercially is the Geulek-Boghaz, or "Cilician Gates," a deep gorge, 3300 feet above sea-level, running about 30 miles north of Tarsus over the Taurus, and connecting Anatolia with North Syria and the Euphrates valley. This famous defile has been followed in all ages by migrating peoples, traders, and conquering hosts. Through it Alexander marched to the overthrow of the Persian Empire, and through it Mehemet Ali in recent times twice penetrated into Anatolia on his march to Constantinople. About 100 miles west of this point the Taurus is crossed by a second pass leading from Karaman southwards to the Gok-su valley, and by a third, 150 miles still farther west, connecting Isbarta southwards with Adalia. The chief openings giving access from the Euxine through the Anti-Taurus to the central plateau are those leading from Ineboli to Kastamuni and Angora, from Sinope to Amasia, from Samsun to the same place, and from Trebizond over the Kolat-dagh to Erzerum.

In the higher regions of the peninsula the chief geological formations seem to be serpentines, granites, and schists, while limestones prevail lower down and almost everywhere in the western provinces. The trachytic formations, which abound in the east, are overlaid towards the centre of the plateau by black volcanic breccia, interspersed with blocks of trachyte. Altogether, igneous formations may be regarded as the dominant feature in the geology of Asia Minor.

3. *Hydrography: Rivers, The Kizil-Irmak, Sakaria, Choruk, Khoja-chai, and others—Lake Tuz-gol.*

The chief Anatolian rivers flow in a north-easterly direction to the Black Sea. Of these rivers, which have not all yet been thoroughly explored, the largest is the Kizil-Irmak (Halys), formed by the junction of two head-streams, one rising in the hills south-west of Tokat and flowing westwards, the other rising farther south on the slopes of Taurus and thence flowing first in a westerly and then in a northerly direction. After pursuing a very winding course of about 800 miles, the Kizil-Irmak discharges its waters through two principal channels into the Euxine below Bafra, and a little to the east of the Gulf of Sinope. Nearly the whole of its lower course below the latitude of Angora, a distance of 280 miles, remained unexplored till 1893, when it was carefully surveyed by Lieutenant Märker and his associates. North of Kalejik, where the stream plunges into a wild rocky gorge 6 miles long, the explorers discovered several rock-tombs, which date from pre-Hellenic times, and are traditionally attributed to the ancient Paphlagonian rulers of the land. Similar remains occur elsewhere in this region, attesting a far denser population than at present. The tombs, which nearly always occupy commanding positions, are cut in the live rock usually at a height of 20 to 50 feet above the alluvial soil. They have always a vestibule, whence a low, wide passage leads to the tomb proper, which contains from one to four chambers. At Osmanjik, farther north, the river is spanned by a fine old stone bridge of fifteen arches, built by Bayazid II., and leading to Haji Hamza and the Devrezchai affluent. Below Cheltek, where it is crossed by a wooden bridge, the Kizil-Irmak pierces the coast range through some rocky gorges, and beyond the thriving town of Bafra it enters the lowlands, where its banks are fringed by a dense belt of woodlands forming overhanging avenues like those of the White Nile tributaries. In summer the stream varies from 60 to 330 yards in width, with a velocity of at least 4½ miles an hour, and when in flood it is said to be accessible to barges from Bafra to the Kaisarieh district. At other times it is unnavigable by the smallest craft, owing to the numerous rapids below the Elmadaghchai confluence (*Berlin Geographical Society*, January 1894). Some 50 miles farther east the Yeshil-Irmak (Iris) enters the Black Sea, about 16 miles to the east of Samsun, after flowing by Tokat and Amasia during a tortuous course of nearly 240 miles.

But next in importance to the Halys is the Sakaria (Sangarius), which rises near Angora on the tableland, and reaches the Black Sea at a point some 80 miles east of the Bosphorus. The Choruk or Joruk (Bathys), the north-eastern frontier river, crosses Armenian territory and falls into the Euxine just south of Batum. The affluents of the Ægean Sea are important historically rather than geographically. While all are of small size, most of them are renowned in song and legend. Especially famous are the Gediz-chai (Hermus), flowing to the Gulf of Smyrna, and formerly noted for its auriferous sands; the Bakir-chai (Caicus ?), reaching the coast below Pergamus; the Khojah-chai (Granicus), flowing from the slopes of Ida, and the scene of Alexander's first victory over Darius in 354 B.C.; the Meinder-su, "called Xanthus by the gods and Scamander by men," which with its tributary, the equally famous Simois, traverses the Troas and joins the Ægean at the mouth of the Dardanelles; lastly, the Buyuk Meinder (Mæander), which flows for 250 miles through wild mountain gorges and rich alluvial plains to the coast near Miletus, and whose remarkable windings have given a familiar word to the English tongue. Most of these streams bring down much alluvial matter, which has during the historic period choked up many of the old harbours of the Ionian seaboard. Of less consequence are the rivers running south to the Mediterranean, two only of which, the Jihun-chai (Pyramus) and Sihun-chai (Carus), are of any considerable size. The "Silver Cydnus," associated with the names of Antony and Cleopatra, reaches this coast close to the mouth of the Sihun.

Some miles west of the Cydnus follow the two coast streams, Lamas and Kalykadnos (Ghiuk-su), which descend from the southern slopes of Taurus, and flow through a district strewn with innumerable ruins of towns, palaces, and temples. Here was the famous Cilician kingdom of Olba, at one time ruled by a dynasty of priest-kings who were reduced by the Romans. Olba,

ROCK CARVINGS NEAR OLBA.

the capital, the site of which was rediscovered by Mr. Theodore Bent in 1889, stands on the Jebel Hissar (5850 feet), which is still crowned by a ruined fortress four stories high, with five chambers on each story.

Close by is the great temple of Jupiter with thirty Corinthian columns still *in situ*, altogether one of the finest ruins in Asia Minor. The Lamas River, whose upper course encloses the Jebel Hissar on the north and east, plunges lower down into an extremely romantic gorge nowhere more than half a mile across, "the stupendous walls of which for miles offer on either side sheer precipices, reaching to the elevation in some places of over 2000 feet" (Bent, *Geo. Proc.*, 1890, p. 450).

A prominent feature of the plateau consists of its numerous fresh and salt water lakes, of which the largest is the Tuz-gol, or "Salt Lake" (Tatta Palus), lying about 60 miles north of Konia (Iconium). It is nearly 50 miles long by 10 to 12 wide; its waters are very brackish, and the saline incrustations on its banks are rich enough to supply the surrounding districts with salt. It is very shallow, and its area is much diminished by evaporation during the summer months. Of the fresh-water lakes the largest is the Egerdir, which lies 2800 feet above the sea, between the Sultan-dagh and the northern spurs of Taurus, and which is 30 miles long by 9 to 10 broad at its widest point. In the north-west the Isnik-gol, near Brussa, is 50 miles round, and drains to the Sea of Marmora.

4. *Natural and Political Divisions—Islands.*

The Anatolian peninsula forms in reality as well as in name a miniature of the whole continent. Both consist mainly of extensive central plateaux, with an inland and seaward drainage, and both are skirted by lofty ranges, behind which most of the streams have their source, which find their way to the coast. But in Asia Minor the alluvial plains developed by these rivers cannot compare in relative extent with those of the greater

Asia. The escarpments of the plateaux approach everywhere so near to the sea that no space is left for great lowland plains such as those of Siberia and China. There are a few low-lying and somewhat marshy tracts about the lower course of the Yeshil-Irmak, Kizil-Irmak, and Sakaria on the Black Sea, along the banks of the Meinder below Smyrna, and about Adalia and Mersina on the south coast. But with these and a few other unimportant exceptions, the whole peninsula may be broadly divided into two main natural divisions—the central plateau, here and there intersected by transverse ridges, and the encircling ranges. This disposition of the surface has largely determined the limits of the eight great vilayets or provinces into which Anatolia is divided for administrative purposes. Two or three—Angora and Sivas—comprise the greater part of the tableland. Of the six others, Adana, Aydin, Kastamuni, and Trebizond coincide with so many distinct sections of the coast ranges, while Brussa and Konia alone include portions both of the plateau and of the seaboard.

The old historical divisions, which fluctuated considerably with the many political and ethnical vicissitudes of this region, have been almost entirely effaced by the modern administrative changes of the Ottoman rule. Nevertheless, the names of these ancient states have never quite died out of history, and such memorable geographical terms as Phrygia, Lydia, Pamphylia, Paphlagonia, Cilicia, Cappadocia, are still familiar to the ordinary reader. How far all the old divisions correspond with the present administrative departments may be seen in the subjoined comparative table:—

Turkish Vilayets.	Ancient Divisions.
Brussa (Khodavendikiar)	Mysia.
Aydin (Smyrna)	Lydia. Caria.
Konia (Iconium)	Lycia. Pisidia. Pamphylia. Isauria. Lycaonia.
Adana	Cilicia.
Sivas	Part of Cappadocia.
Angora	Phrygia. Galatia. Part of Cappadocia.
Trebizond	Part of Pontus.
Kastamuni	Paphlagonia. Bithynia.

All the islands of the Ægean Sea belonging to Turkey, and collectively known as the Sporades, are grouped together in a separate administrative division called the Vilayet Jezairi Bahr-i-Sefid—that is, the "Vilayet of the White Sea Islands."[1] In this division was included the large island of Cyprus till the year 1878, when its administration was transferred to England. Thasos also is attached to the Egyptian Government, while Samos forms since 1832 a semi-independent tributary Christian State, under the suzerainty of the Porte, by whom its prince is appointed. With these exceptions all the Sporades of the White Sea Vilayet are disposed in five Sanjaks, or "Banners," as under:—

Sanjaks.	Islands.	Population.
Bigha	Tenedos. Lemnos. Samothrace. Imbros.	26,916 houses, of which 10,544 Greek, 10,308 Moslem.
Mytilene	Mytilene (Lesbos).	19,522 houses, of which 16,594 Greek, 2,818 Moslem.

[1] The Ægean, for no apparent reason, is always called the "White Sea" by the Turks and Arabs. See *Das Vilayet der Inselen des Weissen Meeres*, by A. Ritter zur Helle. Vienna, 1878.

Sanjaks.	Islands.	Population.
Sakyss	Khios (Sakyss). Ipsaria (Psara).	10,428 houses, nearly all Greek.
Kos (Istankoi)	Kos. Kalymnos. Patmos. Nisyros. Nicaria (Icaria). Leros.	6,394 houses, of which 6,085 Greek.
Rhodes	Rhodes. Karpathos. Kharki. Kasos.	16,762 houses, of which 10,270 Greek, 1,172 Moslem.

Most of these islands enjoy a delightful climate, and are fertile in oil, wine, silk, honey, corn, figs, oranges, and other fruits. Physically they may be regarded as a continuation of the mainland, belonging mostly to the same geological formation as the opposite coast of Anatolia, and, like it, still subject to violent earthquakes. By one of these Khios (Scio) was nearly ruined in the spring of the year 1881. In the Sporades the Greeks have always maintained a large numerical superiority, and the Turks, still numerous in Lemnos, Tenedos, and a few others, are retrograding like their fellow-countrymen on the mainland. The ownership of the land is rapidly passing from them into the hands of the Greeks, Armenians, and Jews.

Cyprus.

Cyprus, third largest island in the Mediterranean, presents the rough outlines of a bill-hook, with its sharp convex edge facing the Nile delta, and its handle, the Carpas promontory, projecting north-eastwards to within 80 miles of the Gulf of Iskanderun. But Cape Andreas, at the extremity of this headland, is distant only 60 miles from Latakia, the nearest point on the Syrian coast, with which it is now connected by submarine cable. The island is 3584 square miles in extent, with a population

(1891) of 209,000, two-thirds Greeks and one-third Muhammadan Turks. Physically it comprises three distinct sections — (1) The low-lying central Mesaorea Plain, originally a marine bed, as shown by the extensive banks of marine shells, large oysters and others, massed in several places; (2) The Kyrenian, or Northern range, which runs as a single ridge, with scarcely any spurs, close to the shore from Cape Kormakiti to Cape Andreas, at a mean height of about 1500 feet, with three conspicuous peaks—Pentedaktylos (2400 feet) near the centre, Buffavento (3140) a little farther west, and Kornos (3105) towards the eastern extremity; (3) The southern highlands, which cover nearly half of the surface, rising and broadening out westwards to a width of 20 miles, and culminating in Mount Troodos (6406 feet), highest point in the island. East of Troodos follow Mounts Adelphe (5305), Machera (4674), and Santa Croce (2260) above Larnaca.

From Pentedaktylos a copious perennial stream, the Kythrea of the Greeks, rushes down to the central plain, where it supplies power to several mills, and at last runs out in numerous irrigation rills. But all the other watercourses are rather winter torrents than rivers, seldom flowing for more than a few weeks continuously, and usually only for two or three days after heavy downpours. The largest are Pediæus and Idalia, both of which rise on the slopes of Machera, and after traversing the Central Plain, mingle their waters during the freshets at the silted-up port of the ancient Salamis, just north of Famagusta on the east coast.

Cyprus, which has no resources except the produce of the land, suffers from three plagues—long droughts, goats, and locusts. The goats, relatively more numerous than in any other region, have been a potent agency in the destruction of the pine and cypress forests, for which the island was long famous, and which supplied the material for ship-building from Phœnician times down to the late oppressive rule of the Egyptians. The goat nuisance remains unabated; but the locusts, not migratory, but of a species indigenous to the island, have been nearly extirpated by an ingenious device invented by an Italian resident.

The disappearance of the woods, now reduced to about 400 square miles in the southern uplands, has seriously affected agricultural prospects. With the forests went the soil, which "was washed down to the plains, choked the river-beds, and formed malarious swamps; the hills became bare rocks, incapable of growing a blade of grass, and the locust at once took possession of the barren ground, whilst the absence of trees deprived the earth of its annual fertilising agent, leaf-mould. There is now a stony desert at the south-east of the island between Famagusta and Larnaca, where tradition says there was formerly a large forest; and to the east of the Mesaorea, on the now dry and desolate plateau, there are many limekilns now in ruins, which could not have been supplied except by a vegetation that has now altogether disappeared" (Sir R. Biddulph, *Geo. Proc.*, 1889, p. 711).

Nevertheless, since the British occupation agricultural interests have improved, the exports (cereals, wine, cotton, raisins, silk, cocoons, wool) having advanced from £210,000 in 1888 to nearly £300,000 in 1892, the revenue from £149,000 to £190,000 in the same period, and the population from 186,000 to 210,000 in the decade ending 1891. There is no public debt, and the annual grant from imperial funds to revenue has gradually fallen from £55,000 in 1888 to *nil* in 1893, although the resources of the island are still burdened with an annual charge of £92,000, payable to Turkey under the Convention of 1878.

The bulk of the population is necessarily concentrated on the arable lowlands, where are situated all the towns —Nicosia, the capital, on the Pediæus, near the centre of the Mesaorea Plain, 12,500; Limasol and Larnaca, the two chief ports, both on the south coast, each over 7000;

(*From Photo. by Bonfils, Beyrout.*)

LIMASOL, CYPRUS.

Famagusta on the east coast, a place of vast antiquity, which, since the Turkish conquest (1571), has been a city of ruins, with a present population of 3300.

Cyprus was probably first settled by the Phœnicians in prehistoric times; to them succeeded Ancient Egyptians, Assyrians, early Greeks, Persians, later Greeks, Romans, Byzantins, Crusaders, Venetians, Turks, Modern

Egyptians, and English. Hence scarcely any other spot in the world is richer in archæological remains representing successive periods of Eastern and Western culture, following without a break from the remotest ages down to the present day.

5. *Climate.*

Owing to the great diversity in its relief, the climate of Anatolia is so varied that a general description becomes very difficult. In some places the transition from winter to summer may be effected by the traveller within the four-and-twenty hours. Along the west coast, at all times famous for its genial temperature, the thermometer varies in summer from 85° to 98° or 100° F., and here the heavy dews partly compensate for the slight rainfall. On the central plateau the winters are often exceedingly severe, the snow lying deep on the ground for about four months. In Karamania these winters are followed by sultry summers, and here also the rainy days are so few between April and November that the people depend nearly altogether on the tanks and reservoirs for their water. In the mountain passes of the Taurus the winters are excessively severe, and the summers correspondingly oppressive. More favourable is the climate of the north coast, thanks to its mild character and abundant rainfall. But while the interior is generally healthy, malaria, produced by the great heat and moisture, prevails, especially in autumn, near Trebizond and at some other points along the shores of the Euxine.

6. *Flora and Fauna.*

The Anatolian flora forms a transition between those of Persia and Syria in the east, and of Southern Europe in the west. On the south coast we are even reminded of the Nile valley, while the western seaboard strongly resembles that of the Morea. Owing to their abundant moisture the northern shores possess a magnificent forest vegetation, including the oak, beech, box, ash, plane, and other leafy trees. Here we meet with dense groves of the walnut, quince, mulberry, pomegranate, peach, apricot, plum, and cherry, while the valleys of the Kizil-Irmak, Sakaria, and other streams, afford excellent pasturage. Storax and other plants yielding valuable resins flourish on the Karamanian coast, whose flora resembles that of the shores of Syria.

In the Taurus grow several forest trees, especially of the coniferous order. But thousands of stately pines are yearly destroyed by fire, which is recklessly applied to them in order to stimulate the yield of turpentine. In Adana the sugar-cane grows well, but does not ripen sufficiently to cause the sap to crystallise. Large quantities of excellent grapes, olives, and figs are produced in the southern valleys, while the flora in many parts of the west and south rivals that of Spain and Sicily in splendour and luxuriance. In these respects a striking contrast is presented by the bleak upland plateaux of the interior, which produce little more than a stunted growth of brushwood, some saline plants, wormwood, wild sage, a few species of ferns, and in some districts nothing but two kinds of bramble. Amongst the cereals there is a species of bearded wheat; but oats are little cultivated, and barley is used as fodder for horses and other animals. The seeds of a species of thistle (*Gundelia Tournefortia*), which abounds on the Taurus slopes, yield a beverage which forms an excellent substitute for coffee, "a little brighter in colour, a little more bitter and aromatic in taste, but practically the same. It grows also in large quantities in Afghanistan" (Bent).

Like the flora, the fauna is akin to that of Southern Europe, but still more to those of Syria and Mesopotamia. Amongst the beasts of prey, nowhere numerous, are a few bears, wolves, hyenas, birds, several species of the cat, and wild dog. Jackals are met in the more secluded districts, where the gazelle and other varieties of the antelope also abound. Of domestic animals the buffalo is most commonly employed in agriculture, and even its milk generally replaces that of the common cow, which is rarely seen in the country. The camel is the chief beast of burden, although the horses are strong and well built, and had once a high reputation. The asses also are active and above the average size. The famous long-haired Angora or shawl goat, formerly peculiar to this region, but now found also in Persia, thrives in Anatolia only in a tract about 11,000 square miles in extent, stretching westwards from the Kizil-Irmak. Elsewhere the breed soon degenerates and loses the fine fleecy texture of its coating. The indigenous sheep belong mostly to the fat-tailed species, common throughout the East from Syria to the Kirghiz steppes.

Amongst birds the most common are the eagle, falcon, bustard, stork, heron, quail, partridge, besides the ordinary European species. Of butterflies the varieties are endless, and many are noted for their rare and gorgeous colours. The coasts teem with all kinds of fish, amongst which are the dye-producing cuttle-fishes. Land tortoises, lizards, frogs are common, while leeches are exported in considerable quantities to France and Italy.

7. *Inhabitants: Turks, Greeks, Kizil-Bashis.*

Ethnically speaking, Asia Minor is at present the true home of the Turks. It is one of the mainstays of the Ottoman Empire, from which this power continues to draw most of the resources that have hitherto enabled it to preserve its footing in the Balkan peninsula. Hence it is that the true character of this race can best be studied in Anatolia. All the western provinces are inhabited chiefly by Turks, who, however, even here are compelled to maintain the struggle for existence with other nationalities, and especially with the Hellenes. Farther east other races, such as the Armenians, Kurds, and Lazis, take part in the rivalry.

Yet, strange to say, the term "Turk" itself, at one time a proud title from the shores of the Adriatic to the remotest confines of Central Asia, is now carefully eschewed in Anatolia itself, where it has become a byword of reproach, answering somewhat to the English "boor," or "clod-hopper." And the people themselves have become all the more sensitive on the point, inasmuch as the "effendi," or refined "gentleman" from Stambul, regards the terms "Turk" and "Anatolian" as practically synonymous with "uncouth" or "clownish." The stalwart and sinewy figure of the Anatolian peasant, his rough manners, his harsh dialect, so different in its primitive type from the Arabo-Persian jargon that passes for Turkish in the capital, combined with his rude pronunciation, which has been compared to the gobbling of an enraged turkey-cock, afford a constant source of merriment to the dandies from the other side of the Bosphorus.

At the same time the social condition of the people must be regarded as backward and unsatisfactory. Since the days of the Trojan war the cultivation of the land has undergone but little improvement, and even the simple art of maintaining meadow lands is still unknown, so that during the dry summer months the herds must still be driven to the uplands in quest of a sorry pasturage. The fig, the vine, and the olive supply the Turkish peasant with his frugal fare, and enable him to meet his scanty wants. What need, therefore, to trouble himself with refined systems of husbandry?

The Turkish village presents a far from inviting appearance. The uncleanly hovels built of adobe, or sun-baked bricks, and pierced with one or two holes

HOUSE IN THE TAURUS.

for windows, usually comprise two compartments, one for the family, the other for the storage of provisions. The fittings of the interior are extremely simple, the furniture consisting mainly of a straw mat on the floor, a trestle bed with woollen mattress and cotton coverlets in the corner, a rude chest for the linen and best clothes, a few copper vessels and stone water-jars.

Dr. Carl Scherzer, a shrewd observer and a competent judge in Oriental matters, paints the present and the future of the Anatolian Turk in a few pregnant touches:—"The Turk, as a rule, understands his own language only, whereas all the other races in the country speak at least two from their infancy. This is due partly to his pride and contempt for all non-Muhammadan peoples, partly to the lack of enterprise and social rivalry. Earnest, reserved, and perhaps somewhat indolent, the Turk is still gifted with a fair share of intelligence. But though a keen observer of character, he lacks the business habits and the calculating spirit which have enabled the rival races to monopolise nearly all the trade of the country. In the rural districts the Turks are occupied mainly with agriculture and stock-breeding; in the towns they either deal in the local products, or else ply such simple trades as suffice to supply the few wants of their existence. Under proper management they make good seamen, and are also well suited for the caravan trade. They are deficient in the qualities of industry, perseverance in the acquisition of wealth, and the upward tendency towards social improvement, and indolence may be regarded as one of their most salient national failings. The morrow troubles them but little; hence they will often pay an exorbitant interest for the means wherewith to tide over temporary embarrassments, and will freely sell their lands without giving a thought to the consequent decrease of future income.

"In the districts where they are surrounded by Greek and Armenian communities the Turks have fallen greatly behind; but, thanks to the natural resources of the land and their own frugal lives, they are seldom reduced to absolute want. The recruiting system is a heavy burden, to which the Muhammadan populations alone have hitherto been subjected."

The exclusion of the female element from the social life of the Turk helps but to intensify the evil. The continuance of this practice is due mainly to the low state of education, which completely fails to meet the requirements of modern ideas.

It is not perhaps surprising that under such circumstances the energetic, mercurial, and quick-witted Greek should threaten to usurp the inheritance of the

A GREEK OF SMYRNA.

Turk even during his lifetime. Occupied with thoughts of gain, a shrewd calculating man of business, a skilful seafarer, and intelligent husbandman, the Greek outrivals his Moslem neighbour in every pursuit of life. The learned professions he almost entirely monopolises, and the doctor, lawyer, teacher, banker are everywhere sure to be of Hellenic blood. The Greek is invariably the broker who negotiates all business matters for "his Turkish friend," and he has secured the almost exclusive control over the local and export trade. He is at the same time indefatigable in his efforts to promote scientific and literary work, while also fostering a lively sense of Hellenic nationality. Thus Smyrna has already become a Greek city, and Athens has become the centre of an ably-directed movement aiming especially at the improvement of education amongst the Anatolian Hellenes. With his unflagging efforts to better his social and political status the descendant of the old Ionian stock is gradually resuming possession of the western provinces.

Other ethnical elements in Asia Minor are the Armenians, Jews, numerous in the large towns, the Gipsies, the Circassians, Abkhasians, Lazis, and the Yuruks, a nomad Turki race occupying the uplands between Erzerum and the plains of North Syria.

"The many tribes of Yuruks, so called from the Turkish *Yurumek*, to wander, are almost the only inhabitants of vast districts in the highlands to the west and south of Asia Minor. They are a very peaceable, friendly race, quite distinct from the Afshars, Kurds, Circassians, Bosdans, and other tribes which winter on the Cilician plain, and whose summer pastures or *yaïlas* are farther east. The Turks look upon the Yuruks, from their law-abiding tendencies, as the policemen of the mountains, and they are always ready to give information concerning the suspicious characters who visit their mountains." They are a remarkably fine race, distinctly white, with fair skins and long beards, Muhammadans, polygamists, and great camel breeders, "producing that valuable sort of mule camel common in Asia Minor, a cross between the Bactrian and the Syrian, which is excellent for mountaineering purposes" (Bent).

Mention should also be made of the Kizil-Bashis, or "Red Heads," a remarkable race, also of Turki stock, scattered over Anatolia, Persia, and Afghanistan as far east as Kabul. Outwardly devout Muhammadans, the Kizil-Bashis are none the less tenaciously attached to their own peculiar tenets and observances. These they never reveal to strangers, and Mordtmann, who frequently visited Asia Minor, never succeeded in obtaining any trustworthy information regarding them. He, however, agrees with Van Lennep[1] in looking on them as the last survivors of the old pagan communities. But W. Gifford Palgrave, when British Consul at Trebizond in 1868, described them as "a sort of Eastern Mormonites, with a dash of Persian or Shiah superstition."[2] He adds that they are as distinct from the Osmanli as the Saxons are from the Swedes. They call themselves "Eski-Turk"—that is, "Old Turks"—a term often applied to the Anatolian Turkoman tribes, to whom they seem to be closely akin in physique and speech. Although reputed atheists, they are said to be believers in the doctrine of transmigration, are very hospitable, and entirely free from the absurd feelings of jealousy which degrade women to the level of the brute creation in most Muhammadan countries. The fertile plains of Raz Ova and Ard Ova near Tokat, and the villages between Angora and Amasia, and between Kara-Hissar and Tokat, are the central quarters of the Anatolian Red Heads.

The Circassians and Abkhasians who have migrated to Turkey since the reduction of West Caucasia by the Russians have never found suitable homes in Asia Minor, where they have consequently become a serious disturbing element. Mrs. Scott-Stevenson, and other

[1] *Travels in Little Known Parts of Asia Minor*, Lond. 1870.
[2] Official Report on prov. Trebizond, in Blue-book for 1868, part ii.

recent travellers, represent them as a source of constant trouble, hopelessly indolent, given to plundering and hectoring over the people, levying blackmail right and left, and actually laying siege to the provincial towns.

8. *Topography: Chief Towns.*

The interior of Asia Minor is rich in towns whose names have been famous since classic times, a circumstance which is apt to give them far greater importance than they now really possess. Such are Kaisarieh (Cæsarea), at the north foot of Mount Argaios, which, though much fallen from its former greatness, still derives some importance from its position at the junction of several highways of commerce; Sivas (Sebaste) on the Kizil-Irmak, and Tokat on the Yeshil-Irmak, 60 miles north-west of it, both centres of a considerable inland trade. Farther west, Konia (Iconium), on the road between Brussa and Adana, gives its name to a large vilayet; formerly capital of the Seljuk empire, its numerous shrines of "saints" still attract devout Moslem pilgrims. Angora, or Engurieh (Ancyra), in the centre of the Angora plateau, is noted for its silky, long-haired animals—cats, dogs, rabbits, and goats, the wool of the last-mentioned forming the staple of its trade. Afiun-Karahissar, midway between Smyrna and Lake Tuz-gol, is the centre of a large opium trade, whence its name, which means "Black Castle of Opium." On the northern route leading thence to the Sea of Marmora stand Kiutayah, or Kutaieh (Cotyæum), near which are some interesting Phrygian remains, and Brussa at the foot of Mount Olympus, whence its classic name of Prusa ad Olympum. It was formerly the capital of Bithynia, and is at present the chief town of the vilayet of Khodavendikiar. A few miles north-east of it are the once famous towns of Isnik (Nicæa) at the east end of Lake Ascanius, and Ismid, now connected by rail with Scutari, the Asiatic suburb of Constantinople. On the coast of the Euxine are the small ports of Sinope, where the destruction of the Turkish fleet by the Russians precipitated the Crimean war of 1854, and Samsun (Amisus), near the mouth of the Kizil-Irmak. East of it lies the flourishing port of Tarabuzun (Trebizond), the great emporium of the overland trade with Armenia and Persia. Here the Greeks under Xenophon, on their memorable retreat northwards from Cunaxa, first struck the coast and hailed the blue waters of the Euxine with shouts of "Thalatta, Thalatta!"

But the true emporium of the Levantine trade and the real capital of Asia Minor is Ismir (Smyrna), which is conveniently situated at the head of the gulf of like name, a magnificent inlet of the Ægean, over 40 miles long, forming a vast and well-sheltered harbour with deep water right up to the quays of the city.

Originally an Æolian, later an Ionian settlement, Smyrna is one of the oldest places in the Hellenic world. It was one of the seven cities that claimed the honour of having given birth to Homer, and the poet's epithet of *Melesigenes* has reference to the local belief that he saw the light on the banks of the neighbouring river Meles. Smyrna also figures amongst the "seven Churches" mentioned in Revelation, and on Mount Pagus is still shown the tomb of its bishop, Polycarp, who suffered under Marcus Aurelius (166 A.D.). But since 1419 it has been held by the Turks, who, however, have failed to greatly modify its essentially Hellenic character. Of its estimated population of 200,000 in 1894, fully 130,000 are still Greeks, and not more than 40,000 Turks, the remainder comprising 15,000 Jews, 10,000 Armenians, and about the same number of Franks and Levantines (Europeans). All these sections of the community occupy separate quarters, of which the most attractive is that of the Franks, where the French language has become the general medium of

SMYRNA.

intercourse. But there are nowhere any fine public buildings, and the most important institution is the large Greek College, the chief centre of neo-Hellenic culture in Asia Minor. It has long been under the protection of England, and occupies quite a district to itself, comprising a museum of antiquities, a large and valuable library, and other aids to the higher education of the Greek population. The Greeks take the leading part in all municipal affairs, and they have monopolised most of the retail trade of the place. But by far the largest share of the foreign exchanges belongs to Great Britain—about £2,000,000 of the £5,000,000 exports, and £1,400,000 of the £3,000,000 imports in 1891.

Nothing gives us a better idea of the varied natural resources of Anatolia than a glance at the export trade of Smyrna. The tables include such diverse commodities as maize, rice, and other cereals, tobacco, silk, cocoons, opium, madder, valonea, gall nuts, yellow berries, mohair, sponges, besides large quantities of dried figs and raisins of prime quality.

9. *Highways of Communication.*

One of the chief impediments to the development of the resources of Anatolia is the lack of good highways of communication. Railway enterprise, however, has made some progress in recent years, and in 1894 over 920 miles were open for traffic. The longest line (430 miles) runs from Scutari to Angora; but the chief centre of the system is Smyrna, from which radiate lines to Dinair (234 miles), to Alasher (105), to Odemish (68), and to Sevdikeni (9). Mersina is also connected by rail with Adana (42), and Brussa with Mudania (32). But all these may be regarded merely as the first links and branches of that great Asiatic trunk-line which perhaps may some day connect Constantinople and the West with the Indus valley.

Meanwhile trade and intercourse are largely dependent on four main and a number of secondary highways, none of which, except those connecting Trebizond with Erzerum and Samsun with Amasia, would pass for roads in the West. Of the four main routes the longest runs from Scutari through Ismid, Boli, Amasia, and Tokat, right across the northern section of the peninsula to Erzerum and the frontier Russian fortress of Kars. The second, starting from the Euxine at Samsun, strikes the former at Amasia, and again leaves it at Tokat, running thence nearly due south to Sivas. Here it branches off in two directions, south-westwards to Kaisarieh and through the Cilician Gates over the Taurus to the Mediterranean at Mersina, eastwards through Arabkir and Erzinghan to Erzerum. Another branch connects Kaisarieh with the Tigris at Diarbekr. The third main line runs from Trebizond southwards tc Erzerum, where it trends eastwards to Bayazid on the Russian frontier, and thence across the Persian border to Tabriz. This has from time immemorial formed the great highway of communication for Persia with the Euxine and the West. Lastly, the fourth main route runs from the Sea of Marmora south-eastwards through Brussa, Kiutayah, and Koniah, to Erekli, beyond which it crosses the Taurus also by the Cilician passes, winding thence by Adana round Alexandretta Bay to Skanderun (Alexandretta), where it sends off branches eastwards to Aleppo, southwards to Antiochia. This formerly much-frequented route is now little used for through traffic.

CHAPTER V

THE EUPHRATES AND TIGRIS BASIN

1. *Boundaries—Extent—Area.*

NEARLY the whole of the eastern provinces, lying between Anatolia and Syria on the west, and the Russian and Persian empires on the east, are drained through the Rivers Euphrates and Tigris to the Persian Gulf. They consist mainly of two great physical divisions —the Armenian and Kurdistan highlands in the north, the Mesopotamian lowlands in the south. But there are nowhere any sharply-defined natural frontiers. The somewhat arbitrary line marking the limits of Turkey in Asia towards Russia and Persia coincides nearly throughout its entire length with the eastern frontier of this basin, which thus stretches from Lazistan to the Persian Gulf. On the west the northern uplands merge almost imperceptibly in the Anatolian plateau, while the southern lowlands rise very gradually towards the Syrian highlands and the Arabian tableland. Even in the north Turkish Armenia is cut off from the Black Sea by the portion of Lazistan which is still left to the Porte, and which is administratively included in the Anatolian vilayet of Trebizond. In the south alone the Persian Gulf gives for some distance a decided natural limit. In most maps a graceful curve, described almost with the regularity of the compass, and stretching across the Syrian desert from near the Dead Sea to the head of the gulf, is supposed to mark off Turkish territory from independent Arabia. But this line has absolutely no significance at all. In official maps it disappears altogether, or is replaced by a straight line drawn much farther south from about the head of the Gulf of Akaba eastwards to the new vilayet of Basra, which now includes all the Shat-el-Arab district and a large slice of North-East Arabia. The extent of this region will therefore vary enormously according as it is made to include or exclude the Syrian desert and portion of the province of Basra. But taking the southern limit at the 30th parallel, which crosses the head of the Persian Gulf, and the northern at the Lazistan coast range under the 41st parallel, the Mesopotamian basin will have a total length of about 770 miles, with an average breadth of 300 from the Russo-Persian frontier to Anatolia and Syria, and an area of over 300,000 square miles.

2. *Relief of the Land: The Armenian and Kurdistan Uplands—The Mesopotamian Lowlands.*

The northern section of this vast region embraces that portion of the Armenian highlands which still remains under the Ottoman rule. It consists mainly of a lofty plateau 4000 to 7000 feet above sea-level, and culminating with Mount Ararat just on the eastern frontier. Its surface is even more mountainous and irregular than that of Anatolia, for within its narrower limits it is crossed by four main ranges, with many secondary branches, forming connecting links between the Caucasian system on the north, the Anatolian on the west, and the Kurdish on the south. But notwithstanding the great mean elevation of the land, only a few of the peaks rise above the line of perpetual snow, and the chains themselves, which are crossed in several directions by accessible passes, are separated from each other by the deep valleys of the Aras, Choruk, and Euphrates, flowing in three opposite directions to the Caspian, Euxine, and Persian Gulf. The surface of the country between the mountain ranges consists of broad and mostly level steppe-like tablelands at various elevations, and forming a series of terraces one above the other. Deep and narrow valleys, gloomy and occasionally imposing mountain masses, broad and bleak plateaux, a severe climate, with rigorous winters, followed by dry and sultry summers, a marked absence of forest trees, but in the valleys an abundant and even luxurious vegetation, such is the general physical aspect of the Armenian highlands.

The Kars district, ceded to Russia in 1878, forms a rugged tableland, terminating south-westwards with the lofty Soghanli range, from 7000 to 8000 feet high, beyond which stretches the great valley of the upper Aras (Araxes). This valley, which crosses the district of Erzerum from west to east, is everywhere enclosed by high mountains—on the south by the Aghri-dagh (9400 feet), the Bingol-dagh (12,000 feet), and others; on the north by the Shamar-dagh (9227 feet); on the west the Boyun and Palantukan-dagh (7300 feet), close to Erzerum. North of Erzerum the land falls towards the valley of the Choruk, beyond which it again rises to the Lazistan coast range, which attains an elevation of 11,000 feet, and forms the northern frontier of Armenia proper. Eastwards the range is pierced by the Choruk, which here trends northwards through a narrow gorge at Artvin, beyond which it flows through alluvial plains to the Euxine at Batum. Here the new Russo-Turkish frontier line has been shifted a few miles westwards to the coast village of Khopa, whence it runs southwards over the hills to the Choruk, thus leaving the eastern and richest division of Lazistan to Russia. The rest of this region, as already stated, is included in the vilayet of Trebizond, which thus stretches between Armenia and the coast eastwards to the Russian frontier. But the Choruk forms a geological and ethnical, as well as a political parting-line. While chalk and jurassic formations prevail in the south, igneous rocks everywhere crop out in the north, where they form the higher ridges of the coast range. The range itself is also inhabited by the Lazis, a western branch of the Georgian race, and consequently quite distinct from the Armenians, whose northernmost limit is marked by the middle course of the Choruk.

The great central tableland of Erzerum, which stretches eastwards to Ararat, may be said to be limited southwards by the valley of the Murad, or eastern head-stream of the Euphrates. Here rise the Sunderlyk-dagh, the Ala-dagh, the snowy Sipan-dagh, and other mountains, attaining an elevation of from 10,000 to 12,000 feet, beyond which the plateau maintains a mean altitude of 5000 feet eastwards to the frontier town of Bayazid. But it falls southward to the land-locked basin of Lake Van (5360 feet), which may be taken as the southern limit of Armenia proper.

At some former period the whole of this region must have been a centre of great igneous activity. An old crater may still be traced on the summit of the Bingol-dagh; the Sipan-dagh terminates in a solitary peak shaped like a truncated cone; the Nimrud-dagh on the west side of Lake Van has a crater nearly 8 miles in diameter, now containing several tarns, and hot-springs and lava streams, as well as a large crater, may also be detected on the Tendarek-dagh in the north-east. "The basaltic ravines on the slopes of the Bingol and Sipan

Mountains, and the numerous fragments of obsidian which can be picked up close to the Sipan-dagh, are further evidence of this. In the Hartoshi Mountains and the upper basin of the Great Zab River numerous sulphurous springs are to be met with. Some are hot-springs, some give off sulphuretted hydrogen, and in others the pure sulphur can be gathered from incrustations round the edge of the springs. This is much prized by the Kurds for the manufacture of powder."[1] The origin of Lake Van itself has been attributed to a lava stream from the Nimrud-dagh, blocking up the valley to the south of the mountain, and preventing the running waters from escaping to the Tigris basin.

Since 1876 the Van district has been separated from the vilayet of Erzerum, and a line drawn from Mush through the lake eastwards to the frontier town of Kotur, ceded to Persia in 1878, will roughly mark off the Armenian from the Kurdistan highlands. But the delimitation is in every sense arbitrary. The term Kurdistan—that is, "Country of the Kurds"—is so far correct that it is mainly occupied by tribes of Kurdish stock. But, on the other hand, these tribes have spread in almost every direction far beyond its present limits, reaching eastwards to the Bakhtiari highlands in Persia, northwards in isolated communities to the parallels of Batum and Tiflis, and north-eastwards to Khorasan. Physically, also, the long and rugged mountain range forming the backbone of Kurdistan stretches beyond the frontier northwards between Lakes Van and Urmia to the foot of Ararat. From this point a second chain branches off south-westwards, sweeping round Lake Van and rejoining the eastern range at the Erdosh-dagh. The united chain runs thence with many ramifications south-eastwards to beyond the

[1] Captain F. R. Maunsell: "Kurdistan," in *Geograph. Jour.* for February 1894, p. 83.

34th parallel. The main eastern axis thus forms a natural frontier line between Turkey and Persia from Bayazid to Karmanshah, and the whole system encloses an area of nearly 50,000 square miles.

In the north the surface is very rugged and mountainous, but one extensive plateau, from 4000 to 7000 feet high, is developed between the Erdosh-dagh and Jebel-Judi,[1] which, running nearly west and east from Jezireh to Persia, rises from about 2000 feet at its western extremity to upwards of 13,000 in the Jawar and Rowandiz peaks near the Persian frontier. Beyond this range the country is generally level, varied only with a few low ridges culminating with the Jebel-Hamrin, about midway between Mosul and Bagdad. Here Kurdistan and Mesopotamia proper may be said to overlap, for, while the former at this point reaches southwards to the 34° N. lat. beyond the Tigris, the latter stretches between the two rivers northwards to the 37° N. lat.

The prevailing geological formation in the north is limestone, with red sandstones and conglomerate. Here the hills generally present bare crests with rugged slopes partly overgrown with dwarf cedars, junipers, and valonea. Limestones and sandstones also prevail along the southern frontier range, but intermingled with schists, quartz, and granites. Here the bleak brown hills present jagged outlines and steep sides, often deeply scored by the action of the mountain torrents which lower down flow through narrow winding valleys. Copper, lead, and iron ores are said to abound in the west, and in several places amongst the hills of the Euphrates; but the only minerals available for export are salt from Van, sulphur, alum, naphtha, and a little iron.

South of the province of Erzerum and west of Lake Van, the Armenian and Kurdistan highlands slope con-

[1] *Dagh* is the Turkish and *Jebel* the Arabic word for "Mountain."

tinually southwards to the plains of Mesopotamia, and westwards to the Euphrates, which here marks the eastern limits of Anatolia. The tract between the Van district and the Euphrates, east and west, and between the Murad and Khabur Rivers, north and south, is often spoken of as "Kurdistan" in a more restricted sense, and on many maps figures as the Turkish province of Kurdistan; but this use of the word can scarcely be justified. There is no Turkish province of the name, and the country as above limited is mostly comprised in the vilayet of Diarbekr. Most of this vilayet is watered both by the Tigris and the Euphrates, consequently nearly as far north as Diarbekr it belongs geographically to the region commonly designated as Mesopotamia—that is, the Interriverain Country, or what in India would be spoken of as a "Doab," or "Land of Two Waters." It also belongs ethnically to two distinct domains, for the Kurdish and Arab nomad tribes, of Iranian and Semitic stock respectively, here meet on common ground. The term El-Jesireh, or "The Island," as Mesopotamia is always called by the Turks and Arabs, was formerly limited to the land strictly lying between the two rivers southwards to the old wall by which they were connected above Bagdad. The tract from this point to the Persian Gulf (that is, the ancient Babylonia) was and is still known as Irak-Arabi (that is, Irak of the Arabs), to distinguish it from the Irak of Persia. But the whole region from Diarbekr to the Gulf and from Syria to the Persian frontier is now commonly spoken of as Mesopotamia, the two divisions being sometimes distinguished as Upper and Lower Mesopotamia. It has a total area of perhaps 180,000 square miles; but it everywhere presents remarkable uniformity in its physical and ethnical conditions.

In the extreme north the land rises towards the Armenian and Kurdish highlands; but even here the mean elevation is little more than 1500 feet above the sea. The upland tract between Jesireh and Mardin is a stony waste, offering a scanty pasturage to the flocks and herds of the nomads in winter and spring. But the plains stretching farther west towards Urfa and Harran, and southwards to the low Sinjar hills, are well watered

RUINS OF BABYLON.

and very productive. These Sinjar hills form an isolated ridge, 7 miles wide and 40 miles long, midway between the Tigris and Euphrates, about the parallel of Mosul. Farther south the land is nowhere more than 600 feet above sea-level. It may be regarded as a northern extension of the Persian Gulf, which at one time probably reached to within 80 miles of the Mediterranean, but which has been gradually filled in by the alluvia of the great rivers, and by the advancing sands of the desert. Indeed, before the formation of the Syrian coast ranges the Mediterranean and Persian Gulf were possibly connected, thus isolating Arabia from the rest of the continent, and offering a direct water highway from the Atlantic to the Indian Ocean. Owing to this geological origin of Mesopotamia, the soil is found to consist everywhere of a sandy clay, abounding in excellent agricultural properties, and incapable of cultivation only where water fails. Its astounding fertility is sufficiently shown by the fact that it still remains unexhausted after having supported the teeming populatians of the Akkadian, Assyrian, Babylonian, and Persian empires, from the dawn of history down to comparatively recent times. The number and vastness of the ruins scattered over this region from Babylon to Nineveh still bear witness to its former flourishing material condition; and since the cuneiform writings abounding in these ruins have yielded up their secret to the ingenious labours of modern science, we now know that the Mesopotamian plains have been the scene of successive cultures, rivalled in splendour and antiquity by those only of the Nile valley.

3. *Hydrography: The Tigris and Euphrates—Lake Van.*

With the exception of a small area in the extreme north, the whole of this region drains through the Euphrates and Tigris to the Persian Gulf. Since the rectification of the Russo-Turkish frontier in 1878, the valley of the Kur (Cyrus) belongs entirely to the Russian territory of Transcaucasia. But the Choruk and the Aras (Araxes) still flow for a considerable distance through Turkish Armenia before crossing the frontier on their course to the Euxine and the Caspian. The Choruk, which rises in the uplands west of Erzerum, is joined below Baiburt by a tributary from the west, after which it flows along the southern base of the Lazistan coast range eastwards to the Russian frontier. Here it bends northwards altogether within Russian territory, and reaches the sea close to Batum, after a precipitous course of about 200 miles. The Aras rises at the north foot of the Bingol-dagh 30 miles south of Erzerum, and flows north-eastwards to the frontier, which it soon reaches at a point considerably to the west of Kars and Ararat. In Turkish Armenia, of which it drains a very small area, it is little more than a rapid mountain torrent.

All the rest of the Armenian and nearly the whole of the Kurdish highlands belong to the basin of the twin rivers Euphrates and Tigris, which flow mainly in a south-easterly direction across the Mesopotamian plains. Rising on the Armenian terrace lands, they pursue on the whole a parallel course, although often approaching and diverging from each other, until they at last mingle their waters at Kurnah, where the united stream takes the name of the Shat-el-Arab about 120 miles above its delta at the head of the Persian Gulf. Above Kurnah their channels approach nearest to each other at Bagdad, thus nearly separating Irak-Arabi from Upper Mesopotamia.

The upper region of the Euphrates resembles that of the Rhine, while its middle course may be compared with that of the Danube, and its lower with the Nile. The Euphrates proper is formed by the junction of two great head-streams—the Kara-su or western branch, and the Murad or eastern branch, whose sources lie over 120 miles apart, in the very heart of the Armenian highlands. The Kara-su—that is, "Black Water"—rises some 20 miles to the north-east of Erzerum, and flows for 270 miles south-westwards to Keban-Maaden, a few miles west of Kharput. Here it is joined from the east by the Murad, which flows from the Ala-dagh south of Bayazid, and near the Russian frontier, and has a total course through Armenia and Kurdistan of about 300 miles. Some 60 miles south of the junction the Euphrates pierces the Upper Taurus near Arghana, beyond which it trends southwards through the vilayet of Aleppo, here coming within 80 miles of the sea. But about the 36th parallel it turns somewhat abruptly to the south-east, and henceforth retains this direction to the Gulf. It is navigable for over 1100 miles for small steamers to Bir (Birejik), near Urfa, the point where it is crossed by the great caravan route from Syria to Bagdad.

The Tigris is also formed by an eastern and a western head-stream, the former rising close to Bitlis, near the west side of Lake Van, the latter flowing from the neighbourhood of Kharput by Diarbekr to the confluence above Finduk. Beyond this point it pursues a southerly course by Mosul to Bagdad, and between these points is joined on its left bank by the Great and Little Zab, and some other tributaries from the Kurdistan highlands. It is navigable for vessels of light draught to Nimrud, 20 miles below Mosul, and again for 300 miles by rafts from Mosul to Diarbekr. But owing to the rapidity of the current the traffic is all down stream, and is still carried on mainly by a primitive style of craft, which is broken up at Bagdad, and transported by camels back to Mosul. The journey between these points occupies three or four days during the floods, and from twelve to fourteen at other times.

Below Bagdad the main streams are connected by several channels and intermittent watercourses, of which the chief are the Nahr Isa or Saklawiyah canal and the Shat-el-Hai. Higher up the Euphrates is joined on its left bank by the Belik near Rakkah, and by the Khabur at Kerkesia. The latter flows intermittently through the desert from the Karijah-dagh hills, 20 miles west of Mardin, round the western extremity of the Sinjar hills. During the floods it is joined by several streams, which at other times run dry in the sands. Below the junction of the Khabur there stretches a desolate desert tract between the Euphrates and the Tigris, which is overgrown with wormwood, and still haunted, as in the time of Xenophon, by the wild ass, ostrich, and bustard. This region is visited by terrific whirlwinds, such as that which on 21st May 1836 nearly overwhelmed the English Euphrates Expedition under Colonel Chesney.

Below Kurnah the Shat-el-Arab traverses a flat and fertile plain, dotted over with villages, and covered with artificially irrigated meadow-lands and date groves. At Mohammerah (Mohamra), 40 miles above its mouth, and 20 miles below Basra, it is joined by the Karun from Persia, and here properly begins the delta, of which one arm only is navigable. For six months in the year this delta is converted into a swampy lacustrine district by the floods caused by the melting of the snows about the head-streams in spring, and occasionally by the autumn rains. From its mouth to Bagdad the main stream is navigable throughout the year for steamers of considerable size. For some years past an English line plying between Basra and Bagdad has contributed much towards the development of the resources of Mesopotamia.

In the whole of the Mesopotamian basin there is only one body of still water deserving the name of lake. This is the magnificent Lake Van, by far the largest in Asiatic Turkey, renowned alike for its romantic beauty and historic associations. It occupies an irregular triangular space, 80 miles by 30, over 5000 feet above sea-level, on the border-land between the Armenian and Kurdistan highlands. From the snowy Sipan-dagh, towering above its northern shore, it is seen to occupy "the centre of a magnificent valley, surrounded on three sides by densely-wooded mountains, whose forests of firs, chestnut, beech, walnut, and ash merge in the broad belt of gardens and melon-grounds that fringe most of the shore line as far as the eye can reach."[1] Its waters, which are diversified with several lovely islets and teem with fish, are very salt, and have no present outflow.

LAKE VAN, AND THE MOUNTAINS TO THE NORTH.

Dr. R. Sieger has shown that the levels

[1] J. C. M'Coan, *Our New Protectorate*, i. 45.

both of Lake Van and of Urmia are subject to periodical changes, which appear to coincide with the periods of climatic variation which, according to Brückner, occur on the surface of the globe. Thus, during the present century, these lakes reached their periodic maxima (greatest heights) in the years 1810, 1840-50, and 1876-80, while the level of Van has been generally lower in the first than in the second half of the century (*Globus*, vol. l. 1894, p. 73).

4. *Natural and Political Divisions: Turkish Armenia and Kurdistan—Mesopotamia.*

The Mesopotamian basin comprises two natural divisions only—the Armenian and Kurdistan uplands, where all the rivers have their sources, and the alluvial plains of Mesopotamia proper, which may be regarded as the creation of those rivers. To these two natural divisions correspond the Turkish administrative divisions of Erzerum, Aziz, Diarbekr, Bitlis, Van, Mosul, Bagdad, and part of Basra, with total area 190,000 square miles, and population 3,800,000.

The Armenian and Kurdistan highlands, which form a border-land between three empires, possess neither physical, ethnical, nor political unity. Thus their drainage is partly to the Euxine through the Choruk-su, partly to the Caspian through the Kur and Aras, partly inland to the closed basins of Lakes Van and Urmia, but mainly through the Euphrates and Tigris to the Persian Gulf. Ethnically, also, they are occupied by peoples of four distinct stocks—the Lazis, a branch of the Georgian race; the Armenians and Kurds, members of the Iranian family; the Turks and Tatars, of Turki origin; the Arabs, Jews, and so-called "Nestorians," of Semitic blood. Lastly, these highlands, taken in their widest sense, are politically distributed between the empires of Russia, Persia, and Turkey, which here converge round the base of Ararat, their culminating point. The late changes that have taken place on the Russo-Turkish and Turko-Persian frontiers have even increased rather than diminished the difficulty of drawing any clear parting-line between the three states, whose boundaries are here almost everywhere purely conventional and even arbitrary. In Armenia the Russo-Turkish frontier-line is now deflected considerably westwards in the direction of Erzerum, thus leaving Ardahan, Olti, and Kars to Russia, to which power the seaport of Batum on the Black Sea was also ceded by the Berlin Congress of 1878.

In Mesopotamia the northern and southern sections of El-Jesireh and Irak-Arabi differ greatly in their main features. The transition from the elevated plateau of Diarbekr to the alluvial plains is effected by the extensive open tract which maintains an elevation of over 1500 feet between the Tigris and the western bend of the Euphrates. Here the hilly wooded districts in the north are succeeded by grassy steppes or arid wastes, which are converted into highly-productive oases wherever water abounds. Such is the fertile district stretching from Urfa southwards to Harran, where splendid crops of maize, tobacco, and cotton are raised. Below Mosul the date-palm begins to make its appearance, and this plant forms the prevailing feature in the landscape throughout the level alluvial plains of Irak from Bagdad to Basra. In the extreme south the numerous back-waters and channels of the two main streams merge imperceptibly in the lagoons and morasses of the Shat-el-Arab delta. But these magnificent lands, so well suited for agriculture, are now little cultivated. The nomads and even the scantily-settled population rely mainly on the produce of their flocks and herds, and the country shows the same signs of misrule, ruin, and decay that are elsewhere visible in Asiatic Turkey. "Except around Bagdad the traveller now sees hardly a trace of the date-groves, the vineyards, and the gardens which excited the admiration of Xenophon" (M'Coan).

5. *Climate.*

The Armenian climate, pleasant enough in spring and autumn, is excessively severe in winter and summer. During the long winter months from October to May the ground is mostly covered with snow, while the mid-summer heats are most oppressive. These conditions also prevail in Kurdistan, where, however, the variations of temperature are not so great as farther north. Here, also, the winter is of shorter duration, with correspondingly longer springs and autumns. In Mesopotamia the mild but short winters become the pleasantest part of the year. But they are succeeded by sultry summers, during which the plains become scorched and bare. Here the Samiel, or "poison wind," prevails in the same season; and the disease known as the "Aleppo button," or "Bagdad date-mark," is seldom absent from the towns fringing the desert. This mysterious affection, which is probably caused by the bite of a fly, though troublesome is never fatal, usually lasting about a twelvemonth. No cure has yet been discovered, but some doctors claim to possess the power of driving it from one part of the body to another.

Throughout the Mesopotamian basin the annual rainfall is below the average. Summer is everywhere very dry, but much snow falls on the uplands in winter; and in Upper Mesopotamia abundant rains prevail from December to March. Farther south vegetation and husbandry depend largely on artificial irrigation, which has been practised in this region from the remotest times.

6. *Flora and Fauna.*

In Armenia there is a marked absence of forest trees, and so deficient is the supply of wood that in many places cattle-droppings form the staple of fuel. The well-watered valleys abound in fruits and cereals; but the bleak plateaux are generally bare, or covered with a scanty vegetation of grass. Far more varied is the flora of Kurdistan, where the hills are often clothed with forests of oak, ash, walnut, and pine trees. Here also the lower grounds yield rich crops of maize, wheat, pulse, hemp, besides tobacco, cotton, mulberries, grapes, melons, and other southern fruits. In Mesopotamia the vegetation becomes more decidedly tropical, and the Shat-el-Arab district produces some of the finest dates in the world.

Wild animals have almost disappeared from this region. But the towns are infested by packs of pariah dogs, which, while doing the work of the scavenger, are occasionally dangerous to the people. M'Coan tells us that on one occasion he nearly fell a victim to the half-jackal breed of Erzerum.

Their countless flocks of sheep form the chief wealth both of the Kurdish and Arab nomads, and the latter also possess many camels, and perhaps the purest breed of Arab horses in Asia.

7. *Inhabitants: The Armenians, Kurds, Nestorians, and Bedouins.*

Although the seat of some of the earliest human cultures, the Mesopotamian basin is still largely occupied by a nomad population. Its inhabitants belong to four distinct stocks—the *Iranian*, represented in the northern highlands by the Armenians and Kurds; the *Semitic*, represented in the north by the so-called "Nestorians" or Chaldeans, and in the plains by the Arab Bedouins; the *Turki*, which, besides some Tatar tribes, supplies the ruling element found chiefly in the towns; lastly, the *Caucasian*, of which there are two branches—Lazis in the extreme north, and Circassians, many of whom have migrated in recent years from Russian to Turkish territory.

The centre of gravity of the Armenian nationality, which formerly lay about the basin of Lake Van, has been gradually shifted northwards to the neighbourhood of the Ala-goz and the famous monastery of Echmiadzin, both within the Russian frontier. The race, like the country itself, has long lost its political unity, and is now distributed over the Russian, Turkish, and Persian empires. Nevertheless, over one-third of the people still continued to reside under the Ottoman rule before the organised massacres of 1892-96. They are distinguished as much by their features, dress, and social habits as by their distinct Christianity from the surrounding Kurdish and Turkish Muhammadans, by whom they have always been hated, both as Christians and for their success in accumulating wealth. Like the Jews, the Armenians, after the loss of their independence, turned to trade, which till lately was almost entirely in their hands. They owned nearly all the capital of the country, so that the money market was ruled by them. The great influence thus ensured to them naturally caused mutual heart-burnings and rivalries amongst themselves, while against the common enemy they combined together and spared no sacrifice for the general weal. Surpassing others in shrewdness, the main object of the Armenian dealers was

TATAR NOMADS.

to purchase cheap wares of attractive appearance, and then retail them advantageously. Thus they often succeeded in amassing great wealth, which, however, they were always careful to conceal.

Timid and taciturn, they displayed at least an outward obedience to their rulers, whom they inwardly despised. Naturally of a mild disposition, they have scarcely ever sought to recover their independence by force of arms; and even when driven to despair by the butcheries of 1894-96, they made a resolute stand against their oppressors only in the town and district of Zeitun. But here the conditions were somewhat peculiar, and Zeitun till about 1870 still enjoyed a measure of political independence. It occupies a strong position north by east of Aleppo on the slopes of the Cilician Taurus, in a district which was till lately spoken of as an isolated fragment of free Armenia.

Among the Armenians the women are little better off than among the Moslems, being practically the drudges of the household. But while the sensual Turk often becomes the slave of his handmaiden, the Armenian man of business still remains the head of the family. All menial work is performed by the wife, who waits on her husband at his meals, which she never shares with him. Although unveiled indoors, she is never seen by strangers, even at entertainments withdrawing to a room set apart for the purpose. This is usually raised a few feet above the level of the large central hall, and shut off by means of a wooden lattice, whence, without being seen, the women command a view of the banquet below.

The Armenian race, whose national name is Hai, Haik, or Haikan, formerly numbered some 8,000,000, but is now (1896) reduced to little over 2,300,000, distributed as under:—

Caucasia and Russia in Europe . .	850,000
Turkish Armenia	1,000,000
Persian Armenia	150,000
Turkey in Europe	250,000
Elsewhere	60,000
	2,310,000

While the settled and peaceful Armenians have been constantly losing ground, the nomad and lawless Kurds have long spread far beyond the limits of the region to which their progenitors, the Karduchi, seem to have been confined. In classic times Armenia included the whole of the Van district southwards to the 38th parallel, and Sachau[1] has determined the site of Tigranocerta, one of its many capitals, at the village of Tel Ermen, or the "Armenian Hill," a little to the south-west of Mardin, within the limits of Upper Mesopotamia. But all this region is now mainly occupied by the Kurds, some of whose tribes reach far southwards to the vilayet of Diarbekr, while others have encroached upon the Armenian district round about Ararat, and are found as far north as the 41st parallel, about the head-waters of the Kur. Others are scattered over parts of Asia Minor, North Syria, West Persia, and the highlands between Khorasan and the Turkoman country. Semi-independent Kurdish tribes still form a dreaded cordon round about the upland town of Van. Still more formidable is the Hormakli branch, occupying the snowy Bingol-dagh south of Erzerum, between the two forks of the Euphrates.

Although not always so chivalrous as they have been described by the few travellers who have occasionally visited them, they still possess the proud and frank address of independent highland tribes. Nor can it be

[1] *Ueber die Lage von Tigranocerta.* Berlin, 1881. This place was hitherto supposed to lie much farther north, at or near Diarbekr, on the Upper Tigris.

denied that many of their lawless propensities and notorious indifference to the rights of property must be attributed to the maladministration of their Turkish and Persian rulers. Under some of their semi-independent chiefs a general rising took place on the Turko-Persian frontier in 1880-81, during which the most deplorable excesses were committed, and the Urmia district wasted with fire and sword almost up to the very gates of Tabriz.

But the worst qualities of the race have been developed in the Nestorian district of Hakhiari, about the head-waters of the Great Zab. This tract stretches from the Persian border-land westwards to the Jebel-Judi, between the Zab and Tigris. But the Nestorians are also found in the extreme north-west of Persia, about Lake Urmia, and in small communities scattered over Upper Mesopotamia. They may almost be regarded as the last surviving erratic boulders of a formerly powerful Christian sect, at one time widely diffused over the vast region stretching from the Euphrates to Western China. But few travellers have succeeded in penetrating to their present home in the Kurdish highlands, a circumstance probably due as much to the inaccessible nature of this alpine region as to the savage character of its Christian Nestorian and Moslem Kurdish inhabitants. The heart of the country can be reached only by the Zab valley, on either side of which lie the dangerous haunts of the fierce Leihun tribe, the name of whose dreaded chief, Bedr Khan, is still remembered after nearly four generations by the surrounding Christian communities. Feuds and forays are still frequent enough, especially in the Tiyari district, where nestle the stone huts of the Nestorians under the shade of mighty walnut trees in the well-watered valleys, here everywhere encircled by snowy crests.

The Nestorians, who number altogether about 200,000, reject both the name "Nesturi" and the doctrine of Nestorius. The term is probably a corruption of "Nessarani," from Nazareth, commonly applied in the East to the Christians. But however this be, they call themselves *Kaldani,* or Chaldeans, and claim to be the survivors of the old Christian people of Mesopotamia, who were of Chaldean or Assyrian stock. Those of Mosul and others still speak a corrupt form of Assyrian, which they call modern Chaldean, and which is certainly an Aramaic dialect closely allied to Syriac.

Notwithstanding their lawless and predatory habits, the Kurds have developed a few simple industries. They breed a degenerate species of the Angora goat, from the hair of which are woven rugs and carpets which have found their way to the European market. They also produce coarse woollen, silken, and cotton stuffs, besides earthenware, leather work, hardware, and especially arms. The widely-scattered tribes of Kurdish stock number altogether probably about 3,000,000, of whom 1,250,000 live in Turkey.

While the Armenians are certainly intruders in their present domain, the Kurds appear to represent the aboriginal pre-Aryan race, which at a remote period extended almost continuously from the southern slopes of the Caucasus throughout the whole of the present Armenia, Lazistan, and Kurdistan. This race, the Allophylian of Herodotus, spoke an archaic form of the Caucasian language, which still survives amongst the Georgians, Mingrelians, Lazes, and most other South Caucasian peoples. That it was also the primitive speech of the Kurds may be inferred from the cuneiform inscriptions discovered in the Van district, which Professor A. H. Sayce has deciphered by means of the Georgian language. At present, however, the Kurds are of Aryan speech, having adopted a rude Iranian dialect at some unknown period, probably under Persian influences. But the ethnology of this region has still to be worked out, and, as is the case with so many nomad peoples, there appears to be clear evidence of two very distinct races, one of noble type, probably representing a conquering people of Aryan speech, the other more debased, representing the primitive Allophylian element of Georgian speech, now assimilated to the conquerors.

In Upper Mesopotamia the Kurdish and Arab nomads are intermingled. But farther south the bulk of the population beyond the walls of the towns consists of Bedouin tribes, whose subjection to the Porte is of a very loose character, and who may in some respects be regarded as the true masters of the land. Besides, the Ottoman Government is quite incapable of introducing a practical system of culture even into the arable tracts of Irak-Arabi. For many years past the governors, pressed by the Anazeh, Shammar, Montefik, Beni-Laam, and other powerful Bedouin tribes, have been able to do little more than keep things from tumbling to pieces. Here, as elsewhere, the history of the last fifty years has been nothing more than a constant feud, in which the advantage has frequently been on the side of the foes of Ottoman rule. Could the Arab tribes be induced to combine their forces, the Government would find it no easy matter to hold in check the powerful hordes, which often number from 10,000 to 20,000 mounted warriors. Along the Shat-el-Arab there is little more than an outward show of authority, which is to some extent rather endured than obeyed.

8. *Topography: Chief Towns—Erzerum, Van, Nineveh, Bagdad, Kerbela, Basra.*

The constant encroachments of Russia have left to Turkish Armenia no towns of any note, except Erzerum, capital of the vilayet of like name. Even this place is important rather for its strategical position, and as the *entrepôt* of the caravan trade between Persia and the Euxine, than for its size or population. It lies in a fertile district some 30 miles north of the Bingol-dagh, and 100 miles south-east of Trebizond on the great commercial highway leading from that town over the plateau to the Persian frontier. But, like most fortified towns, it is irregularly built, its narrow dirty streets, flanked by mean houses, being crowded together in the small space enclosed by its lofty walls. Here the Moslem largely prevails over the Christian element, although Erzerum is the metropolis of the Armenian Church in union with Rome, as Echmiadzin is of the Orthodox or Independent Armenian Christians. Its mosques are very numerous, and it is a chief halting-place for Persian pilgrims *en route* for Mecca.

A more interesting place is Van, which, though the chief town in East Kurdistan, is inhabited mostly by Armenians. It is picturesquely situated on the east side of the lake, above which rises an isolated rock crowned with its citadel. Van has suffered much both from earthquakes and from the turbulent Kurdish nomads of the surrounding district. Some time back these marauders took advantage of a fire in the bazaar to plunder the Armenian shops and houses, and since then its trade has greatly declined.

In Mesopotamia nearly all the large towns are situated, not on the Euphrates, but on the Tigris. Of

ERZERUM.

these the northernmost is Diarbekr, capital of a vilayet, and lying on the western head-stream of the Tigris in a debatable land, where the Kurdish, Armenian, Syrian, and Arab races meet on common ground. It is the seat of a Chaldean patriarch, and does a considerable trade by river and caravan.

In the Shakh district near Jezireh-ibn-Omar, on the left bank of the Upper Tigris, are some extensive ruins, visited in 1892 by Captain F. R. Maunsell, and by him identified with the ancient city of Bezabde. "For some two miles to the east were scattered extensive remains of masonry walls, with towers at intervals. The Kurds have many traditions about this place, and the guide said that seven distinct walls could be traced which used to encircle the town. At the top of the ascent we entered the village of Shakh through a gateway in a strong masonry wall, which evidently formed part of the line of fortifications on this side. Running north is a side valley lined with cliffs, down which comes a large stream. I ascended this for nearly 1½ mile, and found a number of chambers cut in the side of the cliff, which had apparently been used for dwellings. One of these measured 20 feet long, 15 feet broad, and 10 feet high, with a door and window, all cut in the hard limestone, and still in excellent preservation. Higher up the cliff to the east was a large arched opening leading into a chamber 60 feet long, 30 feet broad, and 30 feet high. This the Kurds called the council-chamber. Along the base of the cliff could be traced the line of an aqueduct, 3 feet wide and 2 feet deep, cut in the rock for nearly 1½ mile. In former times this must have been a place of considerable importance, and I think it may be taken as the site of the Roman city of Bezabde, capital of the Zabicene province. This has never been definitely fixed; but hitherto it has been

DIARBEKR.

supposed to be in the neighbourhood of Fenduk, higher up the Tigris" (Maunsell, *loc cit.*)

Lower down the river, and in the heart of the ancient Assyria, stands the town of Mosul, once noted for its fine cotton fabrics, which from this place are still known as *muslins*. Here the Tigris breaks through its southern mountain barrier, which forms a natural boundary between the Kurdistan highlands and the Mesopotamian plains. Although a poverty-stricken and decaying place, Mosul must always remain a hallowed spot in the eyes of the antiquarian, thanks to the neighbouring ruins of Nineveh, which have of late years been so successfully explored. Eastwards there stretches an extensive cultivated tract, limited on the north by the steep walls of an irregular limestone range, and extending beyond the horizon southwards to the confluence of the Great Zab, where the right bank of the main stream is already fringed by the Mesopotamian steppe. The small plateau thus circumscribed is broken only by low hills crowned with numerous hamlets, generally associated with those mysterious artificial mounds or barrows which are found scattered over Western Asia, the Balkan peninsula, Russia, and as far west as the Pomeranian and Mecklenburg marsh-lands. Close to these countless tumuli stand the villages of the agricultural Kurds, while the whitewashed tombs of Moslem "saints" are dotted over the boundless grassy plains. On this plateau the ruins of Nineveh cover a space about 18 miles in length along the river, and extend nearly 12 miles from its left bank, thus occupying an area of over 200 square miles, or rather more than that of London. The famous mound of Kuyunjik, where the excavations were begun in 1841, faces Mosul, while those of Nimrud occupy the angle formed by the confluence of the Tigris and Great Zab, 18 miles farther south. Here Layard discovered the colossal winged bulls, lions with human heads, and winged sphinxes placed as guardians at the entrances of the royal palaces, and now preserved in the British Museum. Since then all the European collections have been enriched by the artistic treasures brought to light in the intervening space. The arrow-headed writings of the brick libraries, which are now

SUPPOSED TOMB OF JONAH, NINEVEH.

deciphered, show that Nineveh was the centre of an Assyrian or Semitic civilisation of great antiquity, but still modern compared with that of the Akkads of Babylonia, whose ethnical affinities have not yet been determined.

Below the ruins of Nimrud the Tigris is joined on its left bank by the Great Zab, in whose lower valley is Arbil, which preserves the name and memory of Alexander's decisive victory over Darius (331 B.C.); higher up stands the frowning stronghold of Rowandiz in the very heart of the Kurdistan highlands. From Rowandiz the southern track over the hills leads through Kerkuk, the chief place in the Little Zab basin, to Sulaimanieh, on a head-stream of the Diala, which, though of recent origin (1778), has acquired great importance, thanks to its strong strategical position near the Persian frontier. Sulaimanieh, which is also a flourishing centre of trade for the surrounding populations, is a typical Kurdish town of some 2500 houses, "nearly all single-storied with flat roofs. The narrow, winding streets of the bazaar, with the stalls on either hand, are shaded from the sun by an arrangement of branches and leaves stretched across overhead. The articles displayed for sale reflect the tastes of these warlike Kurdish mountaineers, always fond of something bright and showy in their accoutrements, and a good weapon by their side. The principal manufactures are saddles and horse furniture, shoes and leather-work generally, the leather being cleverly dyed in various brilliant colours. A curious assortment of flint-locks and guns, swords, knives, and daggers of all shapes, round shields of bullock hide, belts with a row of pouches for powder and bullets, and felt saddle-cloths embroidered with coloured worsted, made a very interesting display. A fair quantity of Manchester piece-goods, cotton kerchiefs, etc., could be seen; but to a Kurd, a Martini rifle or a good horse, looted from some Arab on the plains, are of more value than very many yards of cloth. Of eatables there was the universal Kurdish drink of curdled milk called *yaurt*, cheese, raisins, and several kinds of dried fruits, chiefly plums and apricots" (Maunsell, *ib.*).

Nearly midway between Mosul and the Persian Gulf is situated the famous city of Bagdad, in what was once one of the richest and most productive regions in the world. This city was formerly the most brilliant capital of the Moslem world. Arriving with the Persian caravan from Mandali, we enter the city by the gate

STREET IN BAGDAD.

of Sheikh Omer. The archway has long since fallen in, and the soft-hoofed camels struggle painfully over the breaches formed by time in the dilapidated bastions. In the first purlieus we meet with nothing but piles of rubbish, stagnant waters, and cesspools, while a pack of pariah dogs is scattered in all directions by the shrill voice of the leader of the caravan. Over the city swoops the vulture of the wilderness, and at its very gates flocks of carrion crows settle unmolested on the putrid carcasses strewn about.

East of the river is the district of New Bagdad, containing the Government offices and the chief commercial and public buildings. On the right bank is the old town, enclosed by an extensive tract of orange and date groves. Towards the desert this quarter is protected by a wall with two gates, leaving the part facing the river unenclosed. No other large city of Asiatic Turkey is influenced by the desert to the same extent as is Bagdad; no other stands in such direct contact with Central Arabia. The purest Arabic dialect is here current, and here still prevail the Bedouin manners in the social life of the people, and especially in their intercourse with the non-Muhammadan element. Yet, in spite of their religious fanaticism, their general bearing is preferable to that of most other Asiatic Mussulmans, because of the very sincerity of their belief, combined with the natural dignity and frankness peculiar to the Moslem Arab. The population is of a very motley character, being composed, according to some authorities, of 150,000 Muhammadans of various races, some 18,000 Jews, 2000 "Nestorians," nearly the same number of Latin Christians, several hundred Armenians and Syrians, and scarcely more than 20 Europeans. The collective population was estimated in 1894 at 180,000. Bagdad, though shorn of the greatness for which it was once famed, still possesses importance commercially and politically, which it owes to its situation on the great water highway in a country nearly destitute of land routes.

Up to this point the Tigris is navigable throughout

the year for steamers of considerable size, while from the north there daily arrive the so-called "Kelleks," a sort of craft made of inflated goat-skins, boarded over. On these are floated down quantities of lumber from the Kurdistan uplands, the boatmen returning with the empty skins in company with the caravans. But still more characteristic of Bagdad is the "quffe," or coracle, consisting of a round hull 6 to 8 feet broad, with sides curved inwards, constructed mostly of strong reeds and well pitched on the outside. When the bridge of boats becomes broken, the communication is kept open by means of these frail craft. The type is of great antiquity, being represented on the old Assyrian reliefs, and described by Herodotus, who states that these boats were built in Armenia.

In the untidy soldiers slouching about the streets, in the evil-smelling bazaars and ruined mosques, in the rotten bridge of boats, and the mean dusty post-office, one reads at every step that the curse of Turkish inanition lies as heavily or even more heavily on this once prosperous and magnificent city than even on many other Turkish towns. Pestilences have devoured it; the very Euphrates, to which the country owes its history and former opulence, has helped in the work by bursting its banks, and by rendering the country to the west a marsh, and so bringing fever and ill-health to the ill-fated city. "The glory of the city of the Caliphs has indeed departed" (H. S. Cooper, p. 278).

West of the Euphrates, though at no great distance from Bagdad, lies the village of Kerbela, a spot held in great veneration by the Shiah or Persian Muhammadans. Here is the tomb of Hosein, the Prophet's grandson, and son of Ali, whom the Shiahs regard as his true successor in the Caliphate. They believe that by living or dying here they have nothing to fear in

KERBELA.

the next world, being thereby rendered irresponsible for their conduct in life. So strong is this belief, that many leave instructions in their wills to have their remains brought from great distances and buried in this hallowed place Hence many thousands of bodies are yearly brought from Persia and elsewhere and laid in the ground at Kerbela. The place is also visited by numerous caravans of Shiah pilgrims, all who have performed this pilgrimage henceforth bearing the proud title of "Kerbelai."

A little south of Kerbela, and on the Euphrates, stands the town of Hillah, opposite which are the ruins of Babylon, scattered over a wide tract of country. "Hillah, which may be called the present representative of ancient Babylon, situated as it almost certainly is within the ancient boundaries, and built of bricks gathered from the ruins, is a place of considerable importance, situated on both banks of the Euphrates, which is here less than 200 yards wide, and of a gentle current. The town is prettily set in palm-groves, and is surrounded by a mean brick or mud wall. There are bazaars on both sides of the river, but those on the right bank appear more extensive, and that part of the town is indeed the principal. The population seems to be increasing, as Layard put it down at 8000 or 9000, and Gratton Geary in 1878 states that it was then estimated at 20,000. The population is chiefly Arab, but there is a considerable fraternity of Jews. It is indeed an interesting fact that ever since the Babylonian captivity this remarkable race has dwelt in considerable numbers in the vicinity of ancient Babylon" (H. S. Cooper, p. 334).

Below Hillah the two chief riverside ports are Kornah (Kurnah) at the Tigris-Euphrates confluence, and Basra (Basora) on the Shat-el-Arab, 50 miles in a straight line below the junction. According to a local tradition Kornah occupies the site of the "Garden of Eden"; but despite its position at the converging point of the two great rivers, it is a mere village situated on a narrow point of land covered with luxuriant date-palm groves. Between Kornah and Basra the Shat-el-Arab is a copious stream, from 250 to 300 yards wide, with a depth of from 60 to 90 feet.

The present port of Basra is a modern place lying on the right bank of the river about 2 miles from the famous city of Basra, which was founded by Caliph Omar in the seventh century. But an earlier town existed on the site of old Basra or Zobeir, which stood on a canal supposed to be the old Pallacopas mouth of the Euphrates. "From time immemorial there must of necessity have been a port here, where all the trade of the East would make its way to Babylon and Seleucia in the earlier days, and to Ctesiphon, Kufa, and Bagdad in later times. It was this trade which raised the mediæval Balsora to such importance, and, as Layard has remarked, it was the discovery of the Cape of Good Hope route which ruined its trade" (H. S. Cooper, p. 416). But its position is so favourable for traffic that, despite the surrounding marshy and malarious district, Basra still continues to be an important emporium of Asiatic Turkey for eastern produce. Ships of 500 tons burden reach this point, and since the establishment of the English line of steamers, affording regular communication with Bagdad and the Gulf, its prosperity has considerably revived. The distance from Bagdad to the Gulf is 570 miles by water, and the English steamers generally take about four days on the up trip and two days on the return journey between Basra and Bagdad.

9. *Highways of Communication.*

In the Mesopotamian basin there are scarcely any roads properly so called. The two great arteries of the Tigris and Euphrates still continue to be the chief highways of communication. But the desert is crossed in various directions by caravan tracks, and in the extreme north there is one good road, the already-mentioned route connecting Trebizond through Erzerum and Bayazid with Persia. Erzerum is also connected eastwards by a military road with Kars, and south-eastwards through Yaṇgali with Van. From Van an important route runs southwards through Mosul, and down the Tigris valley to Bagdad, and another west-wards through Mush and Kharput to Anatolia. Of the caravan routes across the desert, by far the most important is that which strikes the Euphrates at Bir (Birejik), here bifurcating through Urfa northwards to Diarbekr, south-eastwards down the Mesopotamian low-lands to Bagdad. Another route runs from Diarbekr along the left bank of the Tigris through Finduk to Mosul, here crossing to the west bank, which it follows to Bagdad. Mosul is also reached from Diarbekr by an alternative route *via* Mardin and Nisibin (Nisibis). But the most direct route between the Mediterranean and the Persian Gulf runs from Alexandretta through Aleppo to Kalaat-Jabar on the Euphrates, thence following the right bank of that river *via* Anah and Hit to Kalat-Ambar. Here it crosses over and pursues a straight course south-eastwards to Bagdad. This route is not essentially different from the line which has been examined and partly surveyed for the project of the Euphrates Valley Railway. The line is proposed to run from Alexandretta to Bagdad, and thence south-east-wards to Basra. At Bagdad it would form a junction with the great South-Asiatic trunk-line, which, starting from Scutari, is intended to connect the Bosphorus with the Persian Gulf through Anatolia and Mesopotamia. There has been speculation also regarding the possibility of carrying a railway from Mesopotamia on to India, through Persia and Afghanistan or Baluchistan.[1] Meantime there are no railways in the Mesopotamian basin, nor is it probable that the projected trunk-line will be undertaken at present.

[1] See the publications on the Euphrates Valley Railway and *India and Her Neighbours*, by Mr. W. P. Andrew, chairman of the Sind, Panjab, and Delhi Railway Company.

CHAPTER VI

SYRIA AND PALESTINE

1. *Boundaries—Extent—Area.*

THE Mesopotamian plains are separated by the great Syrian desert from the Mediterranean coast region which here stretches nearly in a straight line from the Sinai Peninsula northwards to Anatolia. The desert forms a chalk and limestone tableland gradually rising to an altitude of over 2000 feet above the sea, stretching away southwards into the peninsula of Arabia, but on the west sinking abruptly down to the long, deep, and narrow depression of El-Ghor, which forms the eastern limit of the southern section of the coast region known as Palestine or the Holy Land. Farther north the desert merges imperceptibly in the plains of Damascus and Aleppo; consequently Syria, or the northern section of this region, presents no natural well-defined limits towards the east. Elsewhere the boundaries of the whole land are sufficiently clear—the sea on the west, the Amanus (eastern Taurus) on the north, the Euphrates on the north-east, the little river El-Arish on the south-west, Arabia Petræa on the south. This gives a total length north and south of about 430 miles, with a mean breadth of 100, narrowing in the south to 50, expanding northwards to 150. The area is officially given at 115,000 square miles, of which not more than 12,000 are comprised in Palestine. The distinction between these terms has long ceased to be recognised in the East, but is still retained in the West by reason of the religious associations and historical reminiscences with which the southern division is inseparably associated. Palestine is cut off by the Lower Orontes and Mount Hermon from Syria proper, measuring from this point to the southern end of the Dead Sea about 160 miles, with an average breadth of 70.

2. *Relief of the Land: Lebanon and Anti-Lebanon.*

While this strip of coast land serves on the one hand to cut off the desert from the sea, it forms on the other a connecting link between the Anatolian and Arabian tablelands. It is everywhere too mountainous to allow the plateau formation to be clearly developed. But the mass of the land has a mean elevation of probably 3000 feet, above which rise two parallel mountain ranges, clearly marked in the centre, less distinctly defined in the north, and southwards breaking into an irregular upland region, where the hills and low ridges still form two systems west and east of the El-Ghor depression, round which they meet and become interlaced in the Arabian uplands.

The coast line, running nearly due north and south, is varied by but few and unimportant headlands and inlets, the section south of Beyrut forming almost a straight line, broken only by the bold promontory of Mount Carmel nearly midway between Beyrut and Jaffa (Joppa). Throughout its entire length the coast is followed by the outer chain of mountains, leaving but a narrow strip of lowlands between their base and the sea. In Palestine this range is little more than the escarpment of the broad and hilly plateau of Judæa, beyond which the plain of Sharon stretches seawards from Cæsarea southwards to Gaza. Beyond Carmel the hills still recede sufficiently to make room for the less extensive plain of Acre, after which they continue to rise in height and approach constantly nearer to the coast. North of the valley of the Lower Leontes (Nahr-el-Litany) they culminate in the two parallel chains of the Lebanon and Anti-Lebanon, which form the great physical feature of this region. The Anti-Lebanon, or inner range, falls gradually northwards down to the plains of Upper Mesopotamia. But the Lebanon, or Jebel-el-Gharbi—that is, "Western Range"—is continued by the less elevated Jebel-Nusarieh as far as the plain of Antiochia, about the 36th parallel. North of this plain the Jebel-Nusarieh is continued by the Giaour-dagh and Akma-dagh to the Taurus above the Gulf of Alexandretta.

The Lebanon or central coast range runs for about 90 miles south-west, at some points approaching to within 8 or 10 miles of the Mediterranean. Seen from a vessel out at sea it presents the appearance of bare, rocky walls, here and there surmounted by a few snow-clad peaks, of which the highest are the Dhor-el-Khodih (10,200 feet), and the Jebel-Makmel (10,000 feet). From these the range takes the name of the Lebanon or "White Mountains," a name which was already current in the time of Moses (Deut. i. 7), and which has never since dropped out of history. Notwithstanding its rugged aspect seawards, the Lebanon, which is properly limited southwards by the valley of the Lower Leontes, really contains many fertile slopes and valleys, well cultivated and thickly peopled.

Eastwards it is separated by the still more fertile valley of the Bekaa (Cœle-Syria) from the Anti-Lebanon

DISTANT VIEW OF MOUNT HERMON.

or inner range, whose naked rocky walls present far more varied outlines than the coast range. Although with a lower mean elevation, the southern extremity of the Anti-Lebanon rises in the Jebel-es-Sheikh (Mount Hermon) to an altitude of 9,200 feet, the culminating point of the Syrian highlands, some 30 miles south-west of Damascus. Beyond this point it throws off two branches towards the south-west and south-east, thus enclosing the upper sources of the Jordan, and merging eastwards in the rocky uplands of Gilead and Moab.

An expedition was made to North Syria in 1892 by the Rev. Dr. Post and Professor West, for the purpose of extending to the Lebanon region the accurate map of Palestine published by the Palestine Exploration Fund. Excluding the south-eastern flanks of Hermon, both the Lebanon and Anti-Lebanon were found to be of limestone formation. The effects of erosion by water are far more distinct on the Lebanon chain, a fact due to the heavier rainfall to which it is exposed; the romantic wadies which furrow its western slopes find no counterpart in the tame watercourses of the Anti-Lebanon. Lebanon is a single ridge, above which rises a series of commanding peaks, while the Anti-Lebanon is of much more complicated structure. It is rooted in the great ridge of Hermon in the south, from which five ridges diverge northwards like the ribs of a fan. The space included between these ridges is a plateau ranging from 4000 to 5500 feet above sea-level, while the intervening ridge ranges from 7000 to 8557 feet in absolute height. The conical summit of Halaim, at the northern extremity of the second ridge going westwards, is characterised by a more distinctive flora than the rest of the system.

3. *Hydrography: Jordan—Dead Sea.*

Syria and Palestine are still sometimes represented as being intersected in their entire length by a deep depression called in the north El-Bekaa, in the south El-Ghor. But more accurate recent surveys have shown that this view is entirely erroneous, and that El-Bekaa and El-Ghor are totally distinct formations. Although the term El-Bekaa means a "deep plain," the tract in question, answering to the ancient Cœle-Syria—that is, "Hollow Syria"—is only "deep" or "hollow" relatively to the two lofty ranges of the Lebanon and Anti-Lebanon, between which it lies. In itself the Bekaa is not a depression at all, but a plateau at an average elevation of no less than 2000 feet above the sea. On the other hand, the Ghor is not only a true depression, but the very deepest in the earth's crust, falling in the basin of the Dead Sea to a depth of 1292 feet below the Mediterranean, or over 4000 feet below the Bekaa. Nevertheless, these two features of the country are still to some extent connected by its hydrography, which they largely regulate. At the famous ruins of Baalbek, under the 34th parallel and about midway between Antiochia and the Dead Sea, the Bekaa attains its greatest elevation of about 3000 feet above the sea, and here is consequently the chief water-parting of the whole region. Round about Baalbek rise the four main streams—Jordan, Leontes, Orontes, and Abana—which flow in four opposite directions, south to the Dead Sea, south-west and north-west to the Mediterranean, east to the Bahr-el-Ateibeh beyond Damascus. At Lake Merom the Jordan reaches the trough of the Ghor, which it henceforth follows throughout its entire course to the Dead Sea. The Leontes and Orontes traverse the southern and northern sections of the Bekaa respectively, while the Abana pierces through the deep gorges of

Anti-Lebanon down to the smiling plains of Damascus. Of the four rivers, the Jordan and Orontes will here claim a more detailed description.

The Orontes (Nahr-el-Asy) rises with two head-streams on the western slopes of Anti-Lebanon, some 10 miles north of Baalbek, flowing thence northwards to the neighbourhood of Homs (Emessa), where it expands into the lakelet of Kades, 6 miles long by 2 broad. Beyond this point it continues its northerly course by Hamah (Epiphania), and through narrow rocky gorges for about 50 miles to the northern extremity of the Nusarieh range, where it trends suddenly westwards and south-westwards through the plains of Antiochia to the coast, which it reaches near Suedia (Seleucia), after a winding course of about 150 miles. At its northern bend it receives on its right bank the Kara-su, its only important tributary, flowing from the Lake of Antioch 4 miles off.

The Jordan (Sheriat-el-Kebir) is formed by three small head-streams, the farthest of which rises between Baalbek and Mount Hermon. The united stream falls thence over seven low terraces southwards to the muddy little Lake Merom (El-Huleh), which lies at the head of the Ghor in a fertile basin, fringed on the north by an almost impenetrable reedy swamp, and enclosed on the south by a spacious elevated plain. This plain sinks southwards sufficiently to afford an outlet for the Jordan, which now pursues its impetuous course through the deep rocky fissure of the Ghor for 10 miles to the Sea of Galilee (Lake Gennesareth or Tiberias). The fall in this short space is nearly 700 feet, and at this point the trough of the Jordan has already descended to 682 feet below the Mediterranean.

Lake Tiberias is a sheet of clear water, now as of old abounding in fish, and encircled on all sides by lofty mountain walls and hills, which in spring are covered with a soft grassy carpet, but which become parched up during the dry summer months. Although often described as a deep basin with depths of over 800 feet, Tiberias is comparatively speaking a shallow depression. The soundings taken by Lieutenant Molyneux in 1847 nowhere exceeded 156 feet, and the results of this survey are

TIBERIAS.

confirmed by the researches of M. Barrois in 1890 (*Comptes Rendus*, Paris Geo. Soc., 1893). West of the lake stretches the fertile plain of Gennesareth (El-Ghuweir, or "the Little Ghor"); but the Ghor itself continues still to fall for about 200 miles between the Giliad hills and the escarpment of the plateaux of Galilee and Samaria, southwards to the Dead Sea. The total fall in this space amounts to 610 feet, so that at its lowest level the Jordan has descended to a depth of 1292 feet below the Mediterranean through a chasm, which is by far the longest and deepest on the surface of the earth. All further extension of the river southwards is thus rendered impossible, although it will be seen farther on

THE DEAD SEA.

that the Ghor itself continues its southerly course into the Arabian peninsula.

The Dead Sea (Asphaltites Lake, or Bahr-Lut, that is, " Sea of Lot ") is enclosed within a basin formed by naked limestone cliffs, 2500 feet high on its east and 1500 on its west side. It is nearly 50 miles long north and south, with an average width of 8 miles and a mean depth of 1300 feet, but shoaling southwards to the ford between the Lisan promontory and the west shore, which is scarcely more than 3 feet deep. M'Coan tells us that its water is "nearly as clear and blue as that of the Mediterranean, but salt, slimy, and fœtid beyond description; its taste like a mixture of brine and rancid oil; and its buoyancy so great that, as I can personally vouch, the human body will not sink in it, strive as the bather may. Bitumen bubbles up plentifully from the bottom, and with the sulphur, nitre, and rock-salt that abound along most of the shore-line, sufficiently explains the density and the nauseous taste and smell of the water. The old traveller's tale that the water itself and the evaporation from it are alike fatal to animal life is less than half true. The 26 per cent of saline matter precludes indeed the existence of fish; but though its exhalations under a burning sun are thick and fever-inducing, they are in no worse degree poisonous, and birds fly along its shores and over its surface as lively as in the mountains on either side" (i. 103).

At the southern extremity of the lake lies the lofty rock-salt ridge of the Jebel-Usdum, beyond which extends the desolate salt marsh of Es-Sebkah, fed by the Wady-es-Safieh flowing from the Wady-el-Arabah. This now dried-up watercourse forms a southern continuation of the Ghor depression. But it does not extend, as was long supposed, to the Gulf of Akaba at the head of the Red Sea, but only to a water-parting near the Bedouin camping-ground of Arabah, some 500 or 600 feet above the Mediterranean. The Wady-el-Arabah is regarded by Edward Hull as a line of main fault or fracture continuous with the Jordan Valley, which, however, never reached the Gulf of Akabah. Terraces of marl, silt, etc., occurring at 100 feet above the Mediterranean, show that the level of the Dead Sea itself was formerly 1400 feet higher than now, filling a depression 200 miles long north and south (*Mount Seir, Sinai, and Western Palestine*, 1885).

4. *Natural and Political Divisions: Giliad and Moab—Land of Bashan—Trachonitis—Ala District—Plateau of Aleppo—Canaan—The Plains of Sharon—Galilee—Samaria—Judæa.*

Till recently the uplands of Giliad and Moab, whose position beyond the Jordan is indicated by their ancient name of Peræa, were a veritable *terra incognita*. But notwithstanding the lawlessness of their Bedouin inhabitants, their numerous cromlechs, ruins, and other interesting monuments have of late years tempted several European explorers to penetrate into its most secluded retreats.[1]

Seen from the western shores of the Dead Sea, Moab looks like a mountain range, but is in reality merely the verge of a rocky upland plateau about 2500 feet above the sea, or 4000 feet above the level of the lake. This plateau, which is furrowed by deep valleys, stretches eastwards for about 25 miles to a bare limestone range, conventionally regarded as the limit of the land towards Arabia. Moab was formerly a well-peopled region. But the eye everywhere lights on ruined villages. Even now, badly cultivated as it is, the land is rich and fertile, and large tracts of a fine red sandy loam, needing no manure, still produce heavy wheat crops. All the streams flow westwards through deep rocky beds to the Dead Sea.

The Moabite country is continued northwards by the

[1] The Palestine Exploration Fund, after completing the survey of Palestine proper, extended its labours to the region beyond the Jordan, where over 1000 square miles have already been surveyed, and many hundred ruins examined.

volcanic plateau of the Land of Bashan, which attains an elevation of from 4000 to 5600 feet eastwards in the Hauran uplands. Including the three districts of the Leja (Western Trachonitis), Nukrah, and El-Jebel, this region runs 60 miles north and south, and nearly 40 east and west. The Leja is mostly a stony plain; but the Nukrah is a rich tract, containing many small towns and villages, unfortunately exposed to the frequent raids of the Anazeh Bedouins, while the Jebel, or "Highlands," marking the extreme eastern limits of Palestine towards the desert, abound in ruined towns still partly peopled by the Druses.

Between the Hauran and the Oasis of Damascus there stretches a broad expanse of volcanic "tell," covered with recent tertiary and pliocene craters, which, although seemingly scattered about in wild confusion, really lie in three tolerably parallel lines, inclining slightly north and south. This is the Eastern Trachonitis (Tulul-el-Safa), towards the northern verge of which stand the stupendous ruins of Palmyra (Tadmor), under the 35th parallel, in 38° E. long. and 120 miles north-east of Damascus. The ruins cover a space of about 3 square miles, and conspicuous amongst them are the sixty columns still standing of the magnificent Temple of the Sun. This "City of the Palms," as both names mean, dates back to the time of Solomon, and is for ever associated with the sad fate of the hapless Queen Zenobia.

Still more interesting to archaeologists is the Ala region between the vilayets of Damascus and Aleppo. It forms an extensive basaltic upland tract, stretching for many miles east of the Orontes valley. Here are the ruins of many cities, which have evidently been rebuilt over and over again, besides numbers of remarkable tombs and fortified camping-grounds. Few Europeans besides the English travellers Burton and Drake have visited this

PALMYRA.

extraordinary land, within whose limits, though figuring on the maps as a blank space or portion of the Syrian Desert, the Arabs have indicated the sites of no less than 365 ruined cities.

In the extreme north the extensive inland plateaux of Aleppo, Umk, and Aintab occupy all the space between the great bend of the Euphrates and the coast range. Although intersected by several low ridges, they contain many fine and fertile level tracts, thickly peopled by Turkoman and Armenian agriculturists. This region marks the extreme limits of both of these races towards the south-west. In the west of the Umk plateau lies the Bahr-el-Abiad, or Lake of Antioch, a fine sheet of water 8 miles by 6, formed by the junction of several steppe streams, and draining to the Orontes.

Returning southwards and re-crossing the Jordan from Moab, we enter the small territory of Canaan, the "Land of Promise," or Palestine proper, ever venerable as the scene of the history of the "Chosen People," and as the Holy Land of Christianity. This region consists of an irregular hilly plateau falling west of the Jordan down to the level coast lands. This narrow low-lying tract, comprising the ancient land of the Philistines, was at one time studded with large towns and thickly inhabited by a restless warlike population. But at present the only noteworthy places are Gaza, Jaffa, and Ascalon, along a coast stretching in an almost unbroken monotonous line northwards to Cape Carmel. This headland, enclosing the Bay of Acre on the south, forms the northern extremity of the Jebel-Mar-Elias (1800 feet), which runs through the old lands of Manasseh and Asher north-westwards between the plains of Sharon and Acre. The rich plain of Sharon, of which only a small part is now under cultivation, stretches some 15 or 20 miles inland, and skirts the coast from above Cæsarea to Gaza, beyond which its loamy soil gradually mingles with the sands of Arabia Petræa.

The tablelands rising immediately behind Sharon comprise in the north the old land of Galilee lying mainly between the Leontes and Carmel, Samaria in the centre, and Judæa in the south. The regions which fall abruptly eastwards to the El-Ghor depression are generally de-

THE LAKE OF GALILEE.

scribed as of jurassic formation. But Dr. Oscar Fraas has lately shown that they consist rather of chalk deposits with hippurites and other fossil shells. The same formation prevails throughout the land east of Jordan, the Sinai Peninsula north of the zone of primitive rocks, and the Nile valley far beyond Karnak.

Galilee, the northern division of Palestine, is a hilly district from 2000 to 3000 feet above the sea, sinking eastwards abruptly to the Jordan and Lake Gennesareth, and southwards to the rich alluvial plain of Esdraelon (Jezril). Here are many pleasant fertile valleys, varied with bold mountains and splendid woodlands stretching northwards to Mount Hermon.

South of the plain of Esdraelon the plateau again rises to the central district of Samaria, where are also many well-watered and cultivated valleys, producing heavy crops and fruits in abundance. Here the prominent landmarks are the rocky Mounts Ebal (3076 feet) and Gerizim (2849 feet), rising close together about 34 miles due north of Jerusalem.

The southern district of Judaea is traversed by a somewhat ill-defined ridge of bare treeless hills, known collectively as the Mountains of Judah. These hills form a small water-parting between Kedron and other brooks flowing east to the Dead Sea, west to the Mediterranean. But although rich beyond any other land in hallowed memories and stirring events, Judaea is on the whole a somewhat bleak, arid country, far less productive than any other part of Palestine.

5. *Climate.*

In this region climate depends far less on latitude than on the relief of the land. Even in small districts the greatest diversity prevails, according to the varying altitudes. Thus on the exposed upland plateau beyond the Jordan the glass falls at night to 22° F., or 10° below freezing-point, when it stands at 76° F. on the shores of the neighbouring Dead Sea. In general a cold temperature prevails on the higher slopes of the Lebanon and other ranges rising above the snow-line. Here the winters, almost as severe as on the southern shores of the Baltic, are followed by genial springs, summers scarcely warmer than in England, and fresh autumns. Along the west coast and the Jordan valley the summer heat is very oppressive, the winter mild, and rain falls in both seasons. Malaria is prevalent at certain marshy spots along the coast, especially near Tripoli and Alexandretta. Central and South Palestine and the vilayet of Damascus enjoy a warm, dry climate, with mild winters and slight rainfall. Here the hot desert winds prevail in summer, drying up the rivulets, and reducing the land to an arid waste. At Jerusalem the mercury rises to 79° at sunset in midsummer, sinking to 49° in January—hottest and coldest means.

A remarkable feature of the Bekaa is the violent, almost tornado-like wind which prevails, especially in the central districts, where it blows regularly every day for some hours in the afternoon.

6. *Flora and Fauna: The Cedars of Lebanon.*

As a rule, vegetation is much more varied and luxuriant in the north than in the south. In Syria the slopes and many of the coasts are often densely wooded, whereas in Judæa "the hill vegetation is everywhere scanty, and the general aspect of the country east and south of Sharon rugged, desolate, and barren" (M'Coan).

The turpentine tree and the *ballud,* the species of oak which produces the gall-nut of commerce, are common features even beyond the Jordan. The vine, olive, orange, and other Southern fruits, besides the mulberry, cereals, and dates of splendid quality, abound in Sharon, the Damascus district, the sheltered Lebanon valleys, and generally throughout Galilee and Samaria. The tobacco especially of the Latakia district facing Cyprus is noted far and wide for its delicate flavour, and the rose of Sharon still remains more than a reminiscence.

On the other hand, the historic cedars of Lebanon have almost become a thing of the past. At a solitary spot a few miles below Tripoli, and not far from Cannobin, seat of the Maronite patriarch, there still survives all that remains of what must be regarded as undoubtedly the most venerable tree in the whole world—the tree to which the Psalmist compares the vine "brought out of Egypt," the boughs of which "were like the goodly cedars" (Ps. lxxx. 10). In 1875 Fraas counted altogether 377 plants of all sizes, but there remain five only of the gigantic trees, whose trunks measure upwards of 30 feet round. Burton and Drake, who visited the place some years ago, were also greatly disappointed at the appearance of these "Christmas trees on a large scale," which from a distance looked like a clump of enclosed pines, and on a closer inspection were found to consist of a few decayed old stems.[1]

Of wild animals the chief are the Syrian bear, the hyæna, jackal, boar, panther, and ounce. There is a small but hardy breed of horses, but the camel and mule are also employed as beasts of burden, especially for the transit trade between the coast and the interior. Fat-tailed sheep are numerous, but the Angora breed soon degenerates.

7. *Inhabitants: The Syrian Christians — Missionary Work — The Maronites, Druses, Nusarieh, and Fellahin.*

With the exception of a few wandering Kurdish and Turkoman tribes in the extreme north, and of the Turkish officials in the large towns, all the inhabitants of Syria and Palestine belong to the Semitic stock. The modern Syrian, who represents the Aramæan branch of that stock,

[1] *Unexplored Syria.* London, 1872.

is the result of a happy blending of races, in which the Semitic element largely predominates. The natural endowments of the people are displayed in the best light by the Christian section of the community. The Syrian Christians are a highly intelligent people, with a rare capacity for adopting European ideas. The admixture of Greek and Arab blood has evidently in no way impaired the good qualities of their Phœnician and Aramæan forefathers. And Phœnicians, the inhabitants of the coast districts, still remain in their enterprising spirit, commercial skill, and love of travel. In Marseilles, Liverpool, and Manchester there are several Syrian merchants, furthering the interests of their native land, and extending their trading relations even to Scandinavia and North America. The prosperous condition of the Beyrut Christians is the natural result of their intelligent industry. Here are found none of those proletariate classes, who cause so much anxiety in the large European cities. Everybody is either a merchant or else engaged in some settled industry, while still preserving the freshness of the simple patriarchal family life. The women are comely, and, although without much book-learning, good mothers, thrifty housewives, and devotedly attached to their husbands. They associate little with the outer world, passing their days in happy seclusion in the midst of their families. Their reading is limited to their Arabic prayer-books, and the harmless *Beyrut Review*, while novel-reading and piano-strumming are accomplishments which are still rare, except, perhaps, where the superficial French culture has been introduced.

There is no lack of girls' schools, though instruction is here limited mainly to the study of English or French. The "Sisters of Charity," however, have an excellent training school, where woman's work is taught, and where native teachers are trained. The rival houses of the

"Sisters of Nazareth" and of "Prussian Deaconesses" are also highly spoken of. The American missionaries are also doing good work, aiming especially at practical objects. The native Protestant community already numbers several hundred families in Beyrut, where the money flowing in from Great Britain and from beyond the Atlantic has enabled them to build a handsome church, besides supporting several schools and a printing establishment.

Even in the Lebanon, Protestant views are making rapid progress, notwithstanding the existence of some good Roman Catholic institutions. Of these, the most noteworthy is the college of the Melchite Greeks, which is admirably conducted, and already numbers several hundred pupils. The Jesuit College at Ghazir is also efficiently managed, and this is also true of the Lazarist College at Antura. Both are exclusively French establishments, and as most of the young men of Beyrut have been educated at one or other of them the French language has become very general amongst the upper classes. It has already almost entirely superseded Italian, which prevailed in the last generation.

In the year 1862 the district of the Lebanon was detached from the vilayet of Damascus, and formed into a separate pashalik, administered by a Christian governor under the control of the European legations. But the limits of this new government depart considerably from the natural limits of Mount Lebanon, having been laid down solely in accordance with the religious interests of the people. Hence districts where the majority were Muhammadans continue to form part of the Syrian province, while all the Christian communities were included in that of the Lebanon. But Tripoli, Beyrut, and Saida (Sidon), the three most important seaports, were also attached to Syria, so that the boundaries of the modern district of the Lebanon are extremely irregular. It comprises an area of rather over 2000 square miles, with a population of 245,000, mostly Christians.

The Lebanon Christians call themselves Maronites, from the national saint, Maron, a famous recluse supposed to have flourished about the year 400. They are the direct descendants of the orthodox community as constituted in the seventh century, and although united with Rome since the time of the first Crusades, they still retain many local privileges and peculiarities, such as a married clergy, administration of the sacrament under both species, celebration of mass in the Syriac language, but otherwise according to the Latin rite, together with their own hagiology and national feasts. They are devotedly attached to their religion, and are in other respects a brave and energetic people. Their villages, and 200 monasteries, are perched like eyries on the spurs and slopes of the main range, and are often surrounded by corn-fields waving over artificial terraces, so disposed as to prevent the rich loam from being washed away.

Unfriendly neighbours of the Maronites are the mysterious Druses, settled partly in the Acre district south of the Lebanon, partly in the remote Hauran uplands, on the verge of the desert. The origin and peculiar tenets of this half-pagan people have not yet been satisfactorily explained. Though apparently having some affinity in faith to the Mussulmans, they jealously preserve a sort of secret doctrine, said to have been handed down from the ancient Egyptians. In fact, however, they make no outward profession of any religion, although believing in a God. Physically they are a fine race, brave, with something of poetry and heroism, but also fierce, cruel, and treacherous.

Druses and Maronites lived for ages amicably together until bitter feuds sprang up between them during the present century. Sudden raids were followed by sanguinary reprisals, and the restoration of order was frequently attended with much bloodshed. Since 1860 the Druses began to withdraw from the Lebanon and settle in the Hauran uplands, and in the Lebanon district they now (1896) number scarcely more than 35,000. But in Hauran they find themselves opposed to new enemies, the Arab Bedouins and the Circassians who settled in the province of Dámascus after their expulsion from the Caucasus by the Russians. In 1895 the Druses flew to arms, driven to desperation by the raids of these tribes and the oppression of the Turkish pashas. But all these forces acting in concert, they were completely routed towards the end of the year, when twenty of their villages were burnt and all the inhabitants of both sexes put to the sword.

There are some 50,000 Christian Greeks in the Lebanon. Some "Ishmaelites" also dwell here, descended from the murderous sect of "Assassins," who have given a familiar word to most European languages. Here also are some 15,000 Mussulmans on the skirts of the range, and about the same number of Meteollis or Shiah sectaries, who are generally regarded with suspicion by their neighbours.

North of the Lebanon we enter the domain of the mysterious Nusarieh race, which gives its name to the northern coast range, and forms the majority of the population along the whole of these uplands, and even beyond the Amanus mountains, right into Cilicia, as far as Adana and Tarsus. Here dwelt from the remotest times the Nazarini, of whom the ancients seem to have known as little as we do of their direct descendants, the Nusarieh. These highlanders live and die in their mountain homes, which they never willingly leave. Tillage and stock-breeding afford them a sufficient livelihood, but while conducting themselves as true followers of the Prophet in the presence of their Moslem neighbours, they maintain profound secrecy on the subject of their peculiar worship. Their speech is the Arabic dialect elsewhere current in the Syrian highlands. Throughout Syria, where they are called Fellahin, and are said to number from 120,000 to 180,000, they have the reputation of being irreclaimable and desperate highwaymen.

The great bulk of the present population of Palestine, which scarcely exceeds 800,000 altogether, consists of Arabs, partly Bedouin nomads, partly Fellahin, or settled agriculturists. They dwell mostly in wretched mud hovels, or amidst the ruins of old buildings. They all speak Arabic, and are mainly followers of Muhammad. A few Christian communities are found in Nazareth and elsewhere.

But till recently the Jews had almost disappeared from the land of their forefathers. Except in Jerusalem they were scarcely anywhere to be found within the limits of Palestine proper. But since the outbreak of religious persecution in Russia, and the spread of the "Anti-Semitic Movement" in Germany and Austria, large numbers of Israelites have migrated, some to the New World, some to the Holy Land. In his annual report for 1894, Bishop Blyth, the Anglican Bishop of Jerusalem, states that "about 100,000 Jews have entered Palestine during the last few years, of whom 65,000 have come within the last seven years, and the arrival of a vast number is imminent." Lately also a few Protestant enthusiasts, mostly from Wurtemberg, have settled about Mount Carmel, in Jaffa, and a few other places. But it is somewhat premature to speak, as some already do, of the German colonisation of the Holy Land.

8. *Topography: Damascus—Aleppo—Emessa—Beyrut—Nazareth—Jerusalem—Hebron—Jericho.*

In Anatolia and Mesopotamia most of the old cities have either disappeared or sunk to the position of obscure hamlets, whose sites have with difficulty been identified. In Syria, on the contrary, although Tyre, Tadmor, Baalbek, and some other famous places have shared the same fate, many of the most venerable cities in the world, such as Damascus, Aleppo, Emessa, Beyrut, Jerusalem, not only continue to flourish but retain their ancient names in more or less modified forms. This, however, is true only of the region west of the El-Ghor and Bekaa depressions, beyond which hundreds of formerly prosperous towns have been swallowed up in the sands continually advancing westwards from the desert.

Damascus—Aleppo.

Here an almost solitary exception is Damaseus, which claims to be the oldest city in the world, and which, owing to the favourable conditions of the soil and climate, still continues to maintain its political and commercial supremacy almost on the verge of the wilderness. It lies nestled amid gardens and orchards at an elevation of 2300 feet above the sea, in a district which owes its exuberant fertility to the Abana and Pharpar flowing eastwards from the Anti-Lebanon and Mount Hermon. Owing to its thoroughly Oriental aspect it is one of the most interesting cities in the East. The Arabs have corrupted its name to Esh-Sham, which term they have extended to the whole of Syria. But when they wish to speak more particularly of the capital, they lose themselves in raptures about "the breath of heaven," "the mole on the cheek of the earth," "the plumage of the

DAMASCUS.

peacock," "the necklace of beauty," and suchlike Oriental imagery. For them it is one of the four Edens, although the city proper, enclosed within its crumbling walls and projecting towers, is far from corresponding with the favourable impression produced by a more distant prospect. The irregular and narrow streets wind along between high dead walls, broken at long intervals by small grated windows, but nowhere relieved by any touches of art. The monotonous piles of dull stone are varied only by a few ancient gateways, which alone make any attempt at architectural display.

Of great historic interest is the former Church of St. John, now the largest mosque in Islam. But more attractive are the numerous bazaars, in extent and richness surpassing most of those elsewhere met with in Eastern cities. Amongst their motley throngs nearly all the peoples of the East are represented.

Nearly due north of Damascus are Homs (Emessa), Hamah (Epiphania), and Aleppo. Homs is still a considerable place on the right bank of the Orontes. Hamah, on the same river and a little farther north, is an almost exclusively Moslem town, in the neighbourhood of which is the interesting Ala district described farther back. Still farther north, and about midway between the Euphrates and the coast, is Aleppo, second only to Damascus in size and importance. Capital of a vilayet, it does a considerable local and transit trade, and is occupied with some long-established industries. An old aqueduct still supplies it with water from some perennial springs 8 miles off. Aleppo was wasted by a terrific earthquake in 1822, since which time it has never quite recovered its former prosperity.

"The general aspect of the interior of the town is extremely substantial for an Oriental city. The streets are built of excellent freestone, well-fashioned, and the masonry fairly well put together. They are, however, crooked, and of course narrow, although there are one or two by which a carriage can enter the bazaars. The bazaars themselves are extensive, and, as a rule, fairly wide, arched or covered over, and the shops (square niches in the wall, in which sits the merchant among his merchandise) are often larger though less characteristic than those of Cairo. Manchester cotton and prints are seen everywhere, which ugly, but cheap and serviceable, manufactures are cutting out the beautiful old silks of native make. Separate bazaars are devoted to separate wares; and the curious traveller can inspect markets teeming with wool, cotton, or hides, or tramp through long alleys hung with festoons of red slippers or silk kaffiehs.

The citadel is by far the most interesting and remarkable place in the town. It is placed rather towards the east of the centre of the walled city, upon the highest ground within the walls. The great mound upon which the ancient fortifications stand is roughly circular, and surrounded by a wide and deep ditch, close on three-quarters of a mile in circumference. The mound itself is about 200 feet high, and although usually said to be artificial, appears to be only partly so, live rock having been found near the summit. The citadel probably occupied part, if not the whole, of the site of the ancient Beræa, and it may be compared with some of the large mounds which stud the great plains of Northern Syria and Mesopotamia" (H. S. Cooper, *Through Turkish Arabia*, 88). *Beræa* was the Greek name of the old Syrian fortress of *Chalybon*, a name still surviving in the Arab *Haleb*, whence *Aleppo*.

Coast Towns.

All the chief seaports of Syria are still found on the coast of what was formerly the land of the Phœnicians, the most famous navigators of antiquity. Amongst them

BEYRUT.

are Latakia (Laodicea), with a sheltered but shallow harbour ; Tarabulus (Tripoli), at the foot of a spur of the Lebanon, nearly destroyed by the explosion of a powder-magazine in 1864 ; Beyrut (Berytus), 50 miles farther down, next to Smyrna the largest and most flourishing seaport in the Levant. It stands on a noble bay extending in crescent-shape between the spurs of the Lebanon and the sea, and boasts of some fine new quarters and splendid villas, interspersed with shady groves and gardens. Of the population about two-thirds are Syrian Christians.

Nearly all the southern ports—Sidon, Tyre, Acre, Cæsarea, and Ascalon—have gradually lost most of their trade since the stirring days of the Crusades, and are now little more than fishing villages with a small local traffic.

Jaffa (Joppa), however, between Cæsarea and Ascalon, has recovered all its former prosperity, thanks partly to the great development of the orange industry, partly to the opening of the Jerusalem railway, of which it has been chosen as the seaward terminus. The Jaffa oranges, noted for their size and flavour, are now exported in increasing quantities to Europe, America, and even India. In 1892 nearly 1880 acres were planted, new groves are being constantly laid out, and the yearly exports now average 36,000 boxes, while the population has risen from 12,000 in 1878 to 42,000 in 1893. Owing to this trade Jaffa now ranks next to Beyrut in importance amongst the Syrian seaports.

Alexandretta (Iskandrun), in the extreme north, has of late years acquired some importance as the out-port of Aleppo. Here is by far the finest harbour on the whole coast, and, notwithstanding its unhealthy climate, Alexandretta cannot fail to become a flourishing place should the projected railway line ever be executed which is to run from the coast at this point through Aleppo to the main trunk-line in the Euphrates valley.

Nazareth—Jerusalem—Lachish—Gaza.

In Galilee still the most important place is Nazareth (En-Nasirah), west of Mount Tabor, and 1100 feet above the sea. It has now a Christian population of about 7000. The chief place in Samaria is the busy little town of Nablus (Neapolis or Shechem), lying in a fertile and well-watered valley between Mounts Ebal and Gerizim, and on the route from Damascus to the coast. Here still survives a small community of about 200 Samaritans, who, like their forefathers, continue to worship on Holy Gerizim. Amongst them is jealously preserved the precious codex of the Pentateuch in the old Samaritan dialect and in the archaic Hebrew character. Samaria, which gave its name to the land, has dwindled to a hamlet now called Sebastieh, a little to the north-west of Nablus.

From the summit of Gerizim, looking southwards, the eye lights on a limestone plateau, rising 2600 feet above the Mediterranean and nearly 4000 above the Dead Sea, connected northwards with the great tableland of Judæa, and on the three other sides enclosed by rugged gorges. Here stands Jerusalem, to the Christian the most hallowed of all places. It is even by the worshippers of Allah regarded as El-Kuds, or "The Holy Place." Here are still the Holy Places, the Church and Shrine of the Holy Sepulchre, to which are ever turned the footsteps of thousands of pilgrims from the West.

No writer has more vividly described the outward aspect of Jerusalem than Chateaubriand. "In the heart of a mountain range lies a desert basin, enclosed on all sides by yellow, rocky heights. These heights are open only towards the east, thus affording a prospect of the depression of the Dead Sea and the distant hills of Arabia. In the middle of this stony landscape, on an

JERUSALEM.

uneven and inclined plain, encircled by walls that once crumbled beneath the blows of the battering-ram, and are now propped by tottering towers, we behold some scattered heaps of ruins,—ruins overgrown with a few solitary cypresses, aloes, and prickly pears, and overbuilt by Arab huts resembling whitewashed sepulchres,—and such is the mournful picture now presented by Jerusalem. At the first sight of this forsaken spot, the heart is overcome by an overwhelming sense of despondency. But this feeling disappears as we gradually pass from desolation to desolation, and at last reach the boundless open space, which, so far from oppressing, rather inspires us with a certain sense of cheerfulness and buoyancy. Unwonted sights everywhere reveal a land crowded with hallowed memories. The sultry sun, the fierce eagle, the modest hyssop, the stately cedar, the barren fig-tree—here are concentrated all the poetry and all the imagery of Holy Writ. In every name lurks a mystery, every cavern lifts a corner of the veil shrouding the future, every hill-top echoes with the song of the prophet. By these rushing waters God Himself has spoken to man, and their dried-up beds, the rocks rent asunder, the yawning graves, still bear witness to His voice. Still hushed seems the wilderness, awe-stricken, and as if afraid to break the silence; for it has heard the voice of the Everlasting."

The present generation has undertaken with thoughtful piety again to rescue the ancient sites of the Holy Land from the accumulated *débris* of ages, and to determine their identity with the actual spots traditionally bearing their name. Attention has naturally been centred in Jerusalem, and great results have already been achieved, especially by the English "Palestine Exploration Fund," which has been at work since 1875.

Since the opening of the railway to the coast, the

Holy City is rapidly being modernised, and promises soon to become a great trading centre for all the surrounding districts as well as for the land of Moab beyond the Jordan. The city has already far outgrown its former limits; fields and vineyards have been covered with houses, especially in the direction of the west, where an entirely new quarter called "Modern Jerusalem" has sprung up, and the population has risen from 30,000 in 1880 to 80,000 in 1894. A public garden has been opened outside the Jaffa Gate, and the trade in olive oil, olive wood, and mother-of-pearl articles, such as crucifixes and rosaries for the pilgrims to the Holy Places, is rapidly increasing. A company has even been formed to collect the bitumen floating on the Dead Sea, for which there is a great demand in Europe. Sailing boats have been placed on the lake, and it is proposed to establish a steam ferry for the purpose of tapping the resources of Moab, a region abounding in cereals, fruits, and cattle. Kerak, the chief town of Moab, is now held by a Turkish garrison, and the hitherto unruly and predatory Arab tribes have been reduced to order. As soon as direct communications are established it is expected that all the produce of Moab, now forwarded by long caravan routes around the north and south ends of the Dead Sea, will find its way through Jerusalem to the coast (*British Consular Report*, 1894).

Six miles south of Jerusalem is Bethlehem, where the great Church of St. Mary marks the traditional site of the birthplace of the Saviour. Ten miles still farther south is Hebron, one of the oldest places in the world, and traditionally associated with the life and death of Abraham. The wretched village of Eriha (Riha), 18 miles north-east of Jerusalem, and near the north end of the Dead Sea, is supposed to occupy the site of the equally ancient town of Jericho.

A mound standing on a bluff, 60 feet above the Wady-el-Hesy torrent, 16 miles east of Gaza, has recently been explored by Mr. F. J. Bliss of the Palestine Exploration Fund, and found by him to contain the accumulated remains of as many as eleven cities, which here succeeded each other from about 2000 to 400 or 300 B.C. The oldest settlements were certainly pre-Israelitish, perhaps Amoritic, and one of the ruined cities is identified by Mr. Bliss with the Lachish captured by Joshua (x. 32). Here was found a burnt clay tablet bearing the name of Zimrida, probably the Zimridi, Egyptian Governor of Lachish during the eighteenth dynasty, one of whose despatches occurs amongst the documents discovered by Sayce at Tell-el-Amarna in Upper Egypt. Hence this city dates probably from about 1450 B.C.[1] Gaza itself was also at that time a flourishing city of the Philistines, which long held out against the Israelites, and which even still remains a place of some importance on the historic trade route between Palestine and Egypt. But the site has shifted, so to say, with the shifting sands of the desert, on the verge of which it stands; and along its eastward track are found the remains of ancient structures, statues, potsherds, and other objects buried beneath the advancing dunes. About the identity of Gaza (Ghazzeh) there is no doubt, for its record is complete for a period of some 4000 years in Egyptian, Biblical, Greek, Roman, Arabic, mediæval, and recent documents.

Beyond the Jordan there appear to be no inhabited places deserving the name of town. This region, before the survey by the Palestine Exploration Fund, was very little known, and rendered almost inaccessible by the lawless character of its Bedouin inhabitants. When M'Coan visited the Dead Sea in the seventies he was

[1] F. J. Bliss, *A Mound of Many Cities*, 1894.

plundered by some Moab Arabs at the ford of the Jordan, near Jericho. But the marauders seldom extended their raids quite so far west. The district is now pacified, and one or two points held by Turkish garrisons.

9. *Highways of Communication.*

In the north a much-frequented caravan route runs from Alexandretta, the natural port of Aleppo, through that city eastwards to the Euphrates at Bir, here ramifying westwards to Diarbekr and Kurdistan, southwards to Bagdad. As far as Aleppo this route is now accessible to wheeled traffic; above the village of Beilan it crosses the coast range at an altitude of over 2000 feet, thence descending at a comparatively gentle incline down the finely-wooded eastern slopes to the plain of Aleppo. From this point the great caravan and pilgrims' route to Medina and Mecca follows the Orontes valley by Hamah and Homs to Damascus, running thence through the Hauran southwards to Arabia. Damascus itself is connected with Beyrut by a splendid specimen of French engineering, which is carried over the Anti-Lebanon and Lebanon, and across the Bekaa, for a distance of 65 miles. This fine highway, now followed by a narrow gauge railway, constructed for most of the way on the "Abt rack rail" system, gives access to the magnificent ruins of Baalbek (Heliopolis), formerly the chief centre of the worship of the Sun God, whose temple is justly regarded as one of the wonders of the world. Baalbek is also connected with Damascus by another road through the rocky valley of the Wady Yafu'ah, near which it passes the village of Surghaya, 4500 feet above the sea, the highest inhabited point of the Anti-Lebanon. The way lies thence across a stony upland plain to the village of Dumar, where it strikes the French main highway. Another well-known route runs from Damascus across the Upper Jordan valley and through Nablus south-westwards to the coast at Jaffa, where it converges on the main road from the coast to Jerusalem. But the highways are not kept in good repair, and most of the other routes across the country are mere caravan tracks or bridle paths.

From Aleppo three routes run to Bagdad and the Persian Gulf—(1) The Euphrates valley route, by caravan to the river at Meskineh, then down the valley through Hamman, Deir, Anah, and the Hit bitumen springs to Feluja ferry and bridge, whence a short cut leads across El-Jezireh ("The Island") to Bagdad. This route was traversed and carefully described by Mr. H. S. Cooper in 1894 (*Through Turkish Arabia*). (2) By Bir, Urfa, and Mardin to Mosul, and thence by kellek (raft) down the Tigris to Bagdad. (3) By Bir to Diarbekr, and thence by kellek to Mosul and Bagdad.

In Syria railway enterprise has begun with a short line of 54 miles, running from Jaffa to Jerusalem, and opened for traffic in 1892. Since then the far more important Haifa-Damascus line has been taken in hand, and has already made some progress. It starts from two points on the coast, Haifa and Acre, at opposite extremities of the bay, enclosed on the south by Mount Carmel, which is the only inlet on this seaboard suitable for the large ships engaged in modern commerce. The two coast sections converge at the apex of the bay, whence the trunk line runs through the plain of Esdraelon and along the base of Little Hermon to Shunem, leaving Nazareth to the left. Beyond Shunem it runs through Jezreel to the Jordan valley, and after crossing the river skirts the eastern shore of the Sea of Galilee to the Hauran, the ancient Bashan, and thence to the fortress of Gamala and the towns of Kishfin and Nawa, and along the eastern base of Mount Hermon to its present terminus, Damascus. By this route the line, 150 miles long, avoids all mountain ranges, and traverses some of the most fertile districts in Palestine and Syria. The Hauran alone yields over 200,000 tons of cereals annually, and the whole country through which the railway passes is well suited for sericulture and the growth of cotton, wool, fruits, and olives.

The Beyrut - Damascus - Hauran railway already referred to is now (1896) open to Mezerib in the Hauran, a total distance of 140 miles. It is proposed to extend it to Bosrah.

CHAPTER VII

ARABIA

1. *Boundaries—Extent—Area—Coast-line—Islands.*

ALTHOUGH with no very clear limits towards the north, Arabia is on the whole one of the best-defined regions in Asia. In the north it falls, on the one hand, gradually towards the Mesopotamian plains, while on the other merging almost imperceptibly in the uplands of East Palestine and Syria. Here the so-called "Syrian Desert," reaching to about the 35th parallel, might with more propriety be regarded as the "Arabian Desert"; for in its physical and ethnical features it bears a much greater resemblance to the southern peninsula than to the surrounding regions of Syria and Mesopotamia. Like Arabia proper, its watercourses are mere "wadies"; its soil sandy, and in parts destitute of vegetation; its climate dry and almost torrid; and from time immemorial it has been exclusively occupied by nomad tribes of pure Arab stock.[1] Hence many geographers look upon it as merely a northern extension of the peninsula wedged in between the Euphrates and the Syrian highlands, and only in a conventional sense separated from Arabia proper. A convenient line, however, may be

[1] The Sebaa Bedouins, a branch of the great Anazeh family, reach northwards beyond the ruins of Tadmor, and are met even in the neighbourhood of Aleppo.

drawn from El-Arish on the Mediterranean to the Euphrates delta at the head of the Persian Gulf, leaving the vilayets of Damascus and Bagdad on the north, and including on the south all that has at all times and indisputably formed part of Arabia in the strictest sense. Elsewhere the peninsula is surrounded by water —the Persian Gulf and Gulf of Oman on the east, the Arabian Sea on the south, the Red Sea and Suez Canal on the west. Its great axis, running north-west and south-east, measures, as the bird flies, about 1200 miles between the head of the Gulf of Suez (29° 58′ N., 32° 30′ E.) and the Ras-el-Had (22° 23′ N., 60° E.). The mean breadth between the Red Sea and Persian Gulf is about 600 miles, with a total area estimated at rather over one million square miles, and a population of probably not more than five millions.

The shores of Arabia, which stretch from Suez to the Euphrates delta for a total length of nearly 4000 miles, present on the whole a somewhat uniform aspect, and, except in the Persian Gulf, are diversified by few islands or inlets. In the Red Sea the coast is fringed by extensive coral reefs, forming here and there groups of sunken rocks and islets, which render the navigation very dangerous. Between the Straits of Bab-el-Mandeb and Oman, separating the peninsula from Africa and Persia, the coast is generally elevated and rocky, but low and flat thence to the head of the Persian Gulf. The whole coast-line has been admirably surveyed by Moresby, Haines, Elwon, Saunders, Carless, Wellsted, Cruttenden, and other officers of the Anglo-Indian navy, mainly between the years 1819 and 1860.

Of the islands the chief are the small group marking the entrance of the Gulf of Akaba; Farsan, off the Tehama coast; Perim, in the Strait of Bab-el-Mandeb, where the English batteries completely command the approaches of the Red Sea; the Kuria-Muria (Kurian Murian) group and Moseirah, in the Arabian Sea; lastly, in the Persian Gulf the Bahrein Archipelago, centre of an important pearl fishery. The large island of Socotra, although occupied by an Arab population, and politically attached to British India, belongs geographically to Africa, and has been described in the volume of this series devoted to the northern section of that continent. Some of the islands in the Red Sea are volcanic, and one of them, the Jebel-Tir, is still active. Igneous rocks also crop out at Aden and at other points of the coast.

2. *Relief of the Land: Mountains—Plateaux—Lowlands —Deserts—Volcanic Tracts.*

Arabia, taken as a whole, is with good reason regarded as one of the least inviting regions on the face of the globe. The large blank spaces which still meet the eye as it lights on a map of this peninsula bear silent witness to our scanty knowledge of the interior. The glowing and shifting sands of the great southern desert have scarcely ever been visited, and never yet traversed by any European traveller, and fully one-half of this enormous region still remains entirely unexplored. In its general physical aspect, its climatic conditions, fauna, and flora, it so closely resembles the adjacent African mainland that it seems almost more like an eastern extension of this continent than an integral part of Asia.

The bulk of the land consists of a quadrangular mass broadening southwards, and largely covered with arid plains, sandy in the south, gravelly or stony in the north, the whole constituting a vast plateau at a mean elevation of probably 3000 feet above the sea.

The gravelly plain of El-Hamad in the extreme north falls to 2500; but the red sand desert of Nefud, between El-Hamad and the Jebel Shammar, rises to 3000 and 3200, while the land continues to rise thence southwards to 4000 and 5000 feet in the Wahhabi country. Blunt ascertained that from Meshed Ali near the Euphrates in Irak-Arabi (32° N.) there is a regular ascent of 10 feet in the mile to Hail in the Shammar highlands (27° N.); and the whole peninsula may be said to culminate towards the extreme south-west corner, where the Yemen uplands attain an elevation of from 6000 to over 10,000 feet. Thus we see that the tableland is tilted somewhat uniformly towards the north-east and east, so that in a developed water system the drainage would mainly be to the Lower Euphrates and Persian Gulf.

As in the Sahara, the arid tracts are broken by hilly districts and even ranges, where the valleys are watered by short streams or rivulets, and occupied by settled populations residing in small towns and villages. Thus a large portion of the central plateau, comprising the so-called Nejd—that is, "High Land"—consists of fertile hilly tracts everywhere surrounded by uninhabitable wastes and intersected by several ridges running in various directions. The term Nejd is applied to several tracts of this character, hence a certain vagueness inseparable from its use. But the Nejd proper includes, according to Blunt, all the high-lying land enclosed by the Nefuds, or deserts proper. It thus comprises the three provinces of Jebel-Shammar in the north, Kasim in the centre, and Aared or the Wahhabi country in the south, and lies mainly between 24° and 28° N. latitude. It is in no sense a political, but purely a geographical expression, by which may be understood the whole of the interior, bounded on the north by the red sand

Nefud, on the south by the great unexplored Dakhna, or sandy desert, eastwards by desert tracts separating it from the Turkish province of El-Hasa, westwards by the Turkish province of El-Hejas. The arable districts in Nejd, the Hejas, Yemen, and elsewhere are so extensive as to raise the more or less productive lands to about two-thirds of the whole area, leaving not more than one-third of absolutely desert and uninhabitable wastes, lying chiefly in the south. These wastes are variously termed Dakhna (Dahna), Ahkaf, Nefud, or Hamad, according to the greater or less depth or shifting nature of the sands, or the more or less compact character of the soil. The sands, which rest on basalt, limestone, but mainly granitic, beds, have, according to Palgrave, a mean depth of 400 feet, attaining in some places as much as 600 feet. They prevail in the vast unexplored region comprising most of the south, between Nejd and the Hadramaut coast range north and south, and between Yemen and Oman west and east. Here almost absolute sterility is the dominant feature, whereas in the northern Nefud, between El-Hamad and the Jebel-Shammar, not only the hollows but all parts of the plain are well clothed with shrubs. "After a rainy winter I have little doubt that the whole of this Nefud is covered with grass and flowers. Indeed the Nefud explains to me the existence of horses and sheep in Nejd" (W. S. Blunt).

The most clearly developed and best known mountain system is the extensive range skirting the Red Sea at a distance of one to three days' journey from the coast. In the Asir district, south of Mecca, this range attains an altitude of about 8500 feet, and between this point and the Strait of Bab-el-Mandeb it broadens out in the Yemen highlands, where every condition combines to render the south-west corner of the peninsula deserving of the name of Arabia Felix, applied to it by the ancients. These highlands are continued along the south coast by a series of disconnected ridges, which again rise in the extreme south-east to the Jebel-Akhdar, running at an elevation of 6000 to 7000 feet along the Gulf of Oman from the Ras Hadd to the Ras Mussendum. From this point to the head of the Persian Gulf the coast is generally low and flat.

In the interior the Nejd is crossed by several ridges, of which the largest and best known is the Jebel-Shammar, running nearly east and west under the 27th parallel at an altitude of about 6000 feet. Farther south the Jebel-Toweyk attains probably an equal elevation in Aared, on the northern skirts of the Great Desert.

Lowland plains occur chiefly in El-Hasa on the Persian Gulf, and along the shores of the Red Sea. Here the long narrow strip of the Tehama—that is, Low or Hot Land—stretches from Mecca to Mokha, between the coast and the Jebel-Hejas, or "Separating Range," as the term is commonly interpreted.

A conspicuous feature of the peninsula are the so-called *Harra*, or volcanic tracts, strewn with basalt and other igneous rocks. The northern harra south of the land of Bashan is described by Blunt as "a vast plain strewn with volcanic boulders—a black, gloomy region." The harra appear to occupy a wider extent of North Arabia than is generally supposed. In some districts C. M. Doughty found lava beds in the lateral valleys 600 feet deep, with numerous extinct volcanoes and volcanelli. On the route from Kheiber to Hail he crossed a craggy lava-field rising to a height of 6000 feet, and forming the divide between the waters running east to the Wady-er-Rumma and west to the Wady-el-Humth. Signs of volcanic action are still seen in the Harra, smoke issuing from the crevices of the Kheiber, and steam from the summit of the Jebel Ethnan ("Travels in North-West Arabia," in *Proceedings of the Royal Geographical Society*, 1884).

3. *Hydrography: Wadies Sirhan, Dawasir, and Er-Rumma—Coast-Streams.*

Arabia is almost a riverless region, in which the *nahr*, or perennial stream, is mostly replaced by the *wady*, or intermittent and dried-up watercourse. These watercourses, generally dry for nine or ten months in the year, occur everywhere—in the highlands, on the plateaux, in the lowlands, and even in the deserts, and especially in the northern Hamad. Here the great Wady Sirhan runs at an elevation of 1850 feet in a south-easterly direction from the Hauran highlands to the Jof district on the skirts of the Nefud. It is fed by the Wady-er-Rajel in the extreme north-west, and for over 200 miles between Kaf and Jof the wells are plentiful along its whole course. Hence it is much frequented during the summer by marauding tribes, who claim the right of plundering all comers, and acknowledge no authority except that of the tribal chief. Less known is the Wady Dawasir, which receives the Nejran, Bisheh, and other streams on its left bank, and drains all the Asir and Southern Hejas highlands northwards to the Bahr-Salumeh, the only known lake in the whole peninsula. The Aftan, another large wady, runs from the borders of Nejd and the southern desert eastwards to the Persian Gulf. But the most important watercourse in Arabia seems to be the unexplored Wady-er-Rumma, which flows between the Sirhan and the Dawasir from the Hejas coast range right across the peninsula in a north-easterly direction towards the Lower Euphrates, for a total length of nearly

800 miles. With a more abundant rainfall, this would augment the Shat-el-Arab, and give unity to the now disjointed water systems of South-west Asia. As it is, the Wady-er-Rumma, our knowledge of which is mainly due to Wetzstein's studies,[1] receives during the rains a vast quantity of water through countless affluents, some rising apparently in the far south.

Perennial coast-streams occur chiefly in Yemen, where their short courses have been accurately determined by Manzoni.

4. *Natural and Political Divisions: Peninsula of Sinai—West Coast (El-Hejas, Yemen)—South Coast (Belad-Aden, Hadramaut)—Nejd (Jebel-Shammar, Wahhabi Country)—East Coast (El-Hasa, Sultanate of Oman).*

If physically and ethnically one, Arabia is politically a disjointed land. The bulk of the inhabitants being still in the tribal state, there can be no question of a common national sentiment as developed in the west. Hence nearly all the coast lands have fallen to the stranger, while even in the interior Nejd is distracted between the waning Wahhabi and rising Shammar rulers, the only two that here claim sovereign power.

By the ancients the whole peninsula was broadly divided into three great sections, *Arabia Petræa, Deserta,* and *Felix.* The first and last of these answer roughly to the modern divisions of the Peninsula of Sinai in the north-west, and Yemen in the south-west. But Arabia Petræa, which confounded the great central tableland with the surrounding wastes, highlands, and lowlands, must necessarily disappear as the collective name of a

[1] Wetzstein's views receive fresh confirmation from M. Huber, who in December 1880 penetrated to Kheiber.

region which we now know to be composed of several sections differing widely in their physical features. Such are—in the centre the plateau of Nejd, the northern Nefud, and the great Southern Desert; on the west coast El-Hejas; on the south and south-east coasts Hadramaut and Oman; on the east coast El-Hasa or Bahrein. There are no doubt many other geographical expressions of a more or less local character; but these may be taken as the great natural divisions of the land, and they have the convenience of also corresponding on the whole with its political distribution. Thus the coast lands of El-Hasa, Yemen, and Hejas answer to so many Turkish vilayets; Sinai is administered by Egypt; Hadramaut or at least its south-western section bordering on Yemen is controlled by England, firmly entrenched on the rock of Aden. Oman and Nejd are under more or less independent native rule; all the rest is a prey to the Bedouin or the sands.

Sinai Peninsula.

A line drawn from the Dead Sea through the Wady-el-Arabah to the Gulf of Akaba will mark the natural limits of the Sinai Peninsula on the east. From its base on the Mediterranean this triangular section projects with its southern apex far into the Red Sea, thus developing east and west the Gulfs of Akaba and Suez. The triangle will be almost mathematically perfect, if we take the Suez Canal as its north-western limit. But here the conventional frontier between Egypt and Arabia is drawn from Suez through the sands north-eastwards to El-Arish, on the Mediterranean. Hence the mouth of the little River Arish, which gives its name to the port, is the converging point of two continents, and of the three famous lands of Palestine, Arabia, and Egypt.

The Sinai Peninsula forms a rocky limestone plateau

intersected by rugged gorges, and in the north comprising the extensive desert of Et-Tih, which ascends southwards to the alpine region of Sinai proper. This desert waste covers an area of some 10,000 square miles, where a sparse population of perhaps 4000 nomad Bedouins finds a difficulty in procuring sustenance from the arid soil.

RAS SUFSAFEH, A SPUR OF JEBEL-MUSA (THE SUPPOSED SINAI).

Here the land derives its grandeur and peculiar charm from the very nakedness of its rocky heights. In some of the wadies the hillsides are scored by countless seams of the brightest hues, their fantastic designs producing an indescribable pictorial effect. What is seemingly the mere outline of a distant landscape reflects a charming and almost phantasmagoric vista, as if the bare rocks were clothed with woods or vineyards, or their summits capped with eternal snows.

It is remarkable that the scriptural name of Sinai, given to the mountain where Moses communed with Jehovah and received the tables of the law from above, is now unknown in the land. When asked for Mount Sinai, the Bedouin will shake his head or point to the Jebel-Musa (Moses' Mount), one of the highest in the peninsula, where a shrine has been erected to the Jewish lawgiver. A Muhammadan mosque has also been erected there. But we do not know for certain that this is the Sinai of Holy Writ, which many have identified rather with the Jebel-Serbal (6734 feet), lying a two days' journey farther north, while Beke thinks it was the Barghir, or Jebel-en-Nur (Mountain of Light), a peak 5000 feet high in the range bounding the Arabah valley on the east. The view from the granite crest of the Jebel-Musa shows that it is eclipsed by several surrounding peaks, such as the Jebel-Katharine (8536 feet), the more southerly Um-Shaumer (8449 feet), and the Jebel-Gosh (8300 feet), which have rarely been visited by modern explorers. In fact, this alpine region, whose geological formation corresponds with that of the European Alps, and which still bears traces of former glaciers, is still largely an unknown land. It may be added that Professor A. H. Sayce regards the term Sinai Peninsula as altogether a misnomer, tracing the belief that Mount Sinai lay in the peninsula to the Christian anchorites of the second century. He shows from the Old Testament records and the Egyptian monuments that Mount Sinai was certainly not in the peninsular region named from it. The *Yam Suf* of Exodus, translated "Red Sea" in the Bible, he identifies with the Gulf of Akaba, and he is inclined to look for Sinai "on the borders of Midian and Edom, among the ranges of Mount Seir, and in the neighbourhood of the ancient sanctuary of Kadesh-barnea, whose site at 'Ain Kadis has been rediscovered in our own day" (*Asiatic Quarterly Review*, July 1893).

El-Hejas—Yemen.

The west coast of Arabia is comprised in the Turkish vilayets of Hejas and Yemen, which have no well-defined limits towards the interior. Theoretically El-Hejas stretches in the north half across the peninsula, where it is supposed to meet the eastern district of El-Hasa, which is now included in the vilayet of Basra. But between these two provinces lies the powerful state of the Emir of the Shammar, which cannot properly be regarded as forming part of the Ottoman dominions. El-Hejas, however, extends beyond the coast range inland far enough to include the cities of Medina and Mecca and the whole of the El-Asir district bordering southwards on Yemen, which comprises the rest of the south-west coast down to the neighbourhood of Aden. The two vilayets have thus a total length of about 1200 miles, varying in breadth from 60 to 150 between the sea and the western limits of Nejd. El-Hejas consists mainly of the sandy, barren, and torrid plain of the Tehama, varying from 30 to 80 miles in width along the coast, and of the mountain range or highlands with a mean elevation of 3000 feet, separating it from Nejd. The Tehama, which term is by some writers applied to the whole seaboard from above Jidda to Mokha, but by others restricted to the southern section between Yemen and the coast, seems to have formerly formed part of the bed of the sea, from which it has been slowly upheaved. It abounds in marine fossils and saline deposits, and appears to be advancing according as the sea continues to recede. Although everywhere extremely hot and generally very unhealthy, it contains, especially in the south, many well-watered and fertile tracts, affording good pasturage and yielding heavy crops.

But the chief value of Hejas is rather of a political than an economical character, giving to the master of the "holy cities" a great prestige, and perhaps his best title to the Caliphate, or headship of Islam. Yemen is, on the contrary, valuable for its own sake,—a land of fertile and well-watered valleys, rich pastures, and perennial streams, and dotted over with numerous flourishing towns and villages. Fully one-fifth of the entire population is concentrated in this narrow corner of the peninsula, where settled and agricultural communities, elsewhere extremely rare, have existed from the dawn of history. This exceptional position is partly due to the greater mean elevation of the land, partly to its rich soil and happy configuration, calculated to receive from the Indian Ocean and retain in its sheltered valleys an abundance of moisture. Till recently the sovereignty of the Porte in Yemen existed almost more in theory than in reality. But vigorous efforts have been made since 1868 to revive its old claim to absolute sovereignty, and at one time the Turkish and English forces had almost come into collision in the neighbourhood of Aden. The Imam of Sana, formerly the chief potentate in Yemen, is now a mere puppet in the hands of the Turks, and his capital as well as all the other strategical points in Yemen is occupied by a Turkish garrison. Of the other local "Sultans," some are tributary to the Turk, others allied by treaty with the British. "The native chiefs, locally called 'Sultans,' still exercise their old patriarchal sovereignty, and the writ of the Padishah runs little beyond the range of his cannon" (M'Coan).

Owing more perhaps to the jealousy of the Turkish authorities than to the fanaticism of the natives, Yemen continues to be almost as secluded a land as Tibet itself. Few travellers penetrate far inland, and only one or two have traversed the country in an oblique direction from the Gulf of Aden to the Red Sea. This feat, however, was accomplished in 1892 by Mr. Walter B. Harris, although the people were at the time in open revolt against the Turk. Starting from Aden, Mr. Harris passed through Lahej and across the desert to the mountainous Khoreiba district, whence he reached the Turkish frontier town of Kataba. From this point, travelling in disguise, mostly by night, and hiding by day in the jungle, he passed through Yerim and Dhamar to the capital, Sana, taking nearly three weeks to cover the whole distance of nearly 300 miles. Here he was arrested by the Turks, thrown into a pestiferous prison, and ultimately conducted under escort to Hodeidah on the Red Sea, this section of 200 miles being traversed in five days. Mr. Harris speaks of the wonderfully beautiful scenery through which he passed, and confirms the statements of other travellers regarding the surprising fertility and careful cultivation of the plateau, which stands at a mean elevation of from 7000 to 9000 feet above sea-level. There is a good supply of water, and in many districts streams of considerable size.

More definite information is supplied by General F. T. Haig, who in 1887 followed nearly the same route, but in the reversed direction from Hodeidah through Sana to the south coast at Shugra, 60 miles from Aden. After crossing the Tehama, here some 25 miles wide, the route from Hodeidah scaled a series of ridges by exceedingly steep inclines, surmounting the crests at passes which ranged from 4000 to 10,010 feet. This extreme altitude occurred in the district between Suk-el-Khamis and Sana, and as the hills on either side of the path rose about 400 feet higher, it is evident that some of the crests in Yemen must approach, if they do not exceed, 11,000 feet.

General Haig was struck both by the magnificence of the scenery and by the immense development of the terrace system of cultivation along the slopes of the steepest mountains. In one district the whole mountain side, for a height of 6000 feet, was terraced from top to bottom. "The crops had all been removed; only some lines of coffee trees here and there were to be seen, but everywhere, above, below, and all around these, endless flights of terraced walls met the eye. One can hardly realise the enormous amount of labour, toil, and perseverance which these represent. The terraced walls are usually from 5 to 8 feet in height, but towards the top of the mountain they are sometimes as much as 15 and 18 feet. They are built entirely of rough stone laid without mortar. I reckoned on an average that each wall retains a terrace not more than twice its own height in width, and I do not think I saw a single breach in one of them unrepaired" (*Geo. Proc.*, 1887, p. 482).

In Yemen two rainy seasons are distinguished, spring and autumn, when heavy showers of a few hours' duration are of almost daily occurrence. Hence there is generally an abundance of water for irrigating the terraced and other lands, which are very fertile and everywhere carefully tilled, although in a somewhat primitive fashion. The chief crops are coffee, indigo and other dyes, vegetables, such as cabbages and cauliflowers, which grow to an enormous size, grapes also very large and of excellent flavour, and a profusion of other fruits, such as figs, walnuts, peaches, pears, and apricots. Yet despite all this abundance, despite the richness of the soil and the surprising industry of the inhabitants, the bulk of the people themselves are miserably poor, ill-fed, and rudely clothed in undressed sheepskins, worn the woolly side in. This is due to the corrupt and oppressive administration, the heavy taxes usually levied by military force, the venal dispensation of justice, and the many other evils inherent to the Turkish system of government. Such places as Dhamar and Yerim bear the stamp of Turkish rule—decay, poverty, and squalor; and even in the capital, Sana, the bazaars are poor, and the general aspect of the town reflects the social condition of its inhabitants, "an Arab population, intensely hating the few thousand Turks by whom it is held down, heavily taxed, and generally obliged to furnish gratis the supplies required for the large garrison of Turkish soldiers."

Aden—Hadramaut.

The extreme south-west corner from the Strait of Bab-el-Mandeb to Cape Seilan, east of Aden, and reaching inland to the Jaffa range, comprises the so-called Belad-Aden, or Country of Aden, and besides Aden itself includes the Sultanate of Lahej and several of the surrounding tribes, under British protection. The protectorate comprises altogether a tract 200 miles long by 40 broad, with a collective population of about 130,000. "With these tribes we have distinct treaty engagements; we subsidise them so long as they are of good behaviour—that is to say, pay them blackmail to the extent of 12,000 dollars a year, and trouble ourselves no further about them than occasionally to interfere to put down a disturbance, or to decide some disputed question of succession" (Haig, *loc. cit.*).

East of the Belad-Aden the little-known region of Hadramaut stretches between the great desert and the sea eastwards to Oman. The interior of this vast but sparsely-peopled tract was almost a perfect blank until some light was thrown upon it by the travels of A. von Wrede a few years ago. This explorer assures us that the term Hadramaut applied by geographers to the entire south coast is by the natives restricted to its inland or northern section. It is in any case a very old name, for Ptolemy places the Adramitæ in this

HAJARIM, HADRAMAUT.

very region between the Homeritæ of Yemen and the Omanitæ of Oman. The land here rises from the coast in a succession of terraces to the Jebel-Hamra (5284 feet), which is connected north-eastwards with the Jebel-Dahura, probably 8000 feet high. This is the highest of the terraces, and beyond it the land slopes gently northwards. Here Wrede descended by very difficult and dangerous tracks down to the Wady Doan, which flows through the land of the Yssa tribe (Belad-beni-Yssa) northwards, apparently to the verge of the desert. This district is bordered on the west by Belad-el-Hasan, on the east by Belad-Hamum, all three being bounded on the north by Hadramaut proper. But how far this region extends northwards, and whether the sandy desert of El-Akkaf (Bahr-es-Saffi) really begins with the Wady Rakhiya, a branch of the Doan, are points on which Wrede throws no light.

The southern coast lands are on the whole level, and are succeeded by a hilly tract of moderate height, beyond which the upland plains or ranges begin to fall northwards to a depression between the highlands and the vast inland plain of El-Jauf (Gof).[1] A bold attempt to penetrate into the interior from the west coast was made by the French Jew Joseph Halévy in 1870.

Since then little was added to our knowledge of the interior of Hadramaut till the year 1893, when Shibam, which claims to be the capital, or at least the residence of the most powerful Sultan, was reached from the south coast both by Leo Hirsch and Theodore Bent. Starting from Makalla, 280 miles east of Aden, Hirsch made his way up the Wady Howera valley to the Jol or Magad plateau, which forms the water-parting between the streams which flow, when they do flow, south to the Gulf of Aden, and north to the great Wady Masila. The surface of the plateau consists of a thick crystalline limestone resting on a reddish rock of igneous origin, which was exposed by the erosive action of running waters. Crossing this desolate tract in a north-westerly direction, Hirsch struck the Wady Doan, a head-stream of the Wady Masila, opposite the town of Sif, the farthest point reached by Wrede. Thence the route lay north and east through Meshhed Ali, Hora, Shibam, Tariba,

[1] There are several Jaufs or Jofs in Arabia, because the word simply means "low land" in opposition to Nejd, or "high land."

and Terim, all of which places are situated in the fertile and well-cultivated valley of the Masila, which appears to be the main watercourse of Hadramaut. In some districts plantations of the date and dum palm (*Rhamnus nabeca*) extend for miles, and the whole region seemed more like an eastern extension of Yemen than a section of the inhospitable Hadramaut wilderness. Shibam itself, residence of a powerful Sultan, extends some distance along the right side of the valley, with many gardens, much cultivated land, "numberless mosques," and a settled population of about 6000. A highly-prized tobacco is grown in the Wady Riyan valley, which was crossed on the more easterly return journey to Makalla. The neighbouring town of Naga lies in the midst of the most luxuriant vegetation, and coal appears to be plentiful in this district. When Hirsch passed through, the local chief was opening a mine, and had already sunk a shaft 130 feet deep. Elsewhere gold and lead were heard of, and this expedition plainly shows that some parts of Hadramaut possess capabilities and natural resources far beyond what had hitherto been suspected. But the country possesses no political unity, and is still occupied by independent hostile tribes, the most powerful of which appear to be the Kaaitys, with capital, Shibam; their eastern neighbours and bitter enemies the Kathiri, with chief towns, Saiun and Terim; the Sari, Nehds, and others, towards the west and north-west.

The object of Mr. Bent's expedition was mainly archæological, but owing to the hostile attitude of these tribes he was unable to penetrate much beyond Shibam, where he was well received by the reigning Sultan, an enlightened prince, who has resided many years in the East Indies.

Nejd—Shammar and Wahhabi States.

Of all the main divisions of the peninsula the great central tableland of Nejd or Negd is certainly one of the most interesting. It has been fairly well explored by Sadleir, Wallin, Reinaud, Palgrave, Pelly, Guarmani, and most recently by Mr. and Lady Anne Blunt. During the early portion of the century the whole of this region belonged to the powerful and fanatical Wahhabi State, whose capital, originally at Derayeh, is now at Riad. But of late years Mumammad Ibn Rashid, Emir of the Shammar nation in the extreme north, has not only asserted his independence, but is at present by far the most powerful potentate in Nejd. His territory is bordered southwards by the Kasim country, separating it from the Wahhabi State. Northwards his influence extends beyond the Nefud right away to the oases of Kaf and Ittery in the Wady Sirhan in 38° E. long., 31° N. lat., east from the Dead Sea. The inhabitants of these oases acknowledge Ibn Rashid as their suzerain, "paying him a yearly tribute of £4 for each village" (Blunt). The people of the intervening district of Jauf on the northern verge of the Nefud also acknowledge his authority, which reaches westwards to Teyma (27° 30′ N. lat., 37° E. long.), some 80 miles from the Red Sea. He further commands the new pilgrim road from Persia, which formerly passed southwards through Riad, but now runs through Hail, capital of his dominions. This alone brings him in a revenue of £20,000, besides giving him enormous influence throughout the whole of the north from Mecca and Medina to the Lower Euphrates valley. Ibn Rashid's green and purple banner has thus become the symbol of authority in all the land enclosed by Hejas and Palestine on the west, the Syrian Desert on the north,

Irak-Arabi and El-Hasa on the east. Yet he himself, although at present by far the most powerful personage in the peninsula, is content to pay a small annual tribute to the Sherif of Mecca in recognition of the Sultan's suzerainty, such is still the potent influence of the acknowledged head of Islam.

"Although this Richard of the Nejd reached the throne over the murdered body of his young nephew Bender, and by the massacre of sixteen possible future pretenders, he governs his subjects wisely and firmly. His rule is described as mild and just, and the *mejliss* or public court of justice is still daily held in the palace-yard, where the Emir appears, just as Mr. Palgrave describes his predecessors, surrounded by officers of state and a body-guard of 800 soldiers. . . . He is on terms of alliance with all the Bedouins south of the Nefud, and every year brings him in fresh tributaries from among the former dependants of Ibn Saoud [the Wahhabi ruler]. Taxation is light, service in the army voluntary, and Ibn Rashid's government eminently popular. Nowhere in Asia can be found a more prosperous, contented, and peaceable community than in Jebel-Shammar" (Blunt).

And thus has statesmanship succeeded where fanaticism failed. For the once formidable but now almost extinct Wahhabi State had its rise in what was in its origin essentially a religious movement. It aimed at the reform of Islam, but it soon degenerated into a purely political system, upheld by terror and blind fanaticism. It was founded in the middle of the last century by Abd-el-Wahhab, but owed its subsequent expansion to his successor Ibn Saud, in whose family the office of Imam, or spiritual and temporal head, has since remained. By the beginning of the present century the Wahhabi empire had spread over most of the peninsula, and even aimed at creating a united Arabia by the expulsion of the Turks. But it never quite recovered the blow it received in 1819, when the Egyptian troops destroyed Derayeh, and dissipated the Wahhabite dream of universal empire. Their power is now virtually limited to the highland territory of Aared, bounded north-westwards by the independent district of Kasim, and encircled elsewhere mostly by the sands.

In their religious views the Wahhabi are the most rigorous of monotheists, setting their face against all undue veneration of the Prophet, saints, relics, or aught else in the least savouring of idolatry. Their ceremonial is extremely simple, and they carry to a heroic degree the Eastern virtues of hospitality and almsgiving. Their political system is based substantially on the cultivation of the land, and thus was developed a powerful and industrious peasantry, said at one time to have numbered nearly 2,000,000, and capable of raising an army of 60,000 disciplined warriors. But in Central Arabia the seat of power has passed from Riad to Hail; nor can the result be regarded otherwise than as satisfactory. The collapse of the Wahhabite movement, whose influence was at one time felt even in India, lessens the fear of the peace of the world being again threatened by a revival of Moslem fanaticism. The apprehension, however, of disturbance being produced by this cause, from time to time, is not extinguished.

El-Hasa—Bahrein—Oman.

The east coast of Arabia, which is washed by the Persian Gulf, projects almost to a sharp point at the Ras Mussendum, where the Strait of Ormuz separates it from the Persian mainland, and connects the Persian Gulf with that of Oman. The northern section of this coast (El-Ḥasa) is officially included in the Turkish vilayet of Basra, the southern and western frontiers of which are arbitrarily drawn according to the caprice of the Ottoman functionaries. But El-Hasa is naturally limited southwards by the projecting headland of Ras Rekkan, which encloses the group of the Bahrein Archipelago, claimed by Turkey, and valuable for its pearl fishery. Bahrein, as the strip of coast between Capes Rekkan and Mussendum is sometimes called, forms a sort of neutral land between the vilayet of El-Hasa on the north and the independent native State of Oman, which comprises all the rest of the east coast and the whole of the south-east corner of the peninsula.

Although pearl fishing is carried on all along the east coast of Arabia, the industry is chiefly centred in the Bahrein Archipelago. It gives employment to several thousand boats of from four to ten tons burden during the season from May to September, the average annual yield being from £200,000 to £300,000. Nearly all the proceeds go to the speculators, while the unfortunate divers are kept in a state of servitude worse than slavery. The truck system prevails in its worst form, the fishers being compelled to sell their pearls to their employer at his own price, and to hire their boats and purchase their provisions from him at extortionate charges. In the Bahrein Archipelago occurs the somewhat rare phenomenon of fresh-water springs on the bed of the sea, so that the natives dive for their water as they do for their pearls. These pursuits are carried on under great risks, the surrounding waters being infested by sharks and sword-fishes.

The Bahrein group, comprising the large island of Bahrein, 27 miles long by 10 wide, the much smaller horseshoe-shaped Moharek, Sitrah, Sayeh, Arad, and a few other rocky islets, forms since 1784 a little principality ruled by the royal house of El Kalifah, chiefs of the Uttubi Arab tribe from the neighbouring mainland. After some unsuccessful attempts made both by the Persians and Turks to annex the archipelago, the present Sheikh Esau accepted the British protectorate in 1875, and since then peace has been preserved. The group, which was held by the Portuguese for about a century

BARROWS ON BAHREIN ISLAND.

(1520? — 1622) was well known to the ancients, and the remarkable sepulchral mounds of round form explored in 1889 by Mr. Theodore Bent, prove that the islands have been inhabited from remote times. These barrows, covering a great part of the large island, and ranging from a few feet to 40 feet in height, are of unknown origin. But from their peculiar structure and the nature of their contents, Mr. Bent attributes them to the Phœnicians. He even concludes that Bahrein must have been the cradle of the Phœnician race, and in any case he has at least placed beyond doubt that the archipelago was one of the chief seats of that seafaring people long before the dawn of history. Two of the islands were known to Strabo by the familiar Phœnician names of Tyros and Arados, and one member of the group, close to Moharek off the north-east point of Bahrein, still bears the designation of Arad. The present inhabitants are chiefly Arabs from the neighbouring mainland and the African slaves employed in the pearl fisheries. Manameh on Bahrein is the present commercial capital, while Moharek, facing it on the opposite side of the narrow intervening strait, is the seat of Government. Both are relatively large places, each with a population of about 8000.

The Sultan of Oman, formerly more popularly known as the Imam of Maskat,[1] at one time ruled over an extensive territory on the East Coast of Africa. But this was assigned in 1856 to a brother of the reigning Sultan, and now constitutes the protected State of Zanzibar. He, however, still claims jurisdiction over Germansir, a strip of the opposite Persian coast, stretching from about the 28th parallel to the west frontier of Persian Mekran, and including the port of Bundar Abbas (Gombrun) and the large island of Kishm.

On the Arabian mainland the north-east coast is rocky, but well supplied with good natural harbours, while the south coast west of Cape Hadd is flat, and sheltered only by the island of Moseirah. At a distance of 50 or 60 miles from the sea there runs a mountain range, the Jebel-Akhdar, parallel with the crescent-shaped east coast, beyond which the surface is dotted with a number of true oases, abounding in water, incredibly fertile, and covered with an exceedingly dense growth of vegetation.

[1] His proper title is *Sayid*, or "Sovereign." He never assumed the religious dignity of Imam commonly attributed to him.

The Sultan of Oman maintains a small navy to keep down piracy in these waters. He has also long enjoyed the benefit of a close alliance with England, which, while adding to his prestige amongst his own subjects, guarantees him from any overt acts on the part of Turkey or Persia. His States are fairly well governed, justice is efficiently administered, and peace secured within his borders, which merge everywhere inland with the desert.

5. *Climate: Rainless Zone.*

The prevailing climatic conditions are intense heat and dryness. The zone of maximum heat on the surface of the globe in July embraces the whole of the Persian Gulf, the greater part of the Red Sea, and of the intervening Arabian peninsula. This also comprises one of the rainless regions, where rain falls only at intervals of from one to three or four years. Even the periodical wet seasons, to which Yemen and some other favoured tracts are exposed, are occasionally interrupted by counter-atmospheric currents from Africa, and then whole provinces have to depend for two or three years at a time on their wells, tanks, and other reservoirs. It is the vicinity of the African Sahara that prevents Arabia from enjoying, as India does, the full benefit of the moist winds from the Indian Ocean. Hence in summer, when India is often deluged by tropical downpours, the south coast of Arabia swelters under the vertical rays of a fierce sun, and the parched-up land finds no shelter either in an overcast sky or a leafy vegetation. Thus has been developed in the course of ages the great Southern Desert, surpassing the Sahara itself in absolute aridity and barrenness. But as the land rises towards the Yemen highlands the glass naturally falls, the nights become pleasantly cool, and the tanks here freeze in winter. At elevations of 6000 feet storms become frequent, and are at times accompanied by heavy showers. The heat probably reaches its maximum in the low-lying coast district of the Tehama on the Red Sea, and along the west coast of the Persian Gulf. From the bare rocky walls skirting both sides of these land-locked basins the sunbeams are reflected with redoubled strength on the glowing waters, thus producing an enormous evaporation, which converts the surrounding atmosphere into a vapour-bath. For Europeans a trip across the Persian Gulf is considered at these times extremely perilous, and the unhealthy climate of the Tehama has become proverbial. On the other hand, the high central plateau of Nejd enjoys a climate described by Palgrave as one of the most salubrious in the world. Here the pure air, dry atmosphere, and moderate temperature have proved highly favourable to the development of animal life, although the lack of moisture has prevented a corresponding vegetable growth. Altogether, the most favoured region in this respect is Yemen, where the glass, even in July, seldom rises above 90° F. At Sana, Niebuhr found that it did not exceed 85°, while in the neighbouring Tehama it stood at 98° F. in the shade. Here also snow falls occasionally, and it freezes during the three winter months, while at Loheia (Tehama coast) the glass never falls much below 80° F. in January. Such is the astounding difference in temperature produced by the relief of the land in the same district.

Sand-storms prevail very generally, but are not dangerous to travellers, except, perhaps, in the great Southern Desert. On the other hand, the extent of the range of the simooms, or poisonous winds, seems to have been exaggerated.

6. *Flora and Fauna: The Horse and Camel.*

The most valuable plants are the date-palm, of which over 130 varieties are reckoned growing in all the oases, and supplying the chief staple of food; coffee, indigenous in Yemen, and largely exported from Mocha, whence the "Mocha coffee" of commerce; aromatic and medicinal plants, chiefly along the south and west coast, producing frankincense, myrrh, gum-arabic, balsam, senna, which have supplied the markets of the world for ages. The vine is cultivated for its fruit; the peach, apricot, almond, fig, and other fruits of excellent quality, are produced in Yemen, and cotton is cultivated in Oman. Of the few forest trees the chief are the sycamore, the *nebek* or thorny lotus, the cassia, and the manna-yielding ash. Yemen, and some other parts, also yield maize, millet, wheat, barley, durra, lentils, tobacco, madder, indigo. Characteristic of Nejd is the *ghatha*, which grows to 12 or 15 feet high, and yields the purest charcoal in the world. It abounds in the northern half of the Nefud, and is found as far north as Kaf in the Hamad.

Amongst the wild animals are the lion, panther, leopard, wolf, wild boar, jackal, gazelle, fox, monkey, wild cow, or white antelope (Beatrix antelope, genus *Ornyx?*), ibex, webber (marmot?), horned viper, cobra, bustard, buzzard, hawk. The locust abounds in Arabia, but is here rather preyed upon than the spoiler. "It is not generally known how excellent locusts are as food. . . . The red locust, which is, I believe, the female, is the best eating, and should be plain boiled. In taste it resembles green wheat, having a very delicate vegetable flavour. Horses thrive on them, and nearly every animal in the desert devours them. Our dogs caught and ate them greedily. A camel will occasionally munch them in with their pasture, and a hyæna I shot was found to be full of them. Locusts should be gathered in the morning while the dew is still on their wings" (Blunt).

The chief domestic animals are the ass, mule, fat-tailed sheep, and above all the camel and horse. Of the latter there are two classes: the kadishi, of unknown pedigree, used for rough work; and the kokhlani, or koheileh, whose genealogies have been recorded for over 2000 years, and which spring traditionally from Solomon's studs. They are mostly of small size, between 13 and 14 hands high, but symmetrical, hardy, and endowed with extraordinary staying power. The best breed, formerly in the Nejd, is now said to be found amongst the Anazeh and other Bedouin tribes of Mesopotamia. But opinions differ on the point, and while Blunt holds that "the tale of a distinct Nejd breed is entirely fabulous," Rawlinson still considers that the Anazeh is "of much inferior blood to a real Nejd horse." Nejd is supposed to abound in horses, but this would seem to be a mistake. Burckhardt long ago remarked that here they are comparatively rare, and that the Bedouins of the rich Mesopotamian plains own the largest stock. This view is now confirmed by Blunt, who asserts that "horses of any kind are exceedingly rare in Nejd." Here "the camel is the universal means of locomotion with the Bedouins. The townsmen go on foot."

But for the camel the desert would be absolutely uninhabitable. Of this animal there are several species, or rather varieties, abounding especially in Nejd, hence termed *Omm-el-Bel*, or "Mother of Camels." The Nejd "Ship of the Desert" will pass four and even five days in the summer heats without a drop of water; but those most suited for riding are said to come from Oman.

7. *Inhabitants: Bedouin Life.*

Few Asiatic lands can boast of a more homogeneous population than Arabia. The whole peninsula belongs from prehistoric times to the great branch of the Semitic family, who have always called themselves Arabs, a term probably meaning nothing more than "people of the plains." Within this branch there are doubtless many divisions and subdivisions into tribal and sub-tribal groups; but all are essentially one in origin, physique, speech, and religion. The only true distinction that can now be recognised is rather of a social than an ethnical character—that is to say, the distinction between the settled agricultural element residing in towns and villages and the nomad Bedouins of the wilderness. The former are met everywhere in more or less numerous communities, wherever the land is fit for cultivation—in El-Hejas (Mecca, Medina, Taif); in Nejd (Hail, Derayet, Riad); in Oman (Maskat); but especially in Yemen, where the settled political status preponderates over the tribal organisation of camp-life.

Some readers may possibly be surprised at the term "organisation" applied to the social condition of the free children of the desert; but the popular ideas regarding the habits, customs, and usages of the tented Arab are in many respects erroneous. He is usually represented as ceaselessly roaming with his tents and flocks from place to place, whereas there is perhaps no people less given to wandering, or more attached to their homes, than the true Bedouins.[1] Hence Arabic is almost the only language that has a perfect equivalent in the term *watan* (وطن) to the English word *home*. They have their allotted winter and summer camping-grounds, seldom changing

[1] This very word *Bedawi*, plural *Bedawiun*, means rather *pastor*, *stock-breeder*, than *nomad*.

their settlements except when removing from one to the other with the seasons. While *en route* they never pitch their tents, sleeping in the open, wrapped in their flowing garments. Their encampments resemble those of the gipsies, only the occupants are perhaps somewhat wilder and more picturesque in appearance. Women in dark-brown cloaks grinding the corn with primitive hand-mills, or weaving cloth for the tents; children, goats, and dogs, all playing together in happy harmony; the men lounging about smoking, or drinking coffee, form on the whole a not unpleasant scene of homely life.

The Bedouins are often represented as highwaymen and robbers from their birth. Their ideas regarding the rights of property differ seriously from those prevalent in the West; but these very ideas of theirs are based on a keen sense of right, and grow out of the proud spirit which resents the intrusion of strangers or hostile tribes on their domain. It must be allowed that among them there are what may be called marauding tribes by profession; but even these have a certain traditional code of law and honour, strange as the word may sound in such an association, a code which all alike accept and implicitly obey. A curious illustration of this spirit is afforded by the circumstances attending the attack on Mr. Blunt's party in the Wady Sirhan in the year 1878. "Lady Anne and I," he writes, "happened to be separated from the rest of our party, and were sitting under a ghatha bush eating our midday meal of dates, when we suddenly heard the galloping of horse-hoofs in the sand. Looking up, we saw a dozen Bedouins bearing down on us with their lances. . . . Our thick cloaks saved us from the points of the lances, and my Bedouin head-rope saved my head; and when we had cried 'Dahil,' 'I yield,' and given up our mares, they left off knocking us about. It then turned out that our captors were a party of Roala, friends of our own and of Muhammad's, though they knew nothing of us personally; and after we had sworn to our identity, they brought us back our mares and everything that had been dropped in the scuffle."

Jonas Hanway also vindicates the Bedouins from the animadversions of some writers in the last century. "Their skill in horsemanship, and their capacity of bearing the heat of their burning plains, give them a superiority over their enemies. Hence every petty chief considers himself as a sovereign prince, and as such exacts customs from all passengers. When they plunder caravans travelling through their territories they consider it as reprisals on the Turks and Persians, who often make inroads into their country and carry away their corn and their flocks."[1]

Amongst themselves and towards all placed under their protection their sense of honour and trustworthiness are beyond suspicion. Owing to the fearful severity of the custom of blood-revenge, murder is of much rarer occurrence in the wilderness than in more civilised lands. The character of the country and their social habits develop a sort of clannish confederacy amongst the several tribes, as well as a certain common sympathy with all belonging to the Arab race. The Turi or the Maghrabi tribes have now a salutary dread of "the Consul."

In other respects the contrast between the social relations of the Bedouin and those of the "more civilised" inhabitants of the towns and villages is very much to the advantage of the former. Their simple diet and the pure, untainted atmosphere which they breathe render them healthy in mind and body. They are cheerful and even possessed of a fund of humour, and will often endure the greatest hardships without a murmur. Their demeanour is courteous and even refined.

[1] *The Revolutions of Persia*, part v. pp. 221, 222.

As with most Eastern peoples, parents are treated with the greatest respect by their children while under age. But as soon as the young Bedouin is old and strong enough to set up an independent establishment, he considers himself released from this duty, henceforth regarding himself in the light of an equal.

On the men naturally falls the care of supporting the tribe, the means of doing which are often scanty enough. Their chief source of wealth is derived from the camel. The escorting of travellers, pilgrims, and goods is a profitable branch of industry, but restricted to the few tribes recognised as the duly authorised *ghufara*, or "protectors."

A limited trade is also carried on with Suez and Cairo, the Arabs supplying charcoal, millstones, ibex horns, gum-arabic, and the like, in exchange for corn and tobacco. A few inhabiting such fertile districts as the Feiran own a little land, on which they cultivate tobacco, bartering or selling it to the neighbouring tribes. Owners of sheep and goats turn the hair and wool of these animals to account, and use the milk in spring, but seldom kill them except in sacrifices. Another article of trade is the "munn," or manna, a glutinous saccharine substance exuding from the tamarisk tree for about two months, "while the apricot is in bloom."

8. *Topography: Mecca—Medina—Hail—Maskat—Aden.*

In a land of which probably not more than one-tenth is arable, towns cannot be numerous. The largest appear to be Maskat, capital of Oman (60,000); Sana, capital of Yemen, officially given at 50,000, but estimated by Haig at not more than 30,000, or at most 35,000; Mecca 45,000, and its seaport Jidda 30,000. But by far the most important are the two "holy cities" of Hejas, Mecca and Medina, towards which the eyes of one hundred millions of Muhammadans are constantly turned, from the shores of Marocco to the distant islands of the Eastern Archipelago.

Mecca, the Rome of Islam, is an unwalled city situated in a narrow sandy valley enclosed by rocky eminences from 200 to 500 feet high, and about 65 miles from Jidda, its port on the Red Sea. The valley is scarcely 600 yards broad, narrowing southwards to about 300, where it is almost blocked by the Beit-Ullah (God's house), the great mosque enshrining the famous *Kaaba.* The whole building forms a rectangle 250 paces long by 200 broad, of which the north side is formed by four rows of pillars, the other three of three rows each, arched over, and so disposed that each group of four supports a little cupola, making altogether 152 of these structures. Along its entire length suspended from the arches are glass lamps, all of which are kept burning during the Ramadan, or fasting season. The oldest pillars are hewn out of the neighbouring rock; the others, consisting of marble, granite, and porphyry, are mostly offerings of the Faithful, and include some antiques from the old temples of Syria and Egypt.

Within the mosque is the *Kaaba*, or "Holy House," a small, massive building about 40 feet high. Tradition associates this unpretending and curious little structure with a multitude of marvels and legends, one more preposterous than another. On the north side is a doorway leading through steps inlaid with gold and silver to the inner sanctuary. In a corner lies the famous "black stone," supposed to have been given by God to Abraham, but now known to be a meteoric block descended, if not from heaven, at least from the interplanetary spaces. West of the Kaaba is the "golden channel," carrying off

MECCA, AND THE KAABA.

from the flat roof the rain-water, which is reputed to be endowed with miraculous properties.

Access to Mecca is rendered extremely difficult in consequence of the ceremonies imposed on all wishing to visit the birthplace of the Prophet, and expressly designed to exclude unbelievers. Yet the feat has been accomplished during the present century by Burckhardt, Wallin, Palgrave, Burton, Keane (?), and perhaps by others, mostly disguised as pilgrims.

In Mecca resides the great Sherif of Mecca and Medina, a far more important dignitary than the Turkish Vali or Governor of Hejas. As guardian of the holy shrine of Islam he receives a heavy annual stipend from the Porte, in return acknowledging the suzerainty and caliphate of the Sultan.

About seventy miles south-east of Mecca is the small but pleasant town of Taif, to which the pashas condemned for the murder of Sultan Abdul-Aziz were banished. It is one of the most interesting places in Arabia, surrounded by gardens and vineyards, from which Mecca has been supplied with fruits for ages.

Nearly under the same meridian as Mecca, and 240 miles farther north, lies the almost equally venerated city of Medina. Hither fled the Prophet when his obdurate fellow-citizens were deaf to his voice, and from this flight dates the Muhammadan era.[1] Here also is his tomb, a shrine second only in sanctity to the Kaaba itself. Medina lies at an elevation of 3000 feet above the sea, close to a range of hills running north and south between Hejas and Nejd. It is built of solid stone, but the streets are very narrow, and everywhere lined with lodging-houses for the convenience of pilgrims. The great mosque containing the Prophet's tomb is approached by the main street from the gate of Cairo. It is smaller

[1] That is, the *Hijra*, or "Flight."

than that of Mecca, and is supposed to have been really built by Muhammad himself. His coffin is encased in silver, and covered with a heavy marble slab.

By far the most important place in the south-west is Sana, capital of Yemen, and seat of an Imam, who enjoys a nominal jurisdiction under the Turks, real masters of the land. Sana, which is perhaps the finest and best-built city in the whole of Arabia, has been visited of

MEDINA.

late years by Wrede, Halévy, Manzoni, Haig, Harris, and other Englishmen. It lies in a fine, well-cultivated upland valley, 7280, or, according to Haig, 7700 feet above sea-level, about 140 miles north-east of Hodeida, its port on the Red Sea, and 260 from Aden. Its walls, about six miles in circumference, are mounted with cannon, but pointed towards the town, and they enclose two stone palaces of the Imam, besides a great number of highly-ornamental mosques, baths, and caravansaries. As generally throughout the Yemen highlands, the houses, two, three, and even four stories high, are all built of solid stone, whereas those of the Tehama lowlands are usually of mud with thatched roofs. But there is little attempt at architectural display, beyond a peculiar style of ornamental carving on the walls and round the windows. There is a separate Jewish quarter with a population of 5000, about one-twelfth of all the Jews in Yemen, and General Haig was shown the scarcely distinguishable ruins of a Christian church dating from pre-Muhammadan times. The Jews have twenty-three synagogues and twenty schools, attended by 700 boys; female education is almost entirely neglected, whereas all the male population can read Arabic. Numerous ruins are shown over the waste spaces within the enclosures, and there can be no doubt that Sana was formerly a much larger city than at present, with a reputed population of no less than 200,000 so recently as the last century.

Though now eclipsed by Sawakin and Massowah, Hodeidah was a few years ago the busiest seaport in the Red Sea. "There was more life and movement in the streets and crowded bazaars than I had seen elsewhere, more signs of trade and business generally. Being the principal port of Yemen, most of the coffee and hides, which are the staple exports of the country, are shipped from here" (Haig, *ib.* p. 480). The population is estimated at from 25,000 to 30,000, and the market is well supplied with fruits and vegetables; but the water has to be drawn from wells at some distance from the town, and there is no harbour, so that steamers have to ride at anchor in an open roadstead 1½ mile off the shore.

In the interior the only noteworthy places are Hail and Riad, capitals of Jebel-Shammar and Aared respectively.

Hail lies 3500 feet above the sea, not to the south,

as is usually stated, but to the east of the Jebel-Aja, a granite range 6000 feet high, which ends abruptly at this point. In this neighbourhood is the Emir's castle, a formidable stronghold occupying a position of immense natural strength in the Jebel-Aja. Blunt visited this place in 1878, but does not give its exact site, lest the information might be utilised by the Turks under possible future contingencies.

Riad, which has succeeded Dereyah as capital of the Wahhabi State, lies in the heart of the Aared country, enclosed north and south by the Jebel-Toweyk, and about 280 miles south-east of Hail. It is a large place, with a population of probably 30,000; but nothing is known of its present state, as no European has visited it since the time of Palgrave.

On the east coast the only large place is Maskat, capital of Oman, which, although extremely hot [1] and unhealthy, is the centre of nearly all the import and export trade in these waters. For this position it is indebted more to its well-sheltered and convenient harbour than to the fact that it is the seat of government and residence of the Sultan. It is built in a series of terraces rising one above the other on the side of the frowning precipices enclosing its picturesque bay. But though presenting a pleasant prospect from a distance, a nearer view reveals the usual features of large Oriental towns—narrow, dirty, and gloomy streets, tasteless buildings, and masses of dead walls, beyond which stretches a swampy suburb occupied by nomad Arabs and African slaves. The

[1] The heat of Maskat, which is proverbial throughout the East, is thus jocularly referred to by a writer of the fifteenth century quoted by Mr. Curzon: "It was so intense that it burned the marrow in the bones; the sword in its scabbard melted like wax, and the gems which adorned the handle of the dagger were reduced to coal. In the plains the chase became a matter of perfect ease, for the desert was filled with roasted gazelles" (*Persia and the Persian Question*).

MASKAT.

townspeople themselves are a motley mixture of Arabs, Persians, Syrians, Indians, and even Kurds and Afghans, who have either taken refuge here from oppression at home, or else have been attracted to the place by its great facilities for trade.

Politically by far the most important place in the southern section of the peninsula is Aden, occupied since 1838 by the English, who from this stronghold and the neighbouring island of Perim command the whole of the Red and Arabian Seas, and keep open the water highway to British India and the far East. But besides forming one of the most important links in the chain that girdles the eastern hemisphere from London to Hong-kong, this Gibraltar of the Indian Ocean is also a free port, doing a considerable trade with the interior, and with a present population (including Perim) of 34,860. Yet it lies perched on a bare rock in an indescribably barren and desolate coast district, a hotbed of the most deadly diseases, altogether one of the most uninviting and unhealthy spots on the surface of the globe, and in summer sultry almost beyond endurance. The old town lies in the very crater of an extinct volcano, 1775 feet high, whose sides, which have partly fallen in, are crowned with formidable works bristling with cannon.

Aden lies well within the rainless zone, where no rain falls at times for intervals of two or even three years. Hence for its water-supply it is dependent on wells, tanks, condensers, and the magnificent old reservoirs in the neighbourhood, which have been recently restored. Here are two good harbours, formed partly by the adjacent island of Sirah, and, thanks to its convenient and commanding position near the entrance of the Red Sea, Aden has become one of the chief coaling depôts and calling stations for steamers in the Indian waters. It is also an important political centre, barring the further advance of

the Turk in this direction, and guaranteeing the independence and good government of Lahej and the other petty States along the south-west coast.

The settlement, which includes the rocky peninsula, 15 miles in circumference, and extends to the Khor Maksar Creek, two miles north of the defences, is politically subject to the Government of Bombay, and administered by a Resident with two assistants. Since the opening of the Suez Canal the shipping has steadily increased, and a vessel of war is usually stationed at the port, which is in charge of a Conservator and regulated by the Indian Ports Act.

9. *Highways of Communication.*

In Arabia there are scarcely any roads properly so called. But the peninsula is crossed everywhere except in the south by well-trodden caravan routes, whose direction is mainly determined by the greater or less abundance of wells and other reservoirs along their course. There is so little local trade and so much visiting of the holy places from all quarters of Islam, that these routes naturally converge on Mecca and Medina. The two main highways are what might be called the Sunnite and the Shiah *haj*,[1] the former from the north for the convenience of the orthodox Turk, the latter from the east for the heretical Persian.

The northern pilgrim road starts from Damascus and runs nearly due south through the Hauran highlands and the Roala, Sherarat and Harb Bedouin territories between El-Hejas and the Nefud to Medina and Mecca. The chief intermediate stations are Kalaat Belka, east of the

[1] *Haj* means "pilgrimage," whence the "Haji," or pilgrim in a pre-eminent sense, who has visited the holy places, a personage who holds his head very high in the East.

Dead Sea; Maan, east of the Wady-el-Arabah; Tebuk, and Medain Salah, east of the Red Sea. The journey to Medina takes thirty days, and the pilgrim caravan is usually escorted by the governor of Damascus. But this route is not now so much frequented as formerly, the pilgrims from Anatolia and Syria preferring the less fatiguing and more expeditious journey by steamer through the Suez Canal and Red Sea to Jidda, whence they reach Mecca by easy stages in three or four days. Since the opening of the canal the pilgrim traffic of Jidda by sea has increased to from 45,000 to 50,000 yearly. Many also still reach Jidda by land from the Barbary States, Sudan, and Egypt, by a route from Cairo across the Sinai Peninsula and down the coast of Hejas. But this line is also being gradually abandoned in favour of the sea voyage from Suez.

The eastern or rather north-eastern road from Persia runs from Bagdad through Kerbela and Meshed Ali nearly due south, to the wells of Shaibeh (27° 10′ N, 44° E.), here turning west to Hail and thence south-west to Medina. This line traverses the domain of the Montefik and Daffir Bedouins in the north, the Jebel-Shammar State, and the Harb Bedouin country in the west. In lat. 28°-29° and long. 44° 20′ it touches the famous reservoirs built by Zobeyde, wife of Harun-el-Rashid, for the special use of the pilgrims. A caravan route from the Hauran through the Wady Sirhan, the Jauf Oasis, and the Nefud, strikes this line at Hail, and is continued thence south-eastwards through Bereydah and the ruins of Dereyah to Riad. A track from Riad through Yemamah reaches the Persian Gulf at El-Katif above the Bahrein islands. But no certain lines are known to run from this direction southwards. Riad, however, seems to be connected westwards through Taif with Mecca, and this route, if it exists, nearly completes the main lines crossing the peninsula. Blunt denies the existence of a Roman road said to have formerly run from Melakh on the Syrian frontier across the Hamad to Basra on the Shat-el-Arab.

In the foregoing chapters—Asia Minor, Euphrates and Tigris Basin, Syria and Palestine—the headings 10 (Administration) and 11 (Statistics) have not been included, these subjects having been reserved for the present chapter. It was thought best to combine Arabia with Turkey in Asia for this purpose, because some parts of Arabia either belong to, or are claimed by, Turkey, while other parts are independent.

10. *Administration: Turkish System in Asiatic Turkey generally—Social State—Taxation—Justice—Religion—The Ulema—Education.*

In theory the government of Asiatic Turkey is an absolute despotism, limited in practice by many social and religious checks. The Sultan's personal action is now largely controlled by that of the Grand Vizier and Divan (Prime Minister and Cabinet). But he still nominates not only all the members of the Divan but all the provincial governors and lieutenant-governors, whose tenure of office being precarious, the incentive to rapacity often becomes irresistible.

Since 1867 Asiatic Turkey is divided for administrative purposes into vilayets or provinces, sanjaks ("banners"), answering to the French arrondissements, kazas or districts, and nahies or communes. The vilayets, mostly named from the chief town, are governed by valis, ranking as mushirs or pashas of the highest order; the sanjaks by caimacans or lieutenant-governors, ranking as mutessarifs or second-class pashas; the kazas by mudirs, elected in theory by the inhabitants, but in reality by the valis; the nahies by muktars or mayors, ostensibly elected for a year, but really named by the mudir. There are a great number of other officials, of whom it may be affirmed that "not one owes his post to personal merit or qualification, but all to bribery or intrigue. The vali himself buys his appointment from some palace favourite or other patron at the Porte. . . . The same may be said of the cadis (magistrates), of the commandant of the police, and of the directors of the customs" (M'Coan).

Taxation is largely based on the old tithe system, and as the tithes themselves are farmed out, ample scope is given to extortion, the sum raised always far exceeding that imposed by the Treasury. Justice also, although the civil and criminal codes are based on sound principles of equity, is dispensed by servile ministers in such a manner as to become an additional instrument of oppression. In all the courts bribery is a recognised factor, and although the Turk is personally honest and upright, the Turkish official has under this system become the incarnation of servility and corruption. Even the Christian assessors associated with their Moslem rulers would seem to be deeply tainted by the prevailing laxity. M'Coan mentions the case of a Christian member of a civil court waiting on the advocate of some parties in a pending case, and arranging for a bribe of £100 to secure judgment in their favour. This judge is now "president of one of the Stambul courts, a rich and respected functionary."

The real grievance of the Christians is that their testimony carries but little weight, even when not absolutely rejected, in all the courts of the empire. "What we require," said one of them to Captain Burnaby in Smyrna, "is similar treatment for all sects, and that the word of a Christian when given in a court of law should be looked upon as evidence and in the same light as a Muhammadan's statement. If the Caimacans and Cadis were only compelled to do us justice in this respect, we should not have much cause to grumble."

But the maladministration of the Asiatic section of the Turkish empire is essentially, one might say necessarily, of a chronic character, so much so that no real improvement seems possible that depends for its enforcement on the central government. To understand the chaotic state of things which has practically prevailed ever since the displacement of the Christian Byzantine for the Muhammadan Osmanli power, it should be remembered that where Islam rules supreme there can be no equality unless the whole population accept the Mussulman faith. Under this system Church and State are inseparably interwoven, and the prescriptions of the Koran are extended not only to religious matters, but also to the dispensation of justice both civil and criminal, to the rights of inheritance, and to all social institutions. It is as if the States of Christendom were exclusively administered in accordance with the precepts of the Jewish dispensation. Thus two classes only are recognised—the faithful, who enjoy the full rights of citizenship, and the unbeliever, who have in principle no civil status. They are essentially "rayahs"—a subject herd, who pay a poll-tax, but can neither own land nor bear arms, and for whom there is no salvation here or hereafter. Hence the official form of burial certificates granted to the Christians of Turkey runs thus: "We certify to the priest of the Church of —— that the impure, putrid, stinking carcase of ——, damned (*i.e.* deceased) this day, may be concealed underground."[1] Even the Muhammadans themselves, although also suffering from misgovernment, are opposed to any

[1] Canon MacColl, in *The Nineteenth Century*, December 1895, p. 1079.

change that is not in accordance with the immutable principles of the Koran, and that would affect their position as the ruling class. This is perhaps the chief reason that the various projects of reform promulgated or promised during the present century, under pressure of the European powers—the Hatt-i-Humayun, the Hatt-i-sherif of 1853, the agreement of 1878 with Great Britain, the promises made the same year in the Treaty of Berlin (Arts. 61, 62), and the scheme of reform for Anatolia and Armenia announced in 1895—have all proved abortive.

In Armenia the situation has in recent times been aggravated by the religious dissensions of the Christians themselves, as well as by their resistance here and there to intolerable oppression, and by the action of the *Armenian Committee,* which has its headquarters in London, and which aims at the political independence of the Armenian nation. But such aspirations can never be realised, because the nation no longer exists as a compact body still occupying the old Armenian domain, but is scattered in small communities over the empire, and divided into three more or less hostile sections—the Gregorian or Orthodox Christians (75 per cent), the Protestants (15 per cent), and the Uniates or Roman Catholics (10 per cent). In their ancient territory—the "Great Armenia" as it is called, in contradistinction to the "Little Armenia" of Russian Transcaucasia—they are still most numerous, but are here not only subject to Turkish misrule, but also exposed to the systematic raids of the surrounding predatory Kurdish tribes. Hitherto these evils were to some extent endurable, because it was in the interest of the civil authorities to keep the lawless freebooters under some kind of control. But, apparently owing to a belief in the spread of the political propaganda, secret orders were issued about the year 1890, instructing the local governers to act henceforth in concert with the Kurdish raiders. That such a compact was made is evident from the fact that many of these fierce nomads were organised as a corps of irregular cavalry named after the Sultan himself, and it is matter of public notoriety that they co-operated with the regular Turkish troops in the horrible butcheries perpetrated in 1894 in the Sassun district a little west of Lake Van. These atrocities, in which 2000 souls were said to have perished, aroused the indignation of Christendom. But although the worst reports were fully confirmed by an international Commission of Inquiry (1895), and although joint remonstrances were made to the Turkish government by all the great powers acting in apparent concert through their representatives in Constantinople, the massacres were repeated in Erzerum, Kaisarieh, Sivas, Zeitun, Urfa, and many other places, until the Armenians of the Asiatic provinces were driven either to take arms in self-defence or to seek safety in exile, or even embrace Islam to save their lives and property. The object of the Sultan's government in fomenting these disorders was not so much to defy the voice of Christendom as to solve the Armenian question by extirpating the Armenian nation. Despite the protests of the powers, and their insistence on the immediate introduction of the feeble measure of reform extorted from the Sultan, this truly Machiavellian policy continued to be relentlessly pursued to the beginning of the year 1896, when it seemed likely to be crowned with a large measure of success. At that time, according to the extremely defective Turkish returns, nearly 30,000 Armenians had been killed in various towns and rural districts of Asia Minor, and 425,000 reduced to starvation. It is no exaggeration to say that between 1890-96 the Armenian inhabitants of Anatolia were reduced by half a million, partly by butcheries and famine, partly by emigration and absorption in the Muhammadan population.

But in speaking of the Turkish government justice requires that a distinction should be drawn between the "Palace," that is, the Sultan's residence at Yildiz, and the "Sublime Porte," that is, the official heads of the Administration. The situation will be best understood by assuming a spirit of direct antagonism, based on conflicting interests, permanently prevailing between, say, Windsor Castle and Downing Street. In his official report presented to the Sultan in October 1895, and later in other writings,[1] Murad Bey, late Imperial Commissioner to the Public Debt, clearly shows that a great gulf exists between the Cabinet and the Palace, where the Sultan is surrounded by sycophants, favourites, and unscrupulous adventurers. In their hands the Sultan himself, however well intentioned, is helpless, and it is made clear that for a long time the Palace has been a hotbed of intrigue, overruling the action of the Ministers, consulting personal interests alone, and fomenting disorders throughout the provinces, for the express purpose of upholding the present state of affairs and preventing the introduction of the urgently needed reforms. As a remedy for the prevailing chaos, Murad Bey suggests the dismissal of the Sultan's favourites, the suppression of privileges, the establishment of all races and religions on a footing of equality, the constitution of a deliberative Assembly, freedom of the press, a general amnesty, and the limitation of the power of the Palace (that is, of the Sultan's autocracy) to the choice of the Grand Vizier to be appointed by the European Ambassadors. A change in the succession to the Crown and other practical measures in harmony with modern ideas are also proposed,

[1] Especially *Le Palais de Yildiz et le Sublime Porte*, Paris, December 1895.

as the only means of preventing the Empire from breaking to pieces.

Religion—The Ulema.

The Sultan is primarily not so much a temporal sovereign as the accepted Caliph, or spiritual head of Islam. Hence the organic laws of the empire are all based on the Koran, to which the last appeal must be made in all emergencies.

The cardinal doctrine of the Muhammadan religion is pure theism, formulated in the words "There is but one God"; and besides the Prophet, it accepts the divine mission both of Moses and Christ. "The Son of Mary" especially is acknowledged as the Word proceeding from God, as the Messiah of the Jews, Mediator with God in heaven, and the appointed Judge of all. A final judgment, an after state, a heaven and a hell, good and bad spirits, and guardian angels, are amongst the tenets of this religion. Most of its rites, such as punctilious and ceremonious prayer, ablutions, circumcision, pilgrimage, abstinence from alcoholic drinks, are either positively good or at the least harmless; while some parts of its morality, inculcating the virtues of almsgiving, truth, sobriety, mercy to the brute creation, are to be commended.

"Islam," or the Muhammadan faith (literally "submission to God"), differs, in Asiatic Turkey, from most other religions in the absence of a true priesthood. For the Ulema[1]—that is, the "wise" or "learned"—were originally nothing more than a body of interpreters, instituted to study and expound the text of the Koran. But as the Koran contains the secular as well as the religious

[1] **From the Arabic root علم *ilm*, knowledge, science, comes the adjective *álim*, learned, wise, of which the plural is *ulema*.**

code, this body could not fail gradually to usurp a preponderating influence in the councils of the State. This influence it still enjoys and exerts in a spirit hostile not only to Christianity but to all true progress not in accordance with the "letter of law." At present the head of this college is the Sheikh-ul-Islam, or "Head of the Faith"—that is, next after the Caliph, but in purely spiritual matters enjoying a power almost paramount even to that of the Sultan himself.

Education is still in an extremely backward state, and must continue so until emancipated from the control of the Ulema, whose interest it is to restrict its range to the reading and expounding of the Koran. Attempts at reform were made so far back as 1845, when the principle of secularisation was adopted and a new university founded in Constantinople. Primary instruction was soon afterwards made compulsory, but through the influence of the Ulema it was restricted to reading, writing, elementary arithmetic, and the study of the Koran. Even in the *mekteb* or secondary schools, and to a large extent in the *medresseh* or colleges, the teachers are all members of the Ulema, with the inevitable result that education still resolves itself into a training calculated more to fill the mosques and uphold the old system than to produce enlightened and liberal-minded citizens. So much, however, has been secured that the bulk of the people, even in Asiatic Turkey, can now at least read and write.

The above description is applicable to Asiatic Turkey. But Arabia, which is included in this chapter, gave birth to a religion that has extended to several other countries besides the Turkish dominions. The Muhammadans are divided into two sects, the Sunnis and the Shiahs. The Sunnis are usually regarded as the orthodox party. They acknowledge the succession of the four Caliphs who inherited the spiritual and temporal supremacy bequeathed by the Prophet Muhammad himself. Their name indicates those who follow the true tradition. The Shiahs are usually regarded as sectaries, as their name implies. They are considered as heretics by the Sunnis, who formed the dominant party for many generations. In this age, however, they contribute an influential minority. Originally they followed Hasan and Hosen, the grandsons of the Prophet by his daughter Fatima and her husband Ali. The grandsons took up arms against the Caliphs, successors of the Prophet, and were slain in battle. Their memory is revered as that of martyrs. The two religious centres of the sect are Mashhad in the north-east corner of Persia, and Kerbela on the border of Arabia and Mesopotamia. In all the world there is no place more heartily venerated by millions of people than Kerbela. In the main the Shiah country is Persia; but Bokhara, Constantinople, Bagdad, Cairo, Delhi, Kabul, are Sunni. The Mogul emperors of India were Sunni, though, as will be seen hereafter, there are many Shiah sectaries in India. The Sunnis and the Shiahs in India have their respective watch-cries. The Sunnis say "Dam-i-chahár yár"; or "Hail to the four disciples of the Prophet" (that is, the Caliphs). The Shiahs say "Dam-i-panj-tan"; or, "Hail to the five relations of the Prophet"; meaning that the descendants have a prior claim over those who were the disciples only. The Sunnis mean that the disciples were nominated as lawful successors, and that allegiance is therefore due to them.

In Asiatic Turkey the Muhammadan practice at least is understood to be becoming more and more tolerant.

Outside Asiatic Turkey, however, the Muhammadan faith maintains its hold upon the hearts and minds of the influential classes among its adherents. It has priestly classes bearing the names of Mufti, Molavi, Mullah. They are hearty and sincere zealots. Their religious sentiments, originally pure and lofty, often degenerate into bigotry and fanaticism. From time to time, as for instance the Wahhabi movement in Central Arabia, efforts are made to reinvigorate the austere strictness of the Prophet and the Caliphs, his immediate successors. But veneration for the Prophet, his Koran, and his Tradition, never causes the people to forget the attributes of God (Allah), which ever have been, and still are, defined and formulated with extraordinary accuracy and fidelity. The merits of such tenets still infuse potent life into the religion. Though the name of "the most merciful" is constantly invoked, yet something the reverse of mercy and charity, as understood by Christians, is really presented. Almsgiving is indeed proclaimed to be a duty in the loftiest terms. But kindness is really reserved for those within the pale. For all outside the pale, fierce intolerance and an almost sanguinary animosity is declared. These are charged with "unbelief," and the term Kafir, or unbeliever, is still regarded as a severe inculpation. For all that, in countries such as British India, where Muhammadans are brought into contact with Europeans, the common humanity asserts itself, and there many good, faithful, and friendly Muhammadans are to be found.

The Muhammadan nations are retrograding, and the retrogression is in part attributable to their religion. The following sentences are taken from the Rede Lecture delivered at Cambridge in 1881 by Sir William Muir, one of the first authorities on the subject:—

"Some, indeed, dream of an Islam in the future, rationalised and regenerate. All this has been tried already, and has miserably failed. The Koran has so encrusted the religion in a hard unyielding casement of ordinances and social laws, that if the shell be broken the life is gone. A rationalistic Islam would be Islam no longer. The contrast between our own faith and Islam is most remarkable. . . . There are in our Scriptures living germs of truth which consist with civil and religious liberty, and will expand with advancing civilisation. In Islam it is just the reverse. The Koran has no such teaching as with us has abolished polygamy, slavery, and arbitrary divorce, and has elevated woman to her proper place. As a Reformer, Mahomet did advance his people to a certain point, but as a Prophet he left them fixed immovably at that point for all time to come. . . . The tree is of artificial planting; instead of containing within itself the germ of growth and adaptation to the various requirements of time and clime and circumstance, expanding with the genial sunshine and rain from heaven, it remains the same forced and stunted thing as when first planted some twelve centuries ago."

11. *Statistics of Asiatic Turkey and Arabia.*

Areas and Populations.

	Vilayets and Mutessarifats.	Area in sq. miles.	Population [1] (1885-94).
Asia Minor	Ismid	4,296	246,824
	Brussa	26,248	1,300,000
	Bigha	2,895	129,047
	Archipelago	4,963	325,866
	Candia	2,949	294,192
	Smyrna	17,370	1,390,783
	Castamuni	19,300	1,009,460
	Anghora	32,339	892,901
	Konia	35,373	1,088,100
	Adana	14,494	402,439
	Sivas	32,308	996,126
	Trebizond	12,082	1,047,700
	Total	204,618	9,123,432

[1] According to the first census of the Ottoman Empire, begun in 1885 and not yet (1894) completed. Hence some of the figures are still approximate only, while those for Arabia are merely conjectural. The estimates for this region, where no trustworthy data are available, differ enormously.

	Vilayets and Mutessarifats.	Area in sq. miles.	Population (1885-94).
Armenia and Kurdistan	Erzerum	29,614	645,702
	Mamuret-ul-Aziz	14,614	575,314
	Diarbekr	18,074	471,462
	Bitlis	11,522	388,625
	Van	15,440	376,297
	Total	89,264	2,457,400
Mesopotamia	Mosul	29,220	300,280
	Bagdad	54,503	850,000
	Basra	16,482	200,000
	Total	100,205	1,350,280
Syria and Palestine	Aleppo	30,304	994,604
	Zor	38,600	100,000
	Syria	24,009	604,170
	Beyrut	11,773	400,000
	Jerusalem	8,222	339,169
	Lebanon	2,200	245,000
	Total	115,144	2,676,943
Turkish Arabia	Hejaz	96,500	3,500,000 (?)
	Yemen	77,200	2,500,000 (?)
	Total	173,700	6,000,000
	Total Asiatic Turkey	682,931	21,608,055
Independent Arabia	Oman	82,000	1,500,000
	Shammar, Bahrein, etc.	500,000	3,500,000 (?)
Grand Total, Asiatic Turkey and Arabia		1,264,931	26,608,055

	Population.
Turkish territory ceded 1878 to Russia (Batum, Kars, etc.)	415,000
" " to England (Cyprus)	209,000
" " to Persia (Kotur)	5,000
Total	629,000

Approximate Classification by Races and Religions.

Moslem[1] 22,105,000	Turks: Anatolia, Armenia, Syria, etc.	10,500,000
	Arabs: Arabia, Mesopotamia, Syria	7,600,000
	Syrians: Syria	1,750,000
	Kurds: Kurdistan, Armenia, Anatolia	1,500,000
	Circassians and Abkhasians: Anatolia	400,000
	Yuruk Turkomans: Anatolia, Syria	200,000
	Lazis: Lazistan, Anatolia	200,000
	Meteollis: Syria	15,000

1 All Sunnis, except the Meteollis, who are of the Shiah sect, but with peculiar rites.

		Population.
Christian[1] 3,610,000	Greeks: Anatolia, Syria	2,000,000
	Syrians: Syria	350,000
	Armenians: Armenia, Anatolia	760,000
	Maronites: Syria	250,000
	Nestorians: Mesopotamia, Kurdistan	250,000
Sundries 260,000	Druses: Syria, Hauran	90,000
	Jews: Anatolia, Syria, Mesopotamia, Arabia	80,000
	Nusairieh: Syria, Anatolia	50,000
	Kizil-Bashis: Anatolia	30,000
	Yezides: Anatolia, Mesopotamia	10,000
	Ishmaelites: Syria	2,000
		27,550,000

TOWNS WITH UPWARDS OF 4000 INHABITANTS.[2]

	Pop.		Pop.		Pop.
Damascus	200,000	Aden	30,000	Ismid	15,000
Smyrna	200,000	Aintab	30,000	Mardin	15,000
Bagdad	180,000	Angora	30,000	Riad	15,000
Aleppo	120,000	Jidda	30,000	Saida	12,000
Jerusalem	80,000	Kiutayah	30,000	Hail	10,000
Beyrut	75,000	Scutari	30,000	Kurnah	10,000
Brussa	60,000	Tripoli	30,000	Latakia	10,000
Erzerum	60,000	Van	30,000	Nablus	10,000
Kaisarieh	60,000	Amassia	25,000	Tarsus	10,000
Maskat	60,000	Bitlis	25,000	Rowandiz	9,000
Sana	50,000	Chios	25,000	Taif	8,000
Sivas	48,000	Hodeida	25,000	Mokha	7,000
Adana	45,000	Konia	25,000	Nazareth	7,000
Mecca	45,000	Gaza	20,000	Arabkir	6,000
Mosul	45,000	Hillah	20,000	Yanbo	6,000
Trebizond	45,000	Homs	20,000	Acre	5,000
Diarbekr	40,000	Medina	20,000	Bayazid	5,000
Manissa	40,000	Rhodes	20,000	Bethlehem	5,000
Tokat	40,000	Jaffa	16,000	Hebron	5,000
Urfa	40,000	Basra	15,000	Loheia	5,000
Edessa	35,000				

[1] Of these about 800,000 are "Uniates"—that is, in union with the Church of Rome.

[2] These figures are mostly approximative.

CHAPTER VIII

PERSIA

1. *Boundaries—Extent—Area—Iranian Plateau—Coast-line—Islands.*

East of the Persian Gulf and of the Mesopotamian basin, which may be regarded as its northern extension, the land rises abruptly to a vast upland region, occupying the whole space between the Tigris and Indus valleys. From its earliest known inhabitants, the Iranian branch of the Aryan race, this region has received the name of the Iranian plateau. In relation to the general highland system of the eastern hemisphere, it must be regarded as forming the connecting link between the great central and western tablelands. For it is united through the Paropamisus and Hindu-Kush eastwards with the Great Pamir, the focus of the Asiatic system, and through the Armenian highlands westwards with the Anatolian table-land, whence the uplands are continued across the Ægean to the Balkan ranges and the Alps, the focus of the European system.

This vast tableland, which has a total area of about one million square miles, presents the form of a trapeze, enclosed on the south by the Arabian Sea, on the north by the Aralo-Caspian depression, eastwards by the Indus valley, westwards by the Persian Gulf and Mesopotamian basin. It is encircled on all sides by distinct mountain ranges, which descend everywhere abruptly to the surrounding waters and depressions, except in the north-west, where they merge in the still more elevated Kurdistan and Armenian highlands. Through these the plateau is supposed to be connected with the Caucasus range traversing the Ponto-Caspian isthmus. But here there is a deep intervening depression through which the Kur (Cyrus) flows east to the Caspian, while farther west the valley of the Rion (Phasis), draining to the Euxine, forms a less marked line of separation between the two systems.

The Iranian plateau thus forms a clear geographical unit. But ethnically and politically it is a divided land. Although the original home of Aryan peoples, it has for ages been the battlefield of "Iran" and "Turan"—that is, of the rival Caucasic and Mongolo-Tatar races. This struggle, combined with the spread of Islam in the seventh century, has brought about a final rupture of the old Persian Empire, which formerly gave political unity to the land. The eastern section of the plateau is thus at present occupied by the independent States of Afghanistan and Kelat (Baluchistan), the western by all that now remains of the ancient Persian monarchy, which at one time stretched from the Bosphorus to the Indus. And even here the sceptre of the "king of kings" has passed from the old native Persian dynasties to a house of the intruding Turanian race. The usurper Nadir Shah was khan of the Afshar Turki tribe, and the present ruling family belongs to the rival Qajar Turki clan.[1]

Within its present limits, as laid down by various treaties with Russia and Turkey, and by the Sistan and Afghan Boundary Commissions of 1870-72, Persia is

[1] Hence the title of the late Shah—

ناصر الدین شاه قاجار

Násr ud-dín Sháh Qájár.

bounded on the north—1st, by the Russian territory of Transcaucasia, the frontier line here following the River Aras (Araxes) for the greater part of its course to the plain of Mogân and the Lenkoran district on the Caspian, which are included in Transcaucasia; 2nd, by the south coast of the Caspian; 3rd, by the new Russian Transcaspian territory, formerly the Tekke Turkoman country. Here the frontier has not yet been determined by the Russo-Persian Boundary Commission of 1881; but it will probably run from the south-east end of the Caspian, along the Atrak River and Kopet-dagh, through Askabad to Sarakhs on the Tajand River. Westwards, Persia borders on Asiatic Turkey, the limits following a line already laid down at p. 329. But even here the frontier question, referred to an International Commission so far back as 1843, is not yet finally settled, and fierce disputes arise between the conterminous States as to the possession of certain villages, such a dispute having occurred so recently as 1889 (Curzon). On the south-west and south, the Persian Gulf and Arabian Sea form the natural limits, which again become arbitrary, and in some places quite uncertain, on the east towards Baluchistan and Afghanistan. The line has been drawn in the south from Gwattar on the coast to the Maskel River, between which and Lake Sistan it is somewhat vague. Farther north it runs nearly under the 61st meridian to Ghurian, beyond which it follows the Hari-rud to Sarakhs. It will thus be seen that in the south the frontier line should be drawn much farther to the east than is the case in most English maps, so as to include a large slice of Makran (South Baluchistan) and most of Sistan proper, which has always been claimed and is now held by the Shah's government.

With this rectification of the east frontier considerably more than one-half of the Iranian plateau belongs to Persia. The Wazir of Karman has even received the title of Sardar of Baluchistan, and attached to his government are the two large districts of Bampur and West Makran, which are practically Persian territory. Including these outlying tracts, the monarchy forms an irregular triangular mass with a base running from below Mount Ararat for about 1000 miles south-east to the Gulf of Oman, and with nearly equal sides of 700 miles north and east from Ararat to Sarakhs, and thence to the south coast at Gwattar in 61° 30′ E. long. Its contour has been compared to that of a cat on a footstool, and as Persia is specially famous for its cats, the fitness of the resemblance cannot be denied. The total area is about 630,000 square miles, with a population roughly estimated (1893) at 9,000,000, or fourteen to the square mile.

Notwithstanding its extensive oceanic coast-line of over 900 miles from Fao to Gwattar, Persia is almost destitute of islands. In the Arabian Sea scarcely a reef exists, and in the Persian Gulf, besides a few rocks and the small but important islands of Larak and Ormuz, nothing but the Kishm group of islands claimed by the Sultan of Oman. Off the Caspian coast also there is a total absence of islands, and even here the little rock of Ashurada in the south-east corner has been ceded to Russia. The importance of Ashurada as a Russian station is considerable.

2. *Relief of the land: Highlands—Plains—Deserts—The Kavirs.*

Since the surveys of Khanikoff, Lovett, St. John, and others, between the years 1858 and 1876, our notions regarding the extent, direction, and elevation of the Persian mountain systems have been fundamentally modified.

Yet the old ideas still hold their ground in popular treatises, which continue to represent the country as mainly a vast sandy plain fringed on the north and west by continuous escarpments. The truth is that the land is almost everywhere traversed by lofty ranges to such an extent that the strictly highland would seem to prevail greatly over the plateau formation. Even the plains themselves often stand at a considerable elevation above the sea, so that the Persian tableland, taken as a whole, has a mean altitude of from 3000 to 5000 feet. The ranges also, which, with few exceptions, run with surprising regularity in the direction from north-west to south-east, are far higher than was supposed, and so perfect is the parallelism that it actually influences the direction of the atmospheric currents in all the central and western provinces. It is noteworthy that the same direction was followed by the repeated shocks of earthquake felt in September 1881 in the Khoi district, Azairbijan. The disposition of the ranges, especially in the interior, is still far from being perfectly understood; but we now know that some of the ridges run for over 100 miles at mean altitudes of from 8000 to 10,000, rising in some places to 16,000 or 17,000 feet. The most extensive and loftiest seems to be the Kuh-Dinar,[1] traversing the western province of Fars, in the normal direction, at an elevation of perhaps 17,000 or 18,000 feet. Although still unexplored it is perfectly visible from the Persian Gulf at a distance of 130 miles over intervening coast ranges known to be 10,000 feet high. Yet this is about the height given on many old maps to a doubtful Mount Daena, assumed to be the culminating point of the Kuh-Dinar.

[1] *Kuh* or *Koh* (کوه) is the Persian term for mountain, as in *Koh-i-Núr*, "Mountain of Light." Like the Arabic *Jebel* and Turki *Dagh*, it is used also for a continuous chain.

THE ZARD-KUH IN THE BAKHTIARI COUNTRY.

On his journey of 1889 from Shushter across Luristan a total distance of about 250 miles, Mr. Henry

Blosse Lynch had to cross no less than ten mountain ridges by passes ranging in height from over 1000 to 8650 feet. The highest stood farthest inland near Paradomba, 54 miles south by west of Isfahan. Like so many other modern travellers, Mr. Lynch was struck by the highland character of the zone between the coast and the tableland of the interior. "The mountain range of Southern Persia, part of the great system of Europe and Asia, presents a succession of parallel ridges and valleys from north-west to south-east. To reach the Persian plateau from the plains of Khuzistan you cross the grain of the range; but the steep ascents are followed by more gentle declivities; each wall of rock is but a step to higher levels, until, after a tedious march of about 200 miles, you discern the features of the open tableland at an altitude of about 6000 feet above the sea" (*Geo. Proc.*, 1890, p. 534).

The volcanic Damavand, highest peak of the Elburz chain fringing the south coast of the Caspian, usually marked 14,700 feet on the maps, has been fixed by the Russian Caspian Survey at 18,600, while Napier and Wells estimate it at 19,429 feet; and Mount Savalan, between Tabriz and the Caspian, has been raised also by the Caspian Survey from 11,000 to 14,000 feet. Damavand, however, was again reduced to 17,930 feet by the Swedish traveller, Sven Hedin, who reached the summit in 1890. This giant of the Elburz range, which had first been ascended by Sir Taylor Thomson in 1837, and later by Brugsch Pasha, towers to a height of nearly 9000 feet above the sedimentary rocks (lias and jurassic lime and sandstone) of the adjoining parallel chains. The crater, which Sven Hedin found to be of elliptical form, about 1500 yards across, appears to be, if not quite extinct, certainly quiescent and reduced to the condition of a solfatara. Round the edge were large blocks of porphyry and sulphur, and the atmosphere was charged with a sulphurous odour.

The *Kuru-Kuh* range, running south-eastwards to Yazd, maintains for a long way a height of 10,000 feet, and is continued towards the volcanic Kuh-i-Basman (10,000 to 12,000 feet) by the snowy Kuh-Banan and other lofty ridges, culminating with the Kuh-Hazar (14,550 feet). South-east of this point the Kuh-i-Naushada volcano in Sarhad rises to 12,000 or 15,000 feet.

In the Bampur or south-eastern corner of Persia the normal north-westerly direction is broken by the coast ranges, which run either south-west or west and east, parallel with the sea. The only other important exception to the general parallelism occurs in the north, where the eastern section of the Elburz sweeps round the Caspian in a north-easterly direction from Mount Damavand to the valley of the River Gurgan.

In the north-west the separate ranges merge in the general highland systems of Luristan, Kurdistan, and Armenia, where several snowy peaks fall little short of 15,000 feet. In the north-east the Khorasan frontier is usually supposed to be separated from the Turkestan depression by a continuous range running between Afghanistan and the Caspian, and connecting the Hindu-Kush through the Paropamisus and Ghor mountains with the Elburz range. But here also the main direction is south-east and north-west from the Harirud valley to the Great and Little Balkans near Krasnovodsk on the Caspian. Thus the Kuren-dagh,[1] the Kopet-dagh, and the other unsurveyed sections of the north Khorasan highlands, run, not in a continuous line, but rather at an

[1] Although from 8000 to 11,000 feet high, the Kuren-dagh was scarcely known till its rediscovery by V. Baker in 1873 (see *Clouds in the East*, p. 289).

obtuse angle with the north Afghan system, and the break of continuity is well marked by the valley of the Hari-rud and Tajand, giving easy access from Turkestan to Herat. It follows that the northern scarp of Khorasan forms an eastern continuation, not of the Elburz, but of the Caucasus, a fact which has only recently been determined beyond doubt.

The Persian or western section of the Iranian plateau has thus a mean altitude of probably not less than 5000 feet, with a general tendency to rise towards all the seaboards and the western and northern frontiers. Between the coast ranges and the sea there are scarcely any lowland or alluvial plains except those of Khuzistan at the head of the Persian Gulf, and a few strips north of Bushahr and east of Bundar-Abbas. On the Caspian also the only alluvial tract of importance is the delta of the Safid-rud, noted for its great fertility. But in the interior, besides the rich plain of the Urmia basin in the extreme north-west, extensive level tracts everywhere occur between the parallel ridges. Those of Isfahan and Shiraz in the west maintain an altitude of about 5000 feet, rising south-eastwards to perhaps 6000 feet. But eastwards and north-eastwards the land falls continuously to the two great depressions of Sistan and Khorasan, probably not more than 1300 or 1400 feet above sea-level. Here the plains become more extensive, but also more arid, the grassy valleys gradually yielding to sandy and saline swamps and wastes. Eastwards a perpetual struggle for the mastery seems to be going on between the arable tracts and the shifting sands, which have already absorbed even some towns and villages, such as Rhages south-east of Tehran, and which are now threatening to swallow up Yazd in the very heart of the country.

Still farther east the sands themselves yield to those

dreary saline marshy tracts so characteristic of East Persia, which are termed *Kavir* in the north and *Kafeh* in the south. Here all the water from the surrounding streams and from the slight rainfall is collected in the depressions, where it forms a saline efflorescence with a thin whitish crust, beneath which the moisture is retained for a considerable time. Thus are produced those dangerous and impassable slimy quagmires, which in winter are covered with brine, in summer by a thick incrustation of salt.

By far the most extensive of these saline wastes is the Dasht-i-Kavir, or Great Salt Desert of Khorasan, which, with its southern continuation the Desert of Lut in Karman, occupies a great part of East Persia. The northern desert, which is much more salt than the southern, and apparently separated from it by a distinct water-parting, is divided into two great and several minor sections, drained by the Shurab, Kara-su, and other streams, which unite to form the Great Kavir. There are some other large formations of a similar character, north of Kum, west of Yazd, and south of Khaf, while "the ordinary kavirs are innumerable" (St. John).

Such is the general character of the Kavirs, at least in the northern and eastern districts. But Mr. C. E. Biddulph, who in 1891 traversed the western parts, crossing the Darya-i-Namak ("Salt Sea") east of Kum and Kashan, distinguishes between the ordinary saline efflorescences of the Great Kavir and the true salt depressions. The Great Kavir, he remarks, is not accurately described as the "Great Salt Desert," much of the surface being covered with soil quite free from salt and unproductive only because of the absence of water. "The saline efflorescence, known locally by the term 'kavir,' appears only in portions of this desert, and then not necessarily upon low ground, for its appearance is not owing to any deposit of salt left upon the surface of the ground by the evaporation of water containing salt in solution which has covered it, but rather to the presence of salt in the composition of the soil, which, owing to the action of dew or some slight rainfall, has worked itself up from below. This efflorescence is common in Sind, Baluchistan, and Panjab, where it is found in places covering the ground for many miles, but seldom exceeding the thickness of a stout sheet of paper. The Darya-i-Namak, on the contrary, constitutes an extensive tract of hollow ground covered with an incrustation of solid salt several feet in thickness in most places, and in some parts it might be said of almost unknown depth, which it must have taken many centuries to form" (*Geo. Proc.*, 1891, p. 647).

The mean altitude of the Kavirs above the sea appears to be little over 500 or 600 feet, and some authorities have even asserted that parts of the Great Kavir are actually at a lower level than the Caspian. But where it was skirted and partly crossed by Lieutenant Vaughan on his journey through Central Persia in 1889-90, the elevation is stated to be "between 1400 and 2400 feet." The general aspect of the great salt swamp has been described in graphic language by this traveller. "As we quitted the defile a sudden turn in the road presented to our astonished gaze what at first sight looked like a vast frozen sea, stretching away to the right as far as the eye could reach in one vast glistening expanse. A more careful examination proved it to be nothing more than salt formed into one immense sheet of dazzling brilliancy, while here and there upon its surface pools of water, showing up in the most intense blue, were visible. A peculiar haze, perhaps caused by evaporation, hangs over the whole scene, which, though softening the features of the distant hills, does not obliterate their details. This swamp, lying at a low level in the centre of the great desert, receives into its bed the drainage from an immense tract of territory. All the rivers flowing into it are more or less salt, and carry down to it annually a great volume of water. The fierce heat of the desert during the summer months causes a rapid evaporation, the result being that the salt constantly increases in proportion to the water, until at last the ground becomes caked with it" (*Geo. Proc.*, 1890, p. 590).

On the road between Tehran and Kum, a little to the north-west of the Darya-i-Namak, is situated the new lake, which made its appearance in 1883, and of which an account has been published by the Shah of Persia. In this account it is identified with a Lake of Savah, which is mentioned in history, but which dried up about 1300 years ago. Its appearance has been attributed either to waters (springs?) bubbling up in this depression, or more probably to the bursting of a dyke on the left bank of the Kara-Chai tributary of the Kum River. The lake, which has a circuit of about 40 miles, is disposed in two basins communicating through a narrow passage, and its blue waters, clear and brackish, have revived the tradition or legend of a great inland sea, which in pre-Muhammadan times covered a great part of the Dasht-i-Kavir. "There were probably a number of distinct lakes, now patches, of salt desert, which are spoken of in the popular legends of Persia as a vast sea extending from Kazvin on the north to Kerman and Mekran on the south, from Savah on the west to the Sistan depression in the east. These legends, which I have heard at many places on the confines of the desert, not only speak of a great sea, but also mention ships, islands, ports, and lighthouses. The old tower on the hills north-east of Kazvin is popularly called a lighthouse, and the village Barchin, north of Yezd, and not far from Maibud, is called an old seaport and custom-house" (General A. Houtum Schindler, *Geo. Proc.*, 1888, p. 625).

3. *Hydrography: Inland and Seaward Areas of Drainage—The Atrak, Karun, and Tajand Rivers.*

In any case it cannot be doubted that the greater part of the interior has a distinct inland drainage like that of so many other Asiatic tablelands. For while the average elevation of the plateau is about 5000 feet, it rises to 8000 towards the Tigris valley and all the surrounding seas. In fact, the true basin-like character not only of Persia but of the whole Iranian plateau is fully established by a comparison of the inland and outer areas of drainage. Of this plateau about 230,000 square miles drain to the Indian Ocean, and 250,000 to the Aralo-Caspian depression, leaving no less than 550,000 to the inland drainage. Of this area over 200,000 belong to the Helmand or Sistan basin (160,000 in Afghanistan and Baluchistan, 40,000 in East Persia), and the total inland drainage of Persia proper has been estimated at somewhat less than two-thirds of its whole area, as thus:—

		Square Miles.
To the Indian Ocean		130,000
Aral and Caspian		100,000
Inland.	Lake Sistan	50,000
	Lake Urmia	20,000
	Kavirs and other depressions	330,000
Total Area		630,000

The rivers draining the western and south-western uplands to the Indian Ocean diminish in size from north-west to south-east. Thus by far the largest are the Karkhah, Karun, and Jarahi, flowing from the Kurdistan and Luristan highlands through Khuzistan and Arabistan to the Shat-el-Arab at the head of the Persian Gulf. The Karun, which with the Diz forms a stream navigable to the first range of hills, formerly flowed direct to the sea, but now sends most of its waters through an artificial channel to the Shat-el-Arab at Mohammerah, close to the mouth of the Jarahi.

Having been recently thrown open to free navigation, the Karun, which is the only navigable river in Persia, has acquired exceptional importance from the commercial standpoint. Its upper course was crossed several times by Mr. Lynch on his route from Shushter to Isfahan. The Kuh-i-Rung, its farthest headstream, rises in the Zendeh-Kuh range so near the Zendeh-rud, or river of Isfahan, that Shah Abbas the Great attempted to connect the two basins by cutting a passage between the intervening ridge. "The course of the Karun through the mountains is remarkable: where it yields to the laws of nature it defies them. The point where we last crossed it is nearly due east of Shushter, and about 115 miles in a straight line from it. Through the whole distance the range is thrown across the river's path. . : . Beyond Dopulun the Karun turns to south-west, and after a long bend reverses its original direction, and flows north-west to the plains above Shushter. Between Dopulun and Godar-i-Balutak the Karun is fed by two considerable tributaries, the Ab-i-Bazuft and the Ab-i-Bors. The Ab-i-Bazuft flows south-east from the district of Bazuft; the Ab-i-Bors enters the Karun on its left bank, coming from Felat and Sadat, but the point of junction seems only vaguely known. North of Shushter the Shur-i-Labahri taints or tempers the glacier water with its salt stream. The Karun during its course among the mountains is swift and deep; its colour, and that of the Bazuft River, is a grass green" (*loc. cit.*).

At Shushter, where it begins to be navigable, the Karun turns sharply round to the south, and flows for some 50 miles partly through the Abi-i-gargar, originally an artificial canal, now the chief navigable branch, partly in the natural channel of the Abi-i-shateit or Karun proper, to Bund-i-Kir, where it is joined on the right bank by the Ab-i-diz, or river of Dizful, descending from the Zagros mountains in the north-west. About 45 miles below the confluence the main stream is obstructed by the Ahwaz rapids, formed by several red sandstone reefs projecting right across the channel. "Of these reefs five are distinctly perceptible in low water; and it is the water eddying above their summits, or tearing between the gaps by which they are separated, that constitutes the famous rapids of Ahwaz, and creates that practical barrier to continuous navigation which has always impeded, and continues to impede, the mercantile development of the Karun route. On the largest reef are visible the massive remains of the great *bund* or dam that was built across the river, probably in the Sassanian epoch, to hold up the waters for irrigation purposes" (Hon. G. Curzon, *Geo. Proc.*, 1890, p. 519).

Below Ahwaz the Karun, here a stately stream from 300 to 450 yards wide, 12 to 14 feet deep at high water, and with a velocity of 4 to 5 miles an hour, bifurcates some 2 miles above Mohammerah. A portion of its waters continues still to flow for over 40 miles through the Bahmeshir, its natural bed, to the head of the Persian Gulf, a little east of and parallel with the Shat-el-Arab, consequently entirely through Persian territory. But most of its volume was diverted at an unknown date by the artificial Haffar Canal to the Shat-el-Arab, about 40 miles above Fao, at the entrance to the estuary. Of the two branches the Haffar, 400 to 500 yards wide and 20 to 30 feet deep, is alone navigable, so that all the traffic

THE KARUN RIVER AT SHUSHTER.

has at present to pass through Turkish territory in order to reach the interior of Persia. Turkey, apparently jealous of the development of trade by this route, has lately erected fortifications on the Shat-el-Arab, by which the waterway may be blocked at any moment. Hence the Karun can scarcely become a great highway of international commerce until the Bahmeshir branch is rendered navigable by dredgings or embankments, and the Ahwaz rapids removed by blasting or turned by a ship canal. Such operations are not to be contemplated under the present Persian administration.

South of the Karun follows the sluggish river Tab, which has helped to form the Arabistan delta, one of the most extensive and fertile alluvial plains in Persia. But from this point to the Indus not a single navigable stream reaches the coast. Noteworthy is the Minab, which, though scarcely marked on the maps, drains all the extensive plains across the hills to the north-east of Bandar-Abbas.

Of those flowing to the Caspian by far the largest is the Kizil-Uzun (Safid-rud, or "white river"), which drains an area of 25,000 square miles east and south of Lake Urmia. The opposite or south-western coast of the Caspian is reached by the Gurgan and Atrak, the latter of which possesses great political importance, as marking the possible future Russo-Persian frontier line in this direction. In the absence, perhaps, of actual information, its course has been variously laid down, apparently according to the political proclivities of the cartographers. But it is now known to be identical with the Germeh-rud (Germe Khaneh), which rises near Kabushan (Kushan) on the southern slope of the Kuren-dagh, 6000 feet above the sea, about 58° 50′ E. long., and 37° 30′ N. lat. It flows thence mainly north-west through Shirvan and Bujnurd along the southern base of the Kuren-dagh and through the Goklan Turkoman country to the Caspian at Hasan Kuli Bay. Although over 300 miles long, the Atrak is scarcely more than 30 feet broad at its mouth, except during the spring floods, when it overflows its banks to a width of 7000 or 8000 feet. At other times it is nearly exhausted by irrigation canals and evaporation before reaching the Caspian. Near Kabushan also rises the Keshef-rud, which, however, flows south-east past Mashhad to the left bank of the Hari-rud, their junction forming the Tajand. This river, which offers the most accessible approach from Turkestan to Herat, does not end in the sands near Sarakhs, as is generally supposed, but expands into a swamp in the Attok country about 58° E. long. With a sufficiency of moisture it would doubtless reach the Caspian between the Great and Little Balkans.

The inland drainage, notwithstanding its vast extent, receives no rivers of any size, and most of them become brackish before losing themselves in the lakes or the desert. The largest are the Aji-chai and Jaghatu, flowing to the salt lake Urmia; the already-mentioned Kara-su (Hamadan-rud) and Shurab, disappearing in the Great Kavir; the Zainda-rud, watering the Isfahan district, and running dry in the unexplored salt marsh of Gavkhana; the Kur (Bendamir), chief feeder of the salt lake Bakhtegan (Bichegan, or Niriz); lastly, the Mashkel, which filters rather than flows along the Baluchistan frontier northwards to the Hamun-i-Mashkel in the Karan desert, which is separated by a range of hills from Lake Sistan.

4. *Natural and Political Divisions: Irak—Sistan—Khorasan—The Caspian Provinces—Urmia.*

From the foregoing account of its physical constitution, it is evident that the western section of the Iranian

plateau presents two well-marked natural divisions only —the central and lowland plains, and the highlands, by which they are enclosed on every side except towards the Helmand basin. But while the plains are mainly comprised in the two great provinces of Khorasan and Karman, the uplands with the narrow intervening coast strips are, for administrative purposes, subdivided into nine other governments, as shown in the statistical tables at the end of this chapter.

These "Melmeket," as they are called, are grouped round Irak-Ajemi, which forms the political centre of the State. Here are situated both Isfahan and Tehran, the old and new capital. Irak slopes from the Kurdistan highlands eastwards down to the Khorasan wastes, and rises northwards to the Elburz range, separating it from the Caspian. Southwards it reaches to the Kuh-Dinar range, thus including in its general administration the subordinate divisions of Ardalan, Luristan, and Kashan. Here are some rich grassy plains and fertile valleys, which when well watered yield excellent crops of cereals and fruits. But in the east most of the land is waste, and already invaded by the sands continually advancing westwards.

North and south of Irak-Ajemi lie the provinces of Azairbijan, Luristan, and Khuzistan, the latter including the rich alluvial plain of Arabistan at the head of the Persian Gulf. Along the Persian Gulf and Arabian Sea stretch the extensive governments of Farsistan, Karman, and that portion of Makran assigned by the Baluchistan Boundary Commission to Persia. These coast regions consist of lofty highlands rising in terraces rapidly inland, and with their main axes running north-west and south-east everywhere except in Makran, where they run partly south-west and north-east, partly west and east parallel with the coast.

North of Makran, and almost in the very centre of the Iranian plateau, lies the deep depression of Sistan, now partly included in Persian territory, but geographically belonging mainly to the Afghan drainage system. It is an extensive level and low-lying tract on the borders of Afghanistan, Persia, and Baluchistan, partly filled by the Hamun (Sistan) Lake or swamp, which receives the Helmand, Farrah, and other large rivers from the east, but only a few insignificant streams from Persia. The basin does not, however, form a single expanse of water, but is divided into the three depressions fed by the Farrah, Helmand, and Zirreh.[1] The so-called plain of Hamun is generally dry, and the presence of a large lake at this spot, as marked on most maps, can be explained only by supposing that in spring a few pools or tarns are formed in the channels about the river mouths, and occasionally united by floods in a continuous sheet of water some 70 miles long by 25 broad, but scarcely ever more than three or four feet deep. Its margin is covered with a dense growth of reeds, tenanted by numerous herds of wild hogs. The water is fresh about the river mouths, but elsewhere brackish, while its bed seems to be gradually rising, owing to the mass of detritus and mud brought down by its influents. Although fish are scarce, it is much frequented by geese, ducks, and other water-fowl.

Sir Frederic Goldsmid of the English Boundary Commission distinguishes two districts in this region—Sistan proper and Outer Sistan. The former, with an area of perhaps 980 square miles, has a settled population of 35,000, besides 10,000 nomads,—one-third Persians, Baluchis, and Afghans, the rest "Sistanis." The country is generally flat, with a sandy alluvial soil.

[1] The Zirreh, formerly supposed to drain south to the Arabian Sea, flows, on the contrary, north to the Hamun-i-Sistan basin.

growing shrubs, but no trees. There is no lack of irrigation by means of rills and rivulets, and the land is fertile, yielding melons in profusion, besides the staple products, wheat and barley, and excellent pasturage.

Outer Sistan comprises the country stretching along the right bank of the Helmand some 120 miles farther up, and properly forming part of Afghanistan.

The whole of the north-east is comprised in the vast province of Khorasan, which was carefully explored by a Russian expedition conducted by Khanikoff in 1858, and since then visited by Baker, Napier, MacGregor, and several other travellers. Its eastern section, contrary to the general impression, has been found to be very hilly, and the southern portion even bears the name of Kuhistan,[1] or "Highlands." But between the ranges there extend broad tracts of waste lands eastwards to Afghanistan, westwards to Irak-Ajemi.

In this region traces are still everywhere met of the abject fear formerly inspired by the neighbouring Turkoman hordes, whose predatory raids the feeble Persian Government was powerless to resist. But this source of trouble has been removed since the Russian occupation of the Attak in 1881. But so great was at times the distress from hunger that it got the better of the intense fear with which the people regarded their hereditary foe. Bellew tells us that the inhabitants of Mashhad crowded out of the gates of the city in the hope of being seized and carried into captivity by the Turkoman marauders, preferring a

[1] From *Kuh*, mountain, and the ending *stán* or *istán*, so universal in Persian geographical nomenclature. This ending has the general meaning of *country*, as in Farsistan, Afghanistan, Turkestan, etc. It comes from the Aryan root *tan*, as in the Latin *tendo*, with the primary idea of extension, whence a large open space, a plain, and land in general. In this sense it has travelled with the spread of the Aryan race eastwards to *Hindu-stan*, westwards to *Aqui-tania*, and *Bri-tania*.

crust of bread in exile and slavery to a lingering death by starvation at home. The Turkomans spared none but the Arabs, paid no respect to sex or age, and all unable to pay a sufficient ransom were carried off and publicly sold in the slave-market of Khiva before its suppression by the Russians in 1874. But since the Perso-Turkoman has become a Perso-Russian border-land, slave-raiding has everywhere been completely suppressed.

The open country visited by Bellew was found to be dotted over with a peculiar kind of tower, formed by a round mud wall 14 feet high enclosing an empty space open to the sky, and with a low entrance accessible only on all fours. The moment the Turkoman horsemen were detected, the people took refuge with their flocks in these buildings, which offered them a safe if temporary refuge.

The provinces of Ghilan and Mazandaran comprise the wooded northern slopes of the Elburz, besides a more or less extensive strip of flat alluvial coast-land between that range and the Caspian. This tract, often swampy and exposed to deadly fevers, and producing chiefly rice, cotton, silk, and some sugar-cane, is mostly covered with dense forests, like the neighbouring mountains themselves. Herein it presents a striking contrast to the bare, desert, or arid regions to the south of the Elburz—that is to say, to the rest of Persia, which has been caustically described as a land divided into two sections—a salt waste and a saltless waste.

The extreme north-west between the Caspian and Turkey—that is, the "cat's head" in the general contour—is comprised in the province of Azairbijan. It is partly cut off from the Caspian by the Russian district of Lenkoran, and the Russo-Persian frontier is here traced by the River Aras almost from the foot of Mount Ararat nearly to its junction with the Kur. In this upland region, where Mount Savalan attains an altitude of 14,000 feet, the great feature is the remarkable closed basin of Lake Urmia, alike interesting in a geological, ethnical, and economic sense. In this comparatively narrow tract several streams rise almost in close proximity, which nevertheless flow in four opposite directions—the Safid-rud east to the Caspian, the Kara Rud north to the Aras, the Aji-chai west to Lake Urmia, the Zab south-west to the Tigris. Here also, after a lapse of thousands of years, the surrounding antagonistic ethnical elements have hitherto failed to establish an equilibrium—Turkoman, Kurd, Nestorian, Armenian, and Persian still struggling for the supremacy, and apparently unconscious that the shadow of the northern colossus has already fallen on the land.

The lake, which although 4750 feet above the sea is a completely closed basin some 80 miles long by 20 broad, is extremely salt and very shallow. The average depth scarcely exceeds 6 feet, and is nowhere more than 24. It lies in a district of almost unrivalled fertility, covered with vineyards, orchards, gardens, rice grounds, and thickly studded with towns and villages. Urmia, the largest of these, whence the lake and district take their names, is the centre of an American mission, which has for many years worked earnestly and successfully in the cause of true progress and enlightenment.

5. *Climate: Rainfall—Prevailing Winds.*

The climate of Persia is on the whole continental, great dryness being combined with excessive heat, and in many of the uplands with extreme cold. On the northern ranges snow falls as early as November, and it sometimes freezes in Tehran as late as the middle of March. Between these ranges and the Caspian the heat is almost tropical, with an abundance of rain, resulting in the rich and varied vegetation of Ghilan and Mazandaran. The sultry and unhealthy climate of the Persian Gulf seaboard has already been noted. That of Sistan in the extreme east is also very unhealthy, and subject to great extremes of temperature.

North-west and south-east winds prevail throughout the year with great uniformity, their direction being largely determined by the Black and Arabian Seas at these two quarters, and by the remarkable parallelism of the intervening mountain ranges. The atmosphere of the central plateau being rarefied by the great heat of the sun, the cooler currents from the Euxine and Indian Oceans set in to fill up the vacuum; and as the former are the colder of the two, the north-west naturally prevail over the south-west winds. On the south-west coast these two currents often meet, so that a gale from the north-west is often raging at Bushahr in the Persian Gulf when Bandar-Abbas on the Strait of Hormuz is exposed to the fury of a south-easter. As most of the moisture is also brought from the latter quarter, it follows that the prevailing winds are dry, especially as the rain-clouds from the Black Sea and Caspian are mostly arrested by the Armenian and Elburz highlands, while much of the moisture from the Indian Ocean itself is precipitated on the southern and western coast ranges. Hence, excluding the Caspian basin and a few other more favoured tracts, the average annual rainfall on the Persian plateau is probably less than ten inches, and in the eastern Kavir region and Sistan not more than half that quantity. "Were it not that the lofty hills store the moisture in the shape of snow, nine-tenths of Persia would be the arid desert that half of it now is. As it is, cultivation over the greater part of the country is possible only by artificial irrigation, either by canals or by the system of wells connected by underground channels called *kanát* or *kariz*, and peculiar to the Iranian plateau" (St John).

6. *Flora and Fauna: The Camel.*

The results of this deficient rainfall are seen, not only in the undeveloped water system, but also in the vegetation, which is characterised by the absence of trees and even large shrubs almost everywhere except on the outer slopes of the coast ranges. The date-palm flourishes along the sandy shores of the Persian Gulf, but the oaks and other trees of the Bakhtiari and other inner ranges are mostly stunted, and true forests are found only on the northern slopes of the Elburz. Here large tracts are covered with dense plantations of magnificent timber, especially cedars, elms, oaks, the walnut, beech, and the valuable box tree. Wheat and barley are here cultivated to a height of several thousand feet, while the lowlands yield cotton, sugar, silk, grapes, figs, cherries, peaches, plums, and other fruits, in great profusion. Indigo, rice, tobacco, and madder are also cultivated in this region, as well as in the Urmia basin, and on the Isfahan and Shiraz plains, which are almost the only other really fertile tracts in the whole kingdom. Pasture lands are much more extensive, occupying most of the elevated longitudinal valleys and slopes of the parallel ranges in the west and north-east. Hence the Kurdish, Lur, Farsistan, and North Khorasan highlands have been held for ages by nomad pastoral tribes both of Iranian and Turanian stock. The eastern low-lying plains of Khorasan and Karman are almost destitute of vegetation, producing little beyond sands and salt. In Sistan tamarisks and dwarf mimosas are a prevailing feature.

Farsistan is still haunted by the lion, while the tiger, leopard, chitah (used for hunting), hyæna, wolf, lynx, jackal, and some smaller beasts of prey, infest the northern provinces skirting the Caspian seaboard. The *Capra ægagrus*, the supposed ancestor of the domestic goat, is spread over the whole Iranian plateau, and the bustard (*Otis houbara*), here indigenous, is hawked or followed with the gun. Here also the pheasant is indigenous, and the woodlands are enlivened by the song of blackbird, thrush, and bulbul or Eastern nightingale. Fish abound in the Persian Gulf and Caspian, and the sturgeon fisheries of the rivers flowing to the latter sea are very productive. But fresh-water fish are rare, and Urmia and the other lakes are almost destitute of animal life.

With the exception of the Mazandaran black cattle and the familiar "Persian cat," the domestic animals are mostly of inferior stock. The goats, however, of Karman yield a hair equal almost to that of Kashmir, while the fat-tailed sheep supplies the chief staple of animal food. Its wool, also, is of good quality, and either woven into fabrics of various sorts or else left on the skins, which are then cut into garments much affected especially by the nomad Iliat tribes. The chief beast of burden is the mule—a strong, hardy, and sure-footed animal, well adapted to the rude tracks in the highland districts. The camel is also employed for the caravan trade across the sandy plains, and there is a useful breed of small horses, crossed with Arab blood and noted for their speed and shapely forms. The hair of the camel forms the woof and cotton the warp of the camel's hair cloth for which Persia is famous. It is woven very closely so as to be quite waterproof. The camel is not only a very unmanageable beast, but also extremely timid and scared by the least unwonted sound or sight. The jambaz, or riding camel, is, however, an exception, and this breed is also noted for its speed and endurance.

Lieutenant Vaughan describes some remarkable features in the reptile and insect fauna of the great central wilderness traversed by him in 1890. "All the snakes I saw were brown, exactly resembling in colour and

THE DÚMLAK (*Galeodes Araneoides*).

appearance a piece of dead stick. Some of them used to climb bushes, and hitching their tails round a bough would stick their bodies out in imitation of a withered branch, and thus remain motionless for hours. My servant said that they were waiting for a bird to come and perch on them, when they would immediately strike it. There was also a curious spider called the *dúmlak*. He had long hairy legs, formed of shell like those of a crab, while his body was soft and attained the size of a walnut. They were provided with two pairs of curved crab-like claws, which carried a row of teeth like a saw on their inner surface. They spin no web, but run about on the surface of the ground with great velocity, seizing any beetles or insects which fall in their way. The natives say that they are very poisonous, and that all insects living in the desert are so, even though the same species may be found harmless elsewhere" (*loc. cit.*).

7. *Inhabitants: The Modern Persian.*

In its ethnical constituents Persia presents a most striking contrast to the Arabian peninsula. Here all the inhabitants belong to one primeval stock, with scarcely any intrusion of foreign elements. But in Persia not only are the two fundamental Asiatic types fully represented, but several distinct branches of each divide the land into a number of ethnical groups presenting almost as great a variety of races as is found in any other part of the continent. This will be at once evident from the subjoined table of the inhabitants of Persia classed according to their several racial affinities :—

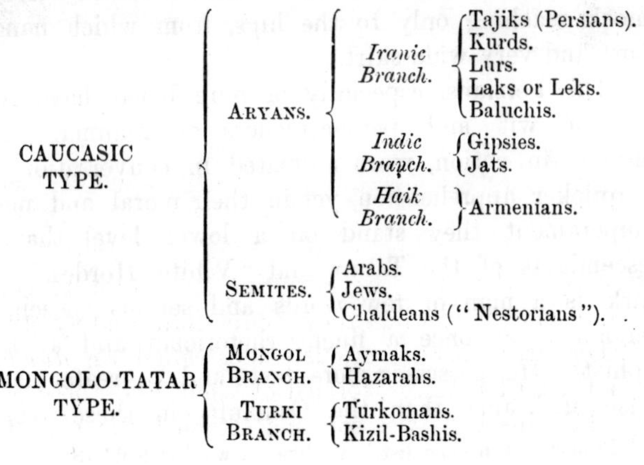

Type	Group	Branch	Peoples
CAUCASIC TYPE.	ARYANS.	*Iranic Branch.*	Tajiks (Persians). Kurds. Lurs. Laks or Leks. Baluchis.
		Indic Branch.	Gipsies. Jats.
		Haik Branch.	Armenians.
	SEMITES.		Arabs. Jews. Chaldeans ("Nestorians").
MONGOLO-TATAR TYPE.	MONGOL BRANCH.		Aymaks. Hazarahs.
	TURKI BRANCH.		Turkomans. Kizil-Bashis.

Nevertheless the bulk of the people still belong to the old indigenous Iranian stock. These western Iranians, or Persians proper, are everywhere throughout Central Asia known exclusively as Tajiks, and in West Irania as Tats, possibly a contracted form of the same word. In the north and east Turan has largely encroached on Iran; but elsewhere the old race has held its ground, hemmed in between the Arabs on the west, the Armenians on the north, and the Turkoman tribes on the north-east and east.

In religion the Persians belong almost exclusively to the Shiah sect, and often harbour feelings of rancour towards their western neighbours the Sunnite Turks. The Persians, often called Qajar,[1] from the tribal name of the reigning dynasty, and usually recognised by the *kulah*, or characteristic black lambskin head-dress, are very extravagant in their dress, the *jube*, or outer garment, often costing from £40 to £50. On the other hand, they seem to be economical in respect to under linen. In fact, when judged by the Persian standard, the Turk himself appears to be a model of cleanliness.

The domestic garb of the women is unattractive, the smock reaching only to the hips, from which hangs a short and very wide skirt.

The Persians, especially of pure blood, have readiness of wit and persuasiveness of manner. More nervous in action, more animated in conversation, and of quicker apprehension, yet in their moral and mental temperament they stand on a lower level than the descendants of the Tatars and "White Hordes." The Turk is a man of few words and serious speech, the Persian is at once a fluent rhetorician and a skilful sophist. He possesses more taste and a greater natural sense of beauty than the Osmanli; in these respects

[1] Pronounced in some districts "like our word cudgel" (St. John).

often betraying a strong likeness both to the Greek and the Jew. And though he is apparently more fanatical than the Sunnite Muhammadan, yet European ideas ought to find more acceptance with him than under the sway of the Crescent. The splendour and the power of the state seem with the Turk to be bound up with his religion, for "the glory of Islam is the glory of the Osmanli." But not so with the Persian, whose forefathers were Persians before the appearance of Islam, and whose nationality had already acquired a recognised political status ages before the days of the Prophet. The Ottoman, again, is a stock-breeder, a husbandman, and a soldier; the Persian, above all, a trader and an artist. And that the natives of Irania descend from an ancient and long-civilised race is agreeably evident to the stranger in the politeness, courteous and even refined demeanour of the people, whether they belong to the urban or rural classes.

But from the old despotic systems they have inherited the taint of cruelty. The savage sentences imposed, especially for murder, theft, and political offences, are carried out in a cold-blooded manner, which implies that some feelings at least have long been deadened to all sentiments of humanity. A frightful story is told of a slave in Shiraz, twelve years old, who had accidentally shot his master's son, and who was sentenced by the governor to be crucified. Here also Bower witnessed the execution of eleven robbers in one morning. The criminals being arranged all in a row and smoking a *kalián*, the executioner walked up, slipped their heads under his left arm, and cut their throats one after the other in the most matter-of-fact way.

Mr. E. G. Browne, author of *A Year Amongst the Persians,* insists on the political importance of the fact that the reigning dynasty is of Turki race. "The whole history of Persia, from the legendary wars between the Kiyanian kings and Afrasiyab down to the present day, is the story of a struggle between the Turkish races whose primitive home is in the region east of the Caspian Sea and north of Khorasan, on the one hand, and the southern Persians, of almost pure Aryan race, on the other. The distinction is well marked even now, and the old antipathy still exists, finding expression in anecdotes illustrative of Turkish stupidity and dulness of wit. Ethnologically, therefore, there is a marked distinction between the 'people' of the north and the people of the south—a distinction which may be most readily apprehended by comparing the sullen, moody, dull-witted, fanatical, violent inhabitants of Azarbaijan with the bright, versatile, clever, sceptical, rather timid townsfolk of Kirman. In Fars, also, good types of the Aryan Persian are met with, but there is a large intermixture of Turkish tribesmen, like the Kashka'is, who have migrated and settled there. Indeed the intermixture has now extended very far, but in general the terms 'northern' and 'southern' may, with reservation, be taken to represent a real and significant difference of type in the inhabitants of Persia. Since the downfall of the Caliphate and the lapse of the Arabian supremacy, the Turkish has generally been the dominant race; for in the physical world it is commonly physical force which wins the day, and dull, dogged courage bears down versatile and subtle wit. Thus it happens to-day the Kajars rule over the kinsmen of Cyrus and Shapur, as ruled in earlier days the Ghaznavids and the Seljuks. But there is no love lost between the two races, as any one will admit who has taken the trouble to find out what the southern peasant thinks of the northern court, or how the Kajars regard the cradle of Persia's ancient greatness" (*A Year Amongst the Persians*, 1893, p. 99).

8 *Topography: Tehran—Kum—Isfahan—Shiraz—Persepolis—Mashhad—Kelat—Tabriz—Seaports.*

From the outward conditions of soil and climate it naturally follows that nearly all the settled population and large towns are found concentrated in the western provinces, where the land contracts between the Caspian and Mesopotamian basin. The desert region east of the 53rd meridian, comprising about two-thirds of the kingdom, contains scarcely any noteworthy places except Mashhad, Tabbas, Yazd, and Karman, which form so many stepping-stones across the saline and sandy wastes from north to south. But west of that line are situated not only the mediæval and modern capitals, but also the ruins of the ancient Persepolis, besides Shiraz, Rasht, Kasvin, Tabriz, Bushahr, Shustar, Karmanshah, Hamadan, Kashan, and several other towns, which either still are or have been important centres of trade and culture during the long annals of the Persian empire.

Tehran, the present capital, lies far to the north, almost at the foot of the Elburz mountains, where they culminate in the majestic Damavand. From the summit of this quiescent volcano the city is perfectly visible, lying in the midst of an arid steppe, apparently one of the most unlikely spots to form the political centre of a large monarchy. Although it has been the capital since the year 1788, Tehran has scarcely a respectable building to show except the quadrangular palace of the Shah, absorbing nearly one-fourth of the enclosed space, and the mansions occupied by some of the nobles and the European legations. The streets are mostly narrow, crooked, and badly paved, and lined with mean houses, whose uninviting exterior corresponds with their miserable internal appearance and fittings. The bazaars, however, contain a good show of the various artistic

objects for which Persia has at all times been famous. Thanks to its political importance, Tehran has considerably increased in size of late years. The old walls, four miles in circumference, have everywhere been encroached upon, and the new quarters have now been enclosed by an outer wall and ditch enclosing a space much larger than the whole of the old town. In summer, when the heat is almost intolerable, the Court, embassies, and wealthy citizens retire to Gulahak and other pleasant retreats on the neighbouring hills. The road to these places passes the Kasr-i-Qajar, or "Palace of the Qajar," which, though now seldom occupied by the Shah, stands on an imposing site in the midst of beautiful grounds that have been compared by Oriental fancy with those of Versailles.

About 85 miles on the road from Tehran to Isfahan lies the town of Kum, which is held next in sanctity to Mashhad "the Holy." Here is the famous shrine of Masuma Fatima, the sister of the Imam Riza, the gilded dome of which has been completed by the present Shah, and which also contains the remains of ten kings and 444 "saints." It is usual to visit this shrine before proceeding to Mashhad or Kerbela, and Kum has become a favourite spot for the interment of the Faithful, whose bodies are brought hither from great distances. But the town itself is mostly in ruins, of its 20,000 houses not more than 6000 being at present occupied. "Its streets and bazaars are deserted, and dangerous from the innumerable holes and pitfalls with which they abound; and its general condition provides an impressive commentary on the state of absolute stagnation which seems to be one of the chief characteristics of the Muhammadan religion" (Major E. Smith).

On the western border of Irak-Ajemi lies Hamadan, the ancient Ecbatana, where the Jews still show the

KUM.

tombs of Esther and Mordocai. Farther south, the apex of an isosceles triangle, whose base connects Hamadan with Tehran, marks the site of Isfahan, the mediæval capital of the kingdom, and the centre of Muhammadan culture in Persia. Isfahan, which was at one time probably larger than any of the old or more recent capitals, lies in a pleasant, well-cultivated plain, almost midway between the Caspian and Persian Gulf, and between Karman and the Turkish frontier south-east and north-west, about 300 miles from all these points, consequently in the most central habitable part of the State. Notwithstanding the many calamities it has suffered and the loss of prestige following the removal of the seat of government northwards, it is still a large place and the centre of many flourishing industries. It is still adorned by several magnificent edifices, dating mostly from its former periods of prosperity, conspicuous amongst which are the large royal palace, the Chhar Bagh, the royal mosque (Masjid-i-Shah), said to be the most sumptuous in the whole Muhammadan world, and the great bazaar of Shah-Abbas. Under Shah-Abbas (1587-1628), who made it his capital, Isfahan reached its greatest splendour, and had at that time no less than 1800 caravansarais, 270 public baths, over 100 large mosques, and a population estimated at 750,000.[1] Even still it ranks with the foremost cities of the East, and, according to the local saying, but for Lahore it would be equal to half the world. It suffered severely during the famine of 1871, but has since then sufficiently revived to give the general impression that it must have fitly represented the regal state and grandeur of modern Persia.

[1] Even when captured by the Afghans under Mir Mahmúd in 1722, "it was esteemed the largest and most magnificent city in Asia, with 600,000 inhabitants" (Jonas Hanway, iii. 122).

Shiraz, capital of Farsistan, occupies one of the most favoured sites in Persia, at an elevation of 4500 feet above the sea, about 220 miles south of Isfahan, and 120 east of Bushahr, its port on the Persian Gulf. Nestling amid rose gardens, vineyards, and cypress groves, Shiraz, although like Isfahan a mere shadow of its former greatness, still retains a certain importance, due largely to its excellent wine, in flavour like the royal Hungarian Tokai, and to its rosewater and attar of roses industries. Its delightful situation has been the everlasting theme of the Persian poets, and the first sight of its soft dark-green vegetation, above which towers the lofty dome of the Shah-Cherak mosque, is naturally calculated to produce some enthusiasm after the traveller's eye has lighted for weeks together on nothing but arid sandy wastes. The abundance of water here produces a flora of tropical luxuriance, and to the charms of a magnificent and varied vegetation are added those of a limpid blue sky and a perennially mild atmosphere. Unfortunately a soft climate, a fertile soil, and an easy life have had an enervating effect on the inhabitants.

South and east of Shiraz are the two salt lakes Mahalu and Bakhtegan (Niriz), and 25 miles to the north-east lie the extensive ruins of Persepolis. Conspicuous amongst them is the so-called palace of Darius, said to have been destroyed by Alexander, and occupying a terrace 1430 by 800 feet, approached by steps cut in the rock. Vast portals and sphinxes, with many still standing pillars and walls covered with sculptures and cuneiform inscriptions, still attest the former magnificence of the royal palace of the Achæmenides.

Some miles south by east of Lar, which gives its name to the province of Laristan, lies the town of Bastak, which, although a place of 5000 inhabitants and capital of a well-peopled district within a few marches of Lingah on the Persian Gulf, was unknown to European geographers before its "discovery" by Lieutenant Vaughan in 1890. It stands on a plain encircled by heights, from which "it presents a most picturesque appearance, being surrounded by green fields and large plantations of date-trees. There is no bazaar, the place being decidedly unsettled, and subject to occasional inroads of hostile tribes. Two months ago the brother of the present khan was murdered in the streets of the town while on his way to the mosque to pray. The murderer was another brother, who wished to become khan himself" (Lieutenant Vaughan, *loc. cit.*). It is noteworthy that the surrounding populations are not Shiah but Sunni Muhammadans.

In the north-east the only noteworthy place is "Mashhad-i-mukaddas," or "Mashhad the Holy," capital of Khorasan, and the religious and trading centre of East Persia. Next to Mecca and Kerbela, this is the most hallowed spot in the Moslem world, for here reposes under a gorgeous gilded dome their most revered saint, the Imam Riza. His shrine, to which no "infidel" is allowed access, is yearly visited by over 100,000 votaries from all parts. Although slumbering in his sumptuous tomb for centuries, Riza is still treated as if he were actually living. "His shrine is enormously rich, possessing land and property in all parts of Persia, and attached to it is a large establishment of officials and servants" (Major E. Smith). This traveller adds: "Holy as Mashhad is said to be, we were struck with the great amount of drunkenness prevalent there amongst the followers of the Prophet."

About 50 miles north of Mashhad, and 60 west of Sarakhs, in 37° N. lat. and 60° E. long., lies the extraordinary natural fortress of Kelat, about 3400 feet above the sea, and close to the new Russo-Persian frontier.

Very little was known of this marvellous place until it was visited by Colonel MacGregor and by Valentine Baker, the latter of whom calls it "one of the wonders of the world," describing it as a gigantic stronghold formed by Nature herself, with very little aid from man. "The walls are mountains of from 800 to 1200 feet high, and with a sheer perpendicular scarp between 300 and 600 feet. It is an irregular oblong about 21 miles long by 5 to 7 broad. There are only five entrances, through narrow natural scarps, and these are fortified. The ground enclosed within is very rich, and it might be a perfect garden, and self-supplying. A stream runs right through the place, in at the southern entrance and out at the northern. There are also several springs within the fortress, and an ample supply of good water could thus be obtained for the cultivation of the whole interior. But everything about it now betokens utter ruin and neglect" (*Clouds in the East*, p. 201). Owing to this neglect, the fortress, where a battalion of troops with cavalry and some guns are maintained, has become so unhealthy that the garrison is often decimated by typhus, and constantly deserting its post.

Near the north-west frontier lies Tabriz (Taurus), the largest city and principal commercial emporium of the kingdom. It stands at the base of the high and rocky Mount Sahend, about 5000 feet above the sea, and on the Aji-chai, 36 miles above its entrance into Lake Urmia. Tabriz, which contains no remarkable buildings except the citadel, originally a mosque, over 600 years old, at one time possessed a large number of khans, splendid mosques, public baths, and a population of 550,000, now reduced to one-third of that number. The neighbourhood is extremely fertile, producing large quantities of magnificent grapes and other fruits.

Of the seaports, the most noteworthy are Rasht and Barfrush on the Caspian; Mohammerah on the Shat-el-Arab; and Bushahr, Bandar-Abbas, and Lingah on the Persian Gulf. Rasht, capital of Ghilan, stands at the head of the shallow lagoon or backwater of Enzeli, where all the shipping stops. It is a thoroughly Persian town, with dirty, close streets, and very unhealthy, as is most of this low-lying, swampy coast. The importance of this

FORTRESS OF TABRIZ.

flourishing seaport is in great measure due to its large export trade in silk, all the raw silk of the province being shipped at Rasht.

Barfrush lies at the mouth of a large sluggish stream 300 feet broad, here crossed by a solid brick bridge. It is surrounded by dense forests, is noted for its numerous schools and colleges, and does a considerable trade in silk and cotton. The population, said at one time to have amounted to 200,000, has now fallen to less than one-fourth of that number.

Although at present a wretched place, built of mud

and bricks, squalid, half-ruined and dirty, Mohammerah seems destined to a prosperous future. It occupies an advantageous position about a mile from the Shat-el-Arab, close to the Turkish frontier on the right bank of the Karun, which here joins the main stream, and which was thrown open to international trade by the Anglo-Persian Convention of 1888. Mohammerah must thus eventually become the chief outlet for the produce of West Persia; but meantime the attempts made to develop the traffic of the country by the Karun water-course have not been very successful. In 1892 the Bombay and Persian Gulf SS. Company had withdrawn their b' ats, and when Mr. Cooper passed down connectior Ahwaz on the Upper Karun was maintained y a small steam-launch. "Persian pigheadedness w.. probably for some time stand in the way of the successful development of trade" (*op. cit.*, p. 425). But behind "Persian pigheadedness" stands the more formidable obstacle of Russian jealousy and intrigue.

Bushahr (Bushire, Abu-Shahr), the chief port on the Persian Gulf, lies 150 miles from the Euphrates delta and about 120 miles west of Shiraz, from which it is separated by the lofty coast range here culminating in the Kuh Hormuj Peak (6500 feet). The anchorage, which is of somewhat difficult access, is distant 2 miles from the town, which stands on a long sandstone ridge separated from the mainland by a saline swamp. Bushahr is a modern place, dating from the time of Nadir Shah; hence the common derivation of its name from *Abu-Shair*, "Father of Cities," cannot be correct (Curzon). It was occupied by the British during the Persian war (1856-57), and is at present the chief emporium of the British and Indian trade with the southern provinces; but although presenting a pleasant appearance from a distance, a nearer approach reveals the usual uninviting features of Persian towns. From this point to the Indus the only port of the least importance is Bundar-Abbas, a small place facing the island of Kishm in Hormuz Strait. It is inhabited chiefly by Arabs, who carry on a considerable coasting trade in fish, salt, and fruits; but the heat is so intense that the place is almost deserted in summer. The Sultan of Oman has long claimed jurisdiction over Bundar-Abbas and the neighbouring strip of coast and adjacent island of Kishm.

Bundar-Abbas has inherited some of the trade, but none of the splendour of the remarkable island of Ormuz (Hormuz), which gives its name to the strait, and which in the seventeenth century had a population of 40,000, now reduced to less than 500. At that time Sir Thomas Herbert (quoted by Curzon) speaks of its "houses furnished with gilded leather, and India and China rarities. Buzzar [bazaar] rich and beautiful, splendid churches [Portuguese], and castle regularly and strongly fortified" (*Persia and the Persian Question*). The remains of water reservoirs, mosques, and numerous other structures are still visible, and excite the astonishment of travellers in an arid island now destitute of vegetation and fresh water, and containing nothing but salt, sulphur, and iron ores. Rock salt and sulphur peaks are visible from the coast, and there are several volcanic cones, besides a remarkable natural curiosity, "a broad stream of water flowing to the sea covered with a dazzling crust of salt, and in the centre, but not mixing with it, a blood-red streamlet tinged with iron ore. A Portuguese fort and lighthouse still exist, the former said to be a wonderful construction of dressed stone" (H. S. Cooper, p. 451). After their expulsion from Persia, the Parsees sojourned some time on Ormuz Island in the eighth century, before finally taking refuge in India from their Muhammadan persecutors.

9. *Highways of Communication.*

In Persia there are one or two good roads of short length—as, for instance, that which runs from Tehran for a few miles to the villas and villages on the southern slope of the Elburz. But all the rest are mere caravan tracks or bridle-paths, whose character depends more on the nature of the land than on the hand of man. The wretched state of these routes is the universal theme of travellers, who are more surprised to find any attempts at repairs than disappointed at the universal neglect. "The absence of roads is the curse of the country. The whole traffic is carried on by mules on the mountains, and camels on the plains, no wheeled carriages existing" (Baker's *Clouds in the East*).

The main highways, such as they are, run in all directions, and even across most of the kavírs between all the large towns and the Russo-Turkish frontiers. Towards Afghanistan and Baluchistan there seem to be scarcely any recognised tracks, and those that formerly existed have been mostly closed and lost through political jealousies. A Persian army could no doubt again find its way from Mashhad to Herat; but for much of the way the route for baggage and artillery would have to be rebuilt. The English Boundary Commission, coming up from the coast to Sistan, was guided in many places more by compass and the stars than by any perceptible paths, and it would probably be impossible to get from Yazd or Karman direct to the Helmand basin. Here the best track, starting from the Hamun swamp, seems to run through Birjand and Kaian, or more to the west through Tun northwards to Mashhad, south-westwards across the Sarhad country to Bam, where it strikes the path running from Bampur north-westwards to Karman, and so on through Yazd, Agda, and Nain to Isfahan, and thence through Kashan and Kum northwards to Tehran. Here it would meet the northern route continued from Mashhad through Sabzawar, Shahrud, Damghan, and Samnan, thus completing the circuit of East Persia. A pilgrims' route from Mashhad to Yazd and Isfahan follows the watershed between the northern and southern deserts, the chief stations being Tun, Tabas, and Gustan, with a branch at Tabas, passing direct either through Ardakan or Nain to Isfahan. The two capitals are connected by trade routes, with Bushahr and Bandar-Abbas on the Persian Gulf, with Rasht and Enzeli on the Caspian, and through Tabriz and Erzerum, with Trebizond on the Euxine.

After the present Shah's return from Europe in 1875 an extensive railway scheme was projected, which began and ended with a small line of a few miles, opened in 1876 at Rasht. But a tolerably complete telegraphic system has been developed under the direction of Sir F. Goldsmid, by which Persia is brought into direct communication with the rest of the world. The lines are laid down in duplicate, one ashore and one submarine, from Karachi, Indus delta, along the coast to Jashak, whence both are submarine to Bushahr. Here they bifurcate, one branch running through Shiraz, Isfahan, Tehran, and Tabriz, to the Russian system at Tiflis, the other crossing the Gulf to Fao, and thence running through Bagdad, Diarbekr, and Constantinople, to the various Western systems.

10. *Administration: Social State—Army—Education.*

The Government of Persia has ever been an absolute despotism in the strictest sense, the head of which bears the title of Shah-in-Shah, or "King of Kings." A revenue is raised of about £1,600,000, a sum which is probably equal to £8,000,000 in England, but which fluctuates considerably with the rise and fall in the price of silver. But it proves often insufficient to meet the requirements of the State. The country suffers from defects in the administration, the administration from faults in its subjects, the subjects from disadvantages of soil and climate,—a vicious circle, from which there seems no escape.

Among the physical disadvantages, the drifting of the sands is prominent. "The sands are in many places visibly gaining on the arable land, and even on the walled towns themselves. It is, in fact, in the process of changing from a series of rocky ridges to one of undulating sandy wastes. . . . You see the sand blowing against the wall, gradually getting higher and higher till it blows over, and then forms a mound in the field beyond, which gradually increases its height till all trace of wall and field is lost, and you have before you a sand-heap. I can quite believe now the story of towns being buried, having myself seen the thing on a small scale" (Col. MacGregor).

To these physical causes of decay must be added the foreign wars and internecine feuds, by which the monarchy was wasted throughout the whole of the last, and for many years during the present century. On the death of Nadir Shah in 1747 it was distracted by a series of fierce dynastic struggles between the rival Afshar and Qajar Turki houses, attended by excesses of every kind, which caused Jonas Hanway almost to despair of its future. "These intestine broils," he exclaims, "have extinguished the glory of Persia. What the fate of that wretched country will be Heaven only knows. But this is evident, that the splendour of their monarchy, all their monuments of art and labour, with all the industry of past ages, are swallowed up by the ravages of war. What numbers of their towns, their cities, their fruitful plains and delicious mountains, are become a dreary waste, and the habitation of wolves!" (iv. 301).

The nominal strength of the army is 105,000 men, of whom perhaps not more than one-third are ever under arms at a time. The rest form a sort of reserve, which, though mostly unarmed and engaged in husbandry, are liable to be called out at any moment. Their arms consist of old English or French muskets, supplemented by a few thousands of home make, and perhaps a hundred available guns of small calibre, with a few Uchatius rifled cannon introduced in 1881. The officers are mostly ignorant and untrained, while the men, with their shabby and tattered uniforms, look more like half-starved mendicants or highwaymen than guardians of the peace. They are drilled after the English fashion, but in a very lax way, and are seldom regularly paid. But their physique, being drawn almost exclusively from the hardy Turkoman, Kurd, and Luri tribes of Azairbijan, Kurdistan, and the Bakhtiari highlands, is magnificent. "It is the concurrent testimony of all who have been connected with the Persian army, that no people in the world present better rough material for soldiers than the Persians" (*Times* Correspondent, Sept. 8, 1881).

Public instruction, which had hitherto been mainly confined to religious teaching, is at present being thoroughly revised and improved. The nucleus of a university was formed in 1881 in Isfahan, where colleges are in course of erection for the teaching of the Oriental and European languages, besides various branches of art and science, mostly under European supervision.

11. *Statistics—Areas and Populations.*

Various Estimates.

	Area in sq. miles.	Population.
Wagner	637,000	6,000,000
Ritter	645,000	5,500,000
Almanac de Gotha	660,000	7,000,000
St. John	610,000	10,000,000
Dickson (1885) . . .	660,000	7,653,000

According to an estimate based on the statistics of the Persian Home Office, and on the observations of travellers, the population exceeded 7,650,000 in 1881, and was estimated in 1891 at about 9,000,000, distributed between the urban, rural, and nomad elements as under:—

Town residents	2,500,000
Village residents (peasantry) . .	4,500,000
Nomad tribes	2,000,000
	9,000,000

APPROXIMATE CLASSIFICATION BY RACES AND RELIGIONS.

Moslem, mostly Shiahs, 6,770,000	IRANIANS.	
	Tajiks and Tats (Persians), all the Towns and Agricultural Districts	5,000,000
	Kurds proper, Persian Kurdistan . . .	300,000
	Mikri Kurds, Azairbijan	50,000
	Shadilu and other Kurds, N. Khorasan ranges	50,000
	Luri proper, Luristan	370,000
	Bakhtiari Luri, Pish-i-Kuh	250,000
	Laks or Leks,[1] Fars, Irak, Mazandaran . .	100,000
	Makrani Baluchis, Sistani Baluchis — Makran, Sistan, Karman .	100,000
	TATARS.	
	Turki Iliats, Irak, Khorasan, etc.	500,000
	Turkomans (Goklans, etc.), Mazandaran, Astrabad	50,000
	MONGOLS.	
	Taemuri Aymaks,[2] South and East Khorasan	250,000
	Hazarahs,[2] towards Herat frontier	50,000
	SEMITES AND SUNDRIES.	
	Arabs,[3] Arabistan, Fars, Laristan, etc. . .	350,000
Christians, 175,000	Armenians, Isfahan, Tehran, Urmia . .	150,000
	Chaldeans ("Nestorians"), Urmia . . .	25,000

1 Many of the Laks, known as "Nasari" and "Ali-Iláhi," reject the Prophet, hence are not regarded as true "Believers."

2 All Sunnis, although the Hazarahs of Afghanistan are Shiahs.

3 Many of these Arabs have become Shiahs, and are in other respects also assimilated to the Persians.

Sundries, 53,000	Jews, the large towns	16,000
	Kizil-Bashis, Khorasan, Karman	12,000
	Ghebrs,[1] chiefly Yazd	7,000
	Gipsies and Jats, Karman, Irak, etc.	? 20,000
	Total	7,650,000

APPROXIMATE AREAS AND POPULATIONS OF THE MEMLEKET (PROVINCES).

		Area in sq. miles.	Population.
North	Astrabad	10,000	150,000
	Mazandaran	8,000	350,000
	Ghilan	6,000	400,000
	Azairbijan	35,000	1,400,000
West	Irak-Ajemi	115,000	1,300,000
	Ardelan	6,000	150,000
	Khuzistan	30,000	600,000
	Luristan	30,000	300,000
	Farsistan	60,000	1,300,000
South	Laristan	20,000	90,000
	Karman, with Kohistan, Makran, and Sistan	150,000	750,000
East	Khorasan	140,000	860,000
	Total	610,000	7,650,000

TOWNS WITH UPWARDS OF 10,000 INHABITANTS.

	Pop.		Pop.
Tehran	210,000	Kashan	30,000
Tabriz	180,000	Dizful	25,000
Isfahan	60,000	Khoi	25,000
Mashhad	60,000	Zenjan	24,000
Barfrush	50,000	Shuster	20,000
Karman	41,000	Burujird	20,000
Yazd	40,000	Kum	20,000
Urmia	40,000	Bandar-Abbas	20,000
Rasht	40,000	Sari	15,000
Kazvin	40,000	Mohammerah	15,000
Hamadan	30,000	Nishapur	11,000
Karmanshah	30,000	Bushahr	10,500
Shiraz	30,000	Astrabad	10,000

REVENUE.

Income.	Expenditure.	Debt (1892).
£1,600,000	£1,900,000	£500,000

[1] Descendants of the old Persian fire-worshippers. Their numbers have been greatly over-estimated. Blackie gives 40,000; but in the town and district of Yazd, where they are chiefly concentrated, Major E. Smith found they had dwindled to 3800 in 1871. They are easily recognised by a uniform turban of a drab or dust colour.

Trade (1893).

Imports.[1]	Exports.[2]
£6,700,000	£6,500,000

Postal Service (1893).

Post Offices	120	Telegraph Offices	100
Letters forwarded	2,050,000	„ Lines, miles	4,150
Receipts	£18,000	„ Wires, „	11,000
		Despatches	800,000

Distances in English Miles.

Tehran to Kum	87	Rasht to Tehran	180
Kum to Isfahan	158	Tabriz to Tehran	360
Isfahan to Yazd	191	Mashhad to Tehran	558
Yazd to Karman	219	Mashhad to Sistan	582
Karman to Bam	136	Sistan to Karman	360
Bam to Bampur	242	Sistan (Nasirabad) to Bam	246
Bampur to Gwadar	220	Bam to Bandar-Abbas	248
Bushahr to Shiraz	120	Shiraz to Isfahan	220

1 Chief imports—Cottons and other woven goods, hardware, glass, paper, metals, tea, sugar.

2 Chief exports—Silk, tobacco, skins, carpets, rugs, opium, gums, rice, dried fruits, pearls, cotton, wool. Trade mainly with England, India, and Russia.

INDEX

THE END